The Microbiology of Respiratory System Infections

The Microbiology
of Respiratory
System Infections

Edited by

Kateryna Kon
Mahendra Rai

AMSTERDAM • BOSTON • HEIDELBERG • LONDON
NEW YORK • OXFORD • PARIS • SAN DIEGO
SAN FRANCISCO • SINGAPORE • SYDNEY • TOKYO

Academic Press is an imprint of Elsevier

Academic Press is an imprint of Elsevier
125 London Wall, London EC2Y 5AS, United Kingdom
525 B Street, Suite 1800, San Diego, CA 92101-4495, United States
50 Hampshire Street, 5th Floor, Cambridge, MA 02139, United States
The Boulevard, Langford Lane, Kidlington, Oxford OX5 1GB, UK

Notices
Knowledge and best practice in this field are constantly changing. As new research and experience broaden our understanding, changes in research methods, professional practices, or medical treatment may become necessary.

Practitioners and researchers must always rely on their own experience and knowledge in evaluating and using any information, methods, compounds, or experiments described herein. In using such information or methods they should be mindful of their own safety and the safety of others, including parties for whom they have a professional responsibility.

To the fullest extent of the law, neither the Publisher nor the authors, contributors, or editors, assume any liability for any injury and/or damage to persons or property as a matter of products liability, negligence or otherwise, or from any use or operation of any methods, products, instructions, or ideas contained in the material herein.

Library of Congress Cataloging-in-Publication Data
A catalog record for this book is available from the Library of Congress

British Library Cataloguing-in-Publication Data
A catalogue record for this book is available from the British Library

ISBN: 978-0-12-804543-5

For information on all Academic Press publications
visit our website at https://www.elsevier.com/

Working together
to grow libraries in
developing countries

www.elsevier.com • www.bookaid.org

Publisher: Sara Tenney
Acquisition Editor: Linda Versteeg-buschman
Editorial Project Manager: Halima Williams
Production Project Manager: Julia Haynes
Designer: Matt Limbert

Typeset by Thomson Digital

Contents

List of Contributors

T. Adam
Universiti Malaysia Perlis, Institute of Nano Electronic Engineering (INEE); Universiti Malaysia Perlis (UniMAP), School of Electronic Engineering Technology, Faculty of Engineering Technology, Kangar, Perlis, Malaysia

J.A. Al-Tawfiq
Speciality Internal Medicine Department, Johns Hopkins Aramco Healthcare, Dhahran, Kingdom of Saudi Arabia; Department of Medicine, Indiana University School of Medicine, Indianapolis, IN, United States

M.K. Md Arshad
Universiti Malaysia Perlis, Institute of Nano Electronic Engineering (INEE), Kangar, Perlis, Malaysia

R.M. Ayub
Universiti Malaysia Perlis, Institute of Nano Electronic Engineering (INEE), Kangar, Perlis, Malaysia

S. Benkouiten
Aix Marseille University, Research Unit on Emerging Infectious and Tropical Diseases (URMITE); University Hospital Institute for Infectious Diseases (Méditerranée Infection), Marseille, France

V. D'Oriano
University of Naples Federico II, Department of Experimental Medicine, Naples, Italy

V. de Carvalho Santos-Ebinuma
São Paulo State University, Araraquara, SP, Brazil

F.S. Del Fiol
University of Sorocaba, Sorocaba, SP, Brazil

T. Dubey
TBON-LAB, Investment Blvd. Hayward, CA, United States

F. Esteves
Department of Genetics, Toxicogenomics & Human Health (ToxOmics), NOVA Medical School/Faculdade de Ciências Médicas, Universidade Nova de Lisboa, Portugal

A. Falanga
University of Naples Federico II, Department of Pharmacy, Naples, Italy

G. Franci
University of Naples Federico II, Department of Experimental Medicine, Naples, Italy

M. Galdiero
University of Naples Federico II, Department of Experimental Medicine, Naples, Italy

S. Galdiero
University of Naples Federico II, Department of Pharmacy, Naples, Italy

P. Gautret
Aix Marseille University, Research Unit on Emerging Infectious and Tropical Diseases (URMITE); University Hospital Institute for Infectious Diseases (Méditerranée Infection), Marseille, France

M. Gerenutti
University of Sorocaba, Sorocaba, SP, Brazil

A.L.S. Gonçalves
MSc at Universidade Federal do Rio Grande do Sul, Porto Alegre, Rio Grande do Sul, Brazil

S.C.B. Gopinath
Universiti Malaysia Perlis, Institute of Nano Electronic Engineering (INEE), Kangar; Universiti Malaysia Perlis, School of Bioprocess Engineering, Arau, Perlis, Malaysia

D. Grotto
University of Sorocaba, Sorocaba, SP, Brazil

J.A. Guisantes
University of The Basque Country, Department of Immunology, Microbiology and Parasitology, Faculty of Pharmacy and Laboratory of Parasitology and Allergy, Research Center Lascaray, Paseo University, Vitoria, Spain

U. Hashim
Universiti Malaysia Perlis, Institute of Nano Electronic Engineering (INEE), Kangar, Perlis, Malaysia

A.P. Ingle
Nanobiotechnology Laboratory, Department of Biotechnology, SGB Amravati University, Amravati, Maharashtra, India

A.F. Jozala
University of Sorocaba, Sorocaba, SP, Brazil

K. Kon
Department of Microbiology, Virology and Immunology, Kharkiv National Medical University, Kharkiv, Ukraine

S.R. Konduri
Wayne State University School of Medicine, Division of Pulmonary, Critical Care and Sleep Medicine, Detroit, Michigan, United States

A. Krishnamurthy
Swinburne University of Technology, Department of Chemistry and Biotechnology, Faculty of Science, Engineering and Technology, Hawthorn, Victoria, Melbourne, Australia

T. Lakshmipriya
Universiti Malaysia Perlis, Institute of Nano Electronic Engineering (INEE), Kangar, Perlis, Malaysia

J. Martínez
University of The Basque Country, Department of Immunology, Microbiology and Parasitology, Faculty of Pharmacy and Laboratory of Parasitology and Allergy, Research Center Lascaray, Paseo University, Vitoria, Spain

O. Matos
Medical Parasitology Unit, Group of Opportunistic Protozoa/HIV and Other Protozoa, Global Health and Tropical Medicine, Instituto de Higiene e Medicina Tropical, Universidade Nova de Lisboa, Portugal

Z.A. Memish
Ministry of Health; Alfaisal University, College of Medicine, Riyadh, Kingdom of Saudi Arabia

L.C.L. Novaes
RWTH Aachen University, Aachen, Germany

L. Palomba
University of Naples Federico II, Department of Experimental Medicine, Naples, Italy

E. Palombo
Swinburne University of Technology, Department of Chemistry and Biotechnology, Faculty of Science, Engineering and Technology, Hawthorn, Victoria, Melbourne, Australia

R. Pandit
Nanobiotechnology Laboratory, Department of Biotechnology, SGB Amravati University, Amravati, Maharashtra, India

P. Paralikar
Nanobiotechnology Laboratory, Department of Biotechnology, SGB Amravati University, Amravati, Maharashtra, India

A. Pasdaran
Phytochemistry Research Center, Shahid Beheshti University of Medical Sciences, Tehran, Iran

A. Pasdaran
Guilan University of Medical Sciences, Department of Pharmacognosy, School of Pharmacy, Research and Development Center of Plants and Medicinal Chemistry, Rasht; Shiraz University of Medical Sciences, Medicinal Plants Processing Research Center, Shiraz; Phytochemistry Research Center, Shahid Beheshti University of Medical Sciences, Tehran, Iran

N. Petrovsky
Vaxine Pty Ltd, Department of Endocrinology, Flinders Medical Centre; Flinders University, Faculty of Medicine, Adelaide, Australia

S.U. Picoli
Universidade Feevale, Novo Hamburgo, Rio Grande do Sul, Brazil

I. Postigo
University of The Basque Country, Department of Immunology, Microbiology and Parasitology, Faculty of Pharmacy and Laboratory of Parasitology and Allergy, Research Center Lascaray, Paseo University, Vitoria, Spain

S. Quereshi
Department of Microbiology and Biotechnology, Indira Priyadarshini College, Chhindwara, Madhya Pradesh, India

M. Rai
Nanobiotechnology Laboratory, Department of Biotechnology, SGB Amravati University, Amravati, Maharashtra, India

R.Y. Ramírez-Rueda
Pedagogical and Technological University of Colombia, Faculty of Health Sciences, School of Nursing, Tunja, Colombia

M. Razzaghi-Abyaneh
Department of Mycology, Pasteur Institute of Iran, Tehran, Iran

L. Rinaldi
Second University of Naples, Internal Medicine of Clinic Hospital of Marcianise, Department of Medicine, Surgery, Neurology, Geriatric and Metabolic Diseases, ASL Caserta, Italy

O. Schildgen
Witten/Herdecke University, Department of Pathology, gGmbH clinics of Cologne, Cologne, Germany

V. Schildgen
Witten/Herdecke University, Department of Pathology, gGmbH clinics of Cologne, Cologne, Germany

D. Sheikhi
Regulations (GCP/ICH), Pharmaceuticals, Denmark

S. Shende
Nanobiotechnology Laboratory, Department of Biotechnology, SGB Amravati University, Amravati, Maharashtra, India

A.O. Soubani
Wayne State University School of Medicine, Division of Pulmonary, Critical Care and Sleep Medicine, Detroit, Michigan, United States

S. Tikar
Nanobiotechnology Laboratory, Department of Biotechnology, SGB Amravati University, Amravati, Maharashtra, India

C. Zannella
University of Naples Federico II, Department of Experimental Medicine, Naples, Italy

Preface

Respiratory infections include a diverse group of bacterial, viral, and fungal infections of upper and lower respiratory systems. Some of them have been known for a long time, such as tuberculosis and influenza, whereas others have recently emerged, such as coronoviral infections SARS, MERS, and human bocaviruses. Despite the presence of advanced hospital techniques and discovery of new antimicrobial drugs, respiratory infections are still associated with significant mortality and morbidity worldwide. The high healthcare importance of this group of diseases makes it necessary to have a well-structured source of up-to-date scientific information, discussing existent clinical and diagnostic guidelines as well as new and perspective trends in the diagnosis, treatment, and prophylaxis of respiratory system infections.

The present book has been divided into three sections according to the types of respiratory pathogens. The first section contain reviews on the most common and epidemiologically important respiratory viruses, such as influenza virus, severe acute respiratory system coronavirus, recently discovered Middle East respiratory syndrome coronavirus and human bocavirus.

The second section is devoted to the respiratory infections caused by bacterial and fungal pathogens, including *Mycobacterium tuberculosis*, multidrug resistant bacteria, such as metallo beta lactamase producing *Pseudomonas aeruginosa*, and fungal pathogens including *Aspergillus* spp., *Pneumocystis jirovecii*, and other fungi. Special attention has been paid to the questions of circulation of respiratory pathogens during mass gatherings, connection between indoor air pollution and respiratory diseases, association of allergic respiratory diseases with the presence of parasites, and to respiratory infections in patients with hematological malignancies.

The third section of this book discusses treatment approaches against different types of bacterial infections of lower respiratory tract. This section reviews classical antimicrobial and phytomedical approaches as well as application of nanotechnology against respiratory pathogens.

This book would be very useful for graduate and postgraduate students, researchers, university teachers, scientists, medical practitioners, and specialists from pharmaceutical and laboratory diagnostic companies.

INFLUENZA VIRUS INFECTIONS: CLINICAL UPDATE, MOLECULAR BIOLOGY, AND THERAPEUTIC OPTIONS

G. Franci*, L. Palomba*, A. Falanga**, C. Zannella*, V. D'Oriano*,
L. Rinaldi[†], S. Galdiero**, M. Galdiero*

*University of Naples Federico II, Department of Experimental Medicine, Naples, Italy;
**University of Naples Federico II, Department of Pharmacy, Naples, Italy; [†]Second University
of Naples, Internal Medicine of Clinic Hospital of Marcianise, Department of Medicine,
Surgery, Neurology, Geriatric and Metabolic Diseases, ASL Caserta, Italy

1 INTRODUCTION

Influenza is an ancient and deadly disease which has sickened and killed millions of people in local epidemics and global pandemics. Nowadays, it is common knowledge that influenza is a highly infectious viral illness, but before the discovery of viruses the etiological factor of influenza was not known and, therefore, we had to relay solely on the clinical picture characterized by a sudden onset of high fever, cough, headache, muscle and joint pain, unwell feeling, sore throat, and runny nose. These symptoms were clearly described by Hippocrates roughly 2400 years ago, but historical data on influenza were of difficult interpretation, since these symptoms can be similar to those of other respiratory diseases, therefore not distinctive enough.

The word *Influenza* originated in the 15th century from the Italian language, meaning "influence" since the disease was ascribed to unfavorable astrological influences. A different origin could be the word "*influsso*" for describing the sweating characteristic of the illness or meaning "influence of the cold." It was not until 1703 when J. Hugger's thesis submitted at the University of Edinburgh and named "*De Catarrho epidemio, vel Influenza, prout in India occidentali sese ostendit*" that the English-spoken world directly associated "influenza" with the disease and its symptoms. After that the name influenza and its shorthand "flu" came into more general use.[1]

The influenza virus was first isolated from pigs in 1930 by Shope and Lewis.[2] This seminal discovery was followed by the isolation in ferrets of influenza A virus by Smith, Andrewes, and Laidlaw.[3] In 1936, Burnet demonstrated that influenza virus could be grown in chicken embryonated eggs,[4] opening the path for the study of the characteristics of the virus.

It is estimated that influenza virus infects every year 5–10% of the adult population worldwide and 20–30% of the children. Even though most patients recover from flu symptoms within a short period

The Microbiology of Respiratory System Infections. http://dx.doi.org/10.1016/B978-0-12-804543-5.00001-4

and without serious sequelae, the estimates indicate from 3–5 million cases of serious illness and over 250,000 deaths per year. Therefore, due to its medical importance, influenza viruses have been the focus of extensive research to decipher the molecular mechanisms that dominate cell invasion and pathogenesis.

2 CLASSIFICATION

Influenza virus belongs to the *Orthomyxoviridae* family, represented by negative-strand RNA viruses whose genome is divided into six to eight individual RNA segments. The orthomyxovirus family name is derived from the Greek words "*orthos*" which means "correct" while "*myxa*" stands for "mucus". The family is subdivided into four genera: *Influenzavirus*, *Isavirus*, *Thogotovirus*, and *Quaranjavirus*. The first genera contain viruses that cause disease in vertebrates, including birds, humans, and other mammals, while isaviruses are fish-infecting viruses (infectious salmon anemia virus).[5] Thogotoviruses have been primarily associated with either hard or soft ticks and they have a wide geographic distribution. Moreover, they can infect several mammals but only a few cases have been reported of human infections and a novel thogotovirus (Bourbon virus) has only recently been associated with a febrile illness and death of a human patient in the United States in 2014.[6] Quaranjaviruses predominantly infect arthropods and birds.[7] Here we focus our attention on the influenza viruses, including influenza A, B, and C, which are of greater medical importance. The most common and also the most medically important of the influenza viruses are those designated type A, which infect humans and a wide array of other mammals, and predominantly birds. In particular, aquatic birds represent the wildlife reservoirs of the virus and play a role of paramount importance in the creation of human epidemic and pandemic influenza strains. Influenza B is not known to give rise to pandemics, and has a more limited host spectrum, in fact it has only recently been found to cause infections in seals, apart from man.[8] Type C influenza viruses infect humans and swine, but only cause mild respiratory illness or no symptoms at all. More importantly, type C influenza viruses are not able to produce epidemics and, therefore, are of limited medical interest.

3 VIRION STRUCTURE AND GENOMIC ARCHITECTURE

Influenza viruses are pleomorphic, spherical particles (about 100 nm in diameter), although filamentous forms can occur. The virions are relatively unstable in the environment and influenza viruses are inactivated by heat, dryness, extremes of pH, and detergents. Virions are enveloped and their lipid membrane is derived from the host cell. The envelope lodges different glycosylated proteins that project from the surface of the virus. These proteins are the hemagglutinin (HA), the neuraminidase (NA), and the M2 ion channel proteins. The morphology of influenza A virus particles is, therefore, characterized by distinctive spikes, with lengths from 10 to 14 nm, which are readily observable in electron micrographs of virus particles. The approximate ratio between HA and NA is 4:1.

The matrix protein (M1) is situated just beneath the envelope, and underlying the M1 layer a helical superstructure, representing the core of the virus particle which is made of the ribonucleoprotein (RNP) complex, is observed. The RNP complex consists of the viral RNA segments, which are coated with the nucleoprotein (NP) and associated with the heterotrimeric polymerase complex (PB1, PB2, and PA) (Fig. 1.1). The chemical composition of virus particles is approximately 1% RNA, 5–8% carbohydrate, 20% lipid, and approximately 70% protein.[9]

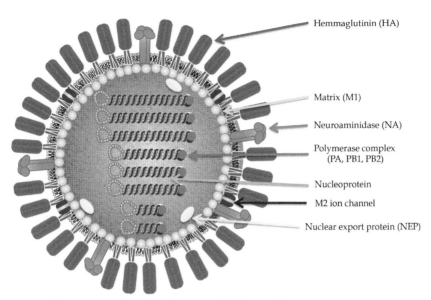

FIGURE 1.1 Schematic Representation of the Structure of Influeanza Virus

The genomes of influenza A and B consist of eight separate negative-sense, single-stranded RNA segments known as viral RNAs. The influenza C genome has only seven segments. Each viral RNA segment exists as a structure known as viral ribonucleoprotein (vRNP) complex in which the RNA strand is wrapped by the NP and forms a helical hairpin that is bound on one end by a single heterotrimeric polymerase complex. Within the RNPs, RNA molecules account for the genetic material of the virus, NP monomers cover and protect it, whereas the polymerase complex is responsible for the transcription and replication.[10] Noncoding sequences are present at both 5′ and 3′ ends of the viral gene segments; partially complementary sequences are located at the extreme termini which are highly conserved between all segments in all influenza viruses. When base-paired, these ends function as the viral promoter that is required for replication and transcription.

Each of the eight segments of the influenza A virus genome encodes for the viral proteins (Table 1.1). The whole influenza genome encodes for 10 major proteins, although alternative protein products have been characterized from several genome segments.[11] The complete coding capacity of the influenza virus genome is only partially known and processes like splicing, and the use of alternative initiation codons or overlapping frames are employed by the virus to augment the variety of viral products that are generated during the infectious cycle. In fact, among the RNAs present, 2 of them (segment 7 and 8) are transcribed to messenger RNAs (mRNAs) which are then spliced into two chains encoding a further gene product each. Additionally, segment 2 presents at the 5′ end an alternative open reading frame that encodes for the 87 aa long PB1-F2 polypeptide. The full expression capacity of the influenza genome is further complicated by the abundant synthesis of small viral noncoding RNAs, whose roles in productive infections are still largely unknown.[12,13]

The three largest RNA segments each encode for one of the viral polymerase subunits, PB2, PB1, and PA.[14] The influenza virus RNA-dependent RNA polymerase is a 250-kDa complex of these three genetic products. Electron microscopy studies have shown a compact structure with the three subunits

Table 1.1 Schematic Representation of the Protein Products of Influenza A Virus Gene Segments

Genome Segment	Name	Protein	Function
1	PB2	PB2 — 759 aa	Component of the RNA polymerase (cap recognition)
	PB2-S1	PB2-S1 — 508 aa (Alternative splicing)	Binds to PB1, inhibits signaling pathways
2	PB1	PB1 — 757	Component of the RNA polymerase (enlongation)
	N40	N40 — 718 aa (Alternative iniziation)	Regulates PB1 and PB1-F2 expression
	PB1-F2	PB1-F2 — 718 aa	Virulence factor (proapoptotic)
3	PA	PA — 716 aa	Component of the RNA polymerase (endonuclease activity)
	PA-X	PA-X — 252 aa (Ribosomal frameshift)	Modulates host response and virulence
	PA-N155	PA-N155 — 568 aa (Alternative iniziation)	Unknown
	PA-N182	PA-N182 — 535 aa	Unknown
4	HA	HA — 580 aa	Surface glycoprotein, binding to receptor and fusion mediator, major antigen
5	NP	NP — 498 aa	RNA-binding and RNA synthesis. RNP nuclear import
6	NA	NA — 465 aa	Surface glycoprotein, neuraminidase activity

Table 1.1 Schematic Representation of the Protein Products of Influenza A Virus Gene Segments (*cont.*)

Genome Segment	Name	Protein	Function
7	M1	M1 — 252 aa	Matrix protein, multiple roles in virion assembly and infection
	M2	M2 — 97 aa — Alternative splicing	Membrane protein, ion channal activity
	M42	M42 — 99 aa — Alternative splicing	Can replace M2 in M2-null viruses
8	NS1	NS1 — 217 aa	Multifunctional protein, INF antagonist activity
	NS2/NEP	NS2/NEP — 121 aa — Alternative	Mediates RNP nuclear export
	NS3	NS3 — 174 aa — Alternative splicing	Associated to new host adaptation ability

tightly associated.[15] PB1, representing the core of the complex and the most conserved of the polymerase subunits, contains the enzymatic motifs needed for RNA polymerization activity.[16] As previously remembered, the second segment also encodes for an accessory protein, PB1-F2, from an alternate open reading frame within the PB1 gene. PB1-F2, which is unique to influenza A, has been found to localize to mitochondria and exhibiting a pro-apoptotic activity. The PB2 subunit recognizes and binds to the host mRNA cap structures generated by the cellular transcription machinery[17,18] and the acidic protein PA has an endonucleolytic activity necessary for the viral cap-snatching process.[19,20]

Segment 4 codes for the HA protein. The mature HA protein is a trimeric type I integral membrane glycoprotein (with a single transmembrane span, N-terminus in the ectodomain and C-terminus in the cytosol) which is found in the lipid envelope of virions and on the surface of infected cells.[21] HA is the most abundant proteic component of the viral envelope and is the major target for neutralizing antibodies. HA undergoes several posttranslational modifications including glycosylation,[22,23] palmitoylation,[24] proteolytic cleavage, disulphide bond formation, and conformational changes. Host cell proteases operate the maturation of the precursor HAO molecule into its HA1 and HA2 subunits (which are linked by a disulphide bridge) (Fig. 1.2, Part 1: HA1 in blue and HA2 in red) and this cleavage mechanism is essential for the fusion activity of HA. Moreover, HA is in charge of the binding of virions to host cell surface receptors (sialic acid).[25]

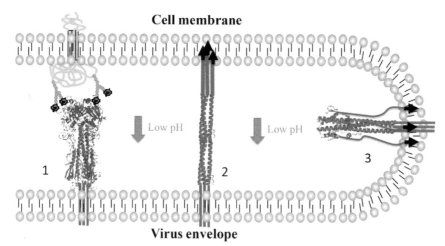

FIGURE 1.2 Influenza Virus HA Mediated Entry Mechanism

(1) Influenza HA binds to sialic acids. (2) Low pH conformational change of HA releases the fusion peptide at the N-terminus of HA2 and a conformational change locates the fusion peptide on the distal part of the extended helix and insertion of the fusion peptide into the cell membrane occurs. The transmembrane domain links the HA2 with the viral envelope. (3) At final low pH, a further conformational change drives a refolding mechanism which leads to the formation of a trimeric coiled-coil (the six-helix bundle) that positionates both the transmembrane domains and the fusion peptides in the same fused membrane.

RNA segment 5 of influenza encodes RNA-binding protein NP.[26] This is a highly basic protein whose main function is encapsidation of the viral RNA (an NP monomer of 56 kD binds 24 bases of RNA)[27] leaving accessibility to the polymerase as a template for transcription.[28]

NP also plays a crucial role in transporting the viral RNPs into the nucleus.[29,30] In fact, RNPs are too large to passively diffuse through the 9 Å nuclear pores.[31] In order to shuttle RNPs in and out of nuclei and from the cytoplasm to the cell periphery, NP makes multiple viral and cellular protein associations involving distinct NP recognition sequences (NLSs, nuclear localization signals, and NES, nuclear export signals). During the late stages of infection, NP also associates with the cytoskeleton[32] and NP has been found to bind to filamentous-actin.[33]

Segment 6 of influenza A encodes the NA protein which is the second major viral glycoprotein. This type II integral membrane protein (single transmembrane span, C-terminus in the ectodomain and N-terminus in the cytosol) is a tetramer and possesses critical sialidase (NA) activity that is needed for a productive release of viral particles from the infected cell.[34]

Segment 7 of influenza A encodes two proteins, the M1 and M2 protein. M1 is expressed from a collinear transcript, while M2 is derived from an alternative spliced mRNA. M1 associates with lipid membranes and plays an essential role in viral budding. It also regulates the movement of RNPs out of the nucleus and inhibits viral RNA synthesis at later stages of viral replication. M2 is a tetrameric type III membrane protein that has ion channel activity. It functions primarily during virus entry where it is responsible for acidifying the core of the particle which triggers dissociation of M1 from the viral RNPs (uncoating).

The shortest RNA (segment 8 in influenza A) encodes the NS1 protein from a collinear transcript and the NEP/NS2 protein from an alternatively spliced transcript. NS1 is a RNA-binding protein that is expressed at high levels in infected cells and is useful to inhibit host antiviral response, besides interfering with the host mRNA processing. The NEP/NS2 protein mediates the nuclear export of newly synthesized RNPs, corresponding with its expression at later times during viral infections.

Recently, several additions have been made to the list of gene products coded by the influenza A genome.[35] As just described earlier in the paragraph, the PB1-F2 protein was discovered in 2001 and is encoded by an alternative open translation initiation sites near the 5′ end of the PB1 gene.[36] Moreover, a third protein is made by the same mechanism by the PB1 gene, named PB1-N40. PB1-N40 is an N-terminal 39 aa polypeptide which is translated from the fifth AUG codon in frame with the PB1 start.[37] In addition, segment 1, coding for PB2, produces a newly discovered viral protein, termed PB2-S1, encoded by a novel spliced mRNA in which the region corresponding to nucleotides 1513–1894 of the PB2 mRNA is deleted.[38] Novel polypeptides are also encoded by genomic segment 3 and are named PA-X, PA-N155, and PA-N182. The first of them is derived from a second open reading frame (X-ORF), accessed via ribosomal frameshifting[39] and modulates host response.[40] The other two (PA-N155 and PA-N182) do not have a polymerase activity, and have been found to be important for virulence and pathogenesis.[41] Expression of specific spliced viral products throughout infection is also applied for two of the smallest segments, M1 and NS1. It was already known for a while that protein products from segment 7 included the matrix (M1) and ion channel (M2) proteins, made from a spliced transcript, but a further protein product with an alternative ectodomain and again by a splicing mechanism, has been recently identified, the M42.[42] Finally, segment 8, besides the nonstructural (NS) protein NS1, also encodes a nuclear export protein NS2/NEP, and NS3 by alternative mRNA splicing.[43]

4 VIRAL REPLICATION

Influenza viruses bind to neuraminic acids (sialic acids) on the surface of cells to initiate the entry process[25] (Fig. 1.3). Notwithstanding the ubiquitous nature of sialic acids, HAs of influenza viruses infecting different animal species show some preference for particular glycosidic linkages of the receptor. Human viruses preferentially bind to sialic acids linked to galactose by alfa 2,6 glycosidic bonds (SAα2,6Gal). On the other hand, avian viruses show a preference for alfa 2,3 glycosidic bonds (SAα2,3Gal).[44,45] The preference of different influenza subtypes for diverse host-species is explained by the fact that SAα2,6Gal is found mostly in human trachea, while SAα2,3Gal is abundant in the gut epithelium of several bird species, but it should be noted that this represents a preference for receptor-antireceptor linkage and not an absolute specificity. In fact, avian and human cells contain sialic acids with both linkages, though differently expressed and distributed. In addition, high viral inoculum or adaptive point mutations in the HA gene may circumvent this problem.[46] HA is able to perform both action of binding and mediation of fusion of juxtaposing membranes. HA is expressed as a trimeric rod-shaped molecule on the virion surface and is produced as a HA0 precursor which is then cleaved into two subunits HA1 (the globular more external part of the molecule) and HA2 (holding a transmembrane domain) by host cell proteases. The cleavage is needed for the full activity of the molecule. The major structural features of the HA trimer are: a long stem of triple-stranded coiled-coil of α-helices and a globular ectodomain exposed to the environment and derived from the HA1 portion of the monomers.[47] The stalk region of HA connects the molecule to the virion envelope by a short hydrophobic

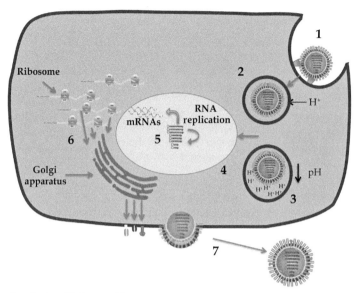

FIGURE 1.3 Replication Cycle of Influenza Virus

(1) Influenza virus enters cells by receptor-mediated endocytosis following the interaction between HA and sialic acids on the cell surface; the virus initially (2) localizes to early endosomes, and then (3) reaches late endosomes where a proton pump generates a gradual acidification of the endosome; HA undergoes sequential conformational changes which finally promotes the fusion of the viral and endosomal membranes. Influenza virus genetic material is released into the cytoplasm and (4) then is transported to the cell nucleus, where (5) replication occurs. (6) Protein prouction and their posttranslational modification happen in the cytoplasm and finally (7) new capsid are assembled and released by membrane budding.

sequence (transmembrane domain),[25] while the globular head is responsible for the receptor binding through a receptor binding pocket located on its distal tip. The interaction between sialic acid and HA is considered to be of low affinity; therefore, in order to increase the total intensity of the interaction, several HA molecules cooperate to link more sialic acids molecules leading to a high-avidity binding on the cell surface.[48] After binding to the receptor (Fig. 1.2, part 1), influenza viruses are taken up into the cell. Early imaging studies revealed that the virus enters the cell by receptor-mediated endocytosis.[49–51] The main mechanism is still recognized to be by clathrin-mediated endocytosis, although a nonclathrin, noncaveolae pathway has been also attributed to influenza entry.[52,53] Regardless of whether the influenza viruses are taken up by endocytosis or macropinocytosis, they end up exploiting the transport system via distinct endosomal stages and consequently changes in pH to release their viral ribonucleoprotein complexes (vRNPs) into the cytoplasm. After internalization by either uptake pathways, the virus initially localizes to early endosomes and then reaches late endosomes. A crucial function is devoted to proton pumps, which mobilitate protons from cytoplasm into the endosomal lumen; therefore, allowing for a gradual acidification of the endosome content.[54,55] Initial lowering of pH in endosomes radically changes HA structure, which is rendered vulnerable to be digested by proteases. Consequently, HA is cleaved into the two subunits, connected together by disulphide bonds. Once the influenza virus is in the acidic environment of the late endosome, HA undergoes further conformational changes that expose the fusion peptide at the N-terminus of the HA2 and locate it toward the endosomal membrane.[56,57] This in turn leads to the interaction of HA with the membranes of the endosome. Following the stepwise conformational changes, the fusion peptide is finally inserted into the target endosomal membrane (Fig. 1.2, part 2) and juxtaposition of the viral and endosomal membranes is reached.[58–60] The crystal structures of pre- and postfusion HA (Fig. 1.2, part 1 and 3, respectively) have been solved, while only models based on such structures are available to predict the intermediate stages of the HA gradual conformational modifications. Furthermore, it has been demonstrated that several HA trimers (at least nine) are needed to execute membrane fusion by concurrent conformational changes to release the required energy.[61] To proceed with fusion between viral and endosomal membranes with the increasing acidification, HA trimers tilt (Fig. 1.2, part 3) at the fusion site and a hemifusion stage is created where the outer leaflets of the interacting membranes are blended together and only the internal leaflets of the two membranes preserve the content from the virus and the endosome to mix together.[62,63] Finally, both membranes fuse and a so-called fusion pore is established.[64–66] As a consequence of these chain of processes, the influenza virus genetic material, in the form of a RNP complex, can be released through the fusion pore into the cytoplasm.[67] Once fusion of viral and endosomal membranes has been achieved, uncoating of the viral capsid is necessary to release the VRNPs into the cytosol. This process appears to require the coordinated action of two viral proteins, namely M1 and M2. Considering influenza virion particle, M1 can be visualized by electron microscopy forming a structured stratum beneath the viral envelope.[68,69] M1 serves to link the viral membrane containing the glycoproteins with the RNPs in the virus core through interactions between the middle domain of M1 with the NP on the RNP.[70] During the uncoating process, the interactions of M1 with either the viral membrane or the vRNPs need to be released for allowing complete uncoating and subsequent transport of the RNPs into the nucleus. This requires the activity of the viral protein M2 which mediates the influx of H$^+$ ions from the endosome into viral particle. This ion channel activity of M2 is regulated by pH, in fact, lowering the pH, the ion channel activity increases.[71] Upon acidification of the endosome, M2 facilitates the proton influx from the endosome into the virion, leading to a pH values drop inside the virus particle.[72] This change in pH inside the virus particle causes conformational changes in M1, therefore, the interaction between the RNPs and M1 is greatly weakened or lost pushing the release of the vRNPs into the cytoplasm.[73,74] Finally the eight segments are transported as one moiety to the nucleus[75] where RNA synthesis of influenza virus occurs. RNP complexes are not able to passively diffuse through nuclear pores (since the small size of the pores); therefore, proteins belonging to the family of α-importins recognize nuclear localization signals (NLS) on the NP protein and play an important role in the transport of RNA complexes.[76] In order for the genome to be transcribed, it first must be converted into a positive sense RNA to serve as a template for the production of viral RNAs. Once in the nucleus, genomic ssRNA of negative polarity serves as a template for the synthesis of two different ssRNAs of positive polarity: (1) messenger RNAs (mRNAs), and (2) full-length complementary copies (cRNAs). Messenger RNAs are the incomplete copies of the template and they are capped and polyadenylated. The viral RNA-dependent RNA polymerase (RdRp) is made up of the three viral proteins: PB1, PB2, and PA. During transcription, viral RdRp interacts with the host polymerase II.[77] PB2 shows endonuclease activity and binds to the 5′ methylated caps of cellular mRNAs cleaving the cellular mRNAs' 10–15 nucleotides 3′ to the cap structure (this phenomenon is called "cap-snatching"). This cellular capped RNA fragment is then used by the viral RdRp to prime viral transcription.[78–81] It is recognized that the three subunits of the polymerase act cooperatively and the synthesis of RNAs is a concerted action of all subunits. The synthesis of full-length complimentary copy (cRNA) of the vRNA does not need capped primer and the cRNA chain is not prematurely terminated and polyadenylated, as is the case of viral mRNA synthesis. The next step of replication of influenza genomic segments is the copying of vRNA on the template of positive strand cRNAs. This process also generates full-length products which assemble with NP and polymerase subunits and form RNP complexes. Such complexes are finally exported into the cytoplasm; the M1 protein and NS2 protein play a vital role in the translocation of these macromolecular structures.[30] Packaging of eight different RNA segments in the form of RNPs into virion shells is a poorly understood phenomenon, though many models have been proposed.[82–84] Some facts suggest the presence of packaging signals at both ends of the genomic segments; however, the precise sequences or structures responsible for specific packaging are still not well defined. All eight segments probably form a supercomplex before arrival at the plasma membrane binding sites.[75] Before being directed into lipid rafts, these two glycoproteins are posttranslationally modified. The modifications take place in the endoplasmic reticulum and in the Golgi apparatus. In the endoplasmic reticulum, these proteins become correctly folded and glycosylated. They are also assembled into oligomers: HA into trimers and NA into tetramers. Capsids get assembled underneath the cell membrane where the inserted viral glycoprotein interacts with the matrix and the RNPs and the envelope virions exit by membrane budding. NA is required to remove sialic acids in order to allow the virus to leave its host cells.

5 CLINICAL FEATURES AND PATHOGENESIS IN HUMANS

Influenza A viruses cause a respiratory illness known as "flu" in humans of all age groups all across the world. Generally, the disease is mild, but in immunocompromised individuals, pregnant women, children, and the aged, it may cause severe symptoms and in some cases it can be life threatening.[85] Direct person-to-person spread during acute infections represents the major mechanism for maintenance of the virus in human populations. Influenza is generally regarded as a seasonal disease for temperate climate regions.[86] In the northern hemisphere, epidemics usually peak between January and April,[87,88] while in the southern hemisphere, major outbreaks hit between May and September. On a global perspective, influenza virus infections are detectable throughout the year.[88] The period of incubation is 1–3 days and the most effective mechanism of spread among humans is by aerosol production.[89,90]

Notwithstanding the fact that influenza is generally considered a mild disease, the collective burden can be significant. In fact, the direct costs include hospitalizations, medical fees, drugs, and testing, while indirect costs such as loss of productivity mainly derive from school and workplace absenteeism. Children younger than 2 years of age and the elderly have the highest hospitalization rates.[91]

The severity of infections depend on the level of preexisting immunity, age of each individual, and virulence of the virus involved, all of these factors can thoroughly vary among individual outbreaks.[92–94]

Reinfection with a closely related variant can occur,[95,96] but symptoms are usually less severe than those resulting from a previous encounter with a similar virus strain. In fact, it was reported that, during the H1N1 pandemic in 2009, older subjects exposed to viruses of previous pandemics (mainly 1918 and 1957) could count on a partial protection against the 2009 strain.[97–100]

As previously noted, influenza infections can, in some cases, lead to a fatal outcome; however, it is difficult to determine the number of influenza virus–related deaths, because of the absence of a laboratory diagnosis which do not allow the certification of influenza as a primary cause of death.

Anyway, fatal cases have been increasing over the last decades, perhaps as a consequence of an ever increasing number of elderly and/or immunocompromised individuals. Mortality can affect all age groups but is principally reported in people older than 65 years (90%).[101,102] Major circumstances that lead to a more severe outcome of the infection in the elderly are poor cardiovascular and pulmonary conditions as well as metabolic or neoplastic diseases.[103]

Human influenza viruses replicate almost exclusively in superficial cells of the respiratory tract, with virus being recoverable from the upper and lower tracts of infected people. The optimal site of growth in the respiratory tract for influenza viruses is, in part, determined by the prevalence of the $SA\alpha2,3Gal$ or $SA\alpha2,6Gal$ receptors. The latter is abundantly expressed on human epithelial cells of the whole respiratory tract, allowing human strains of influenza to bind and infect cells, while $SA\alpha2,3Gal$ is only present on nonciliated cuboidal bronchiolar cells at the junction between the respiratory bronchiole and alveolus, and on cells lining the alveolar wall. This finding offers an explanation not only for the severe pneumonia in humans with avian strains but also for their limited spread in humans.[104,105] Virus replication reaches its peak after about 48 h, then the virus titer starts to slowly decline but virus shedding is still present after days 6–8. A good positive correlation has been found between the amount of virus shed and the symptomatology. Interestingly, children still shed virus after 12–13 days following the onset of symptoms. Therefore, children are major contributors in the spread of influenza higher titers and more prolonged shedding highlights the important role of this population.[106,107]

Influenza A virus induces changes throughout the respiratory tract, but the most clinically important pathology develops in the lower respiratory tract.[108–110] In uncomplicated influenza infections, acute inflammation of the larynx, trachea, and bronchi showing mucosal inflammation and edema are present. Infected cells become vacuolated, edematous, and lose cilia before desquamating. Infiltration of neutrophils and mononuclear cells lead to submucosal edema and hyperemia.[85] In more severe primary viral pneumonia, an interstitial pneumonitis develops with a predominantly mononuclear leukocyte infiltration. Some influenza types have a higher pathogenicity (HPAI) (discussed below) and the pathologic changes associated with HPAI viruses include a hemophagocytic syndrome, renal tubular necrosis, lymphoid depletion, and diffuse alveolar damage with interstitial fibrosis.[111–115] Necrotizing changes may occur with rupture of alveoli and bronchiole walls. At the cellular level, induction of apoptosis represents an additional mechanism of cell destruction.[116] Regeneration of the epithelium begins 3–5 days after the onset of clinical illness. Complete healing of the epithelial damage takes up to one month.

From a clinical point of view, infection with influenza A viruses results in an array of alternatives ranging from asymptomatic infection to primary viral pneumonia that rapidly progresses to a fatal outcome. The typical uncomplicated influenza syndrome is tracheobronchitis with some involvement of small airways.[117] The onset of illness is usually abrupt, with headache, chills, and dry cough, which are rapidly followed by high fever, myalgias, malaise, and anorexia. Fever is the most prominent sign of infection and generally peaks within 24 h with values of 38–40°C.[85] Other respiratory tract symptoms include nasal obstruction, rhinorrhea, sneezing, and pharyngeal inflammation. Sometimes, conjunctival inflammation may occur. As the fever declines, the respiratory signs and symptoms may become more prominent. The cough that produces small amounts of mucoid or purulent sputum often arise after a dry and hacking one, and persists for 1–2 additional weeks. As a consequence of the loss of the mucociliary stratum, a higher temporary predisposition to secondary sinusitis and bacterial pneumonia develops, but little permanent damage in the lung remains.[118]

The clinical manifestations of influenza in children are similar to those in adults, but higher fevers that may be accompanied by febrile convulsions can often be observed. Also otitis media, croup, pneumonia, myositis, and gastrointestinal manifestations are more frequent in children than in adults.[85]

Three distinct syndromes of severe pneumonia can follow influenza infection in children or adults, where the etiology may be viral (direct consequence of viral infection of lungs), secondary bacterial infection (impairment of local defense mechanisms), and mixed (combined viral–bacterial pneumonia).[119]

Primary influenza virus pneumonia may hit predominantly individuals at high risk for the complications of influenza virus infection (ie, the elderly or patients with cardiopulmonary disease),[120] therefore, usually occurs in patients with elevated left atrial pressure but has also been described in patients with chronic lung disease and in pregnant women. The typical case of primary viral pneumonia has a sudden start after the onset of influenza clinical illness and quickly progresses to a severe form of pneumonia with rapid respiration rate, tachycardia, cyanosis, high fever, and hypotension. The illness may progress rapidly to hypoxemia and death in 1–4 days. Survivors can develop diffuse interstitial fibrosis.[119]

Combined pneumonia, due to both viral and bacterial pathogens, is the most common complication in the course of influenza epidemics. Several bacteria can be involved but most often *Streptococcus pneumoniae*, *Staphylococcus aureus*, and *Haemophilus influenzae* are identified.[118] From the clinical point of view, this syndrome may be indistinguishable from primary viral pneumonia, except for the fact that the symptoms of pneumonia, generally, appear after the influenza symptoms. The diagnosis depends solely on the demonstration of bacteria in the sputum or in the pleural fluid.

Secondary bacterial pneumonia can establish on debilitated respiratory tissues. This is a major complication of influenza that contributes dramatically to the increased mortality. The bacterial pathogens most frequently involved are *S. pneumoniae*, *S. aureus*, and *H. influenzae*.[121] The pathogenesis of this disease is due to a correlation between viral and bacterial factors. In fact, influenza viruses favor the colonization of bacteria in the lung by acting on the epithelial cells of the upper respiratory tract, thus removing the physical barrier which opposes the lung bacterial colonization, and reducing the ciliary mucus clearance.

In this syndrome, individuals recovering from a typical influenza illness and a period of improvement of 2–3 days after the acute phase develop shaking chills, pleuritic chest pain, and an increase in cough productive of bloody or purulent sputum with the reappearance of fever and the onset of dyspnea. Often, influenza virus is no longer recoverable and the same bacterial pathogens as the earlier event are involved.

The myositis and rhabdomyolysis have been reported more frequently in children while viremia is highly unusual in influenza virus infections.[122] Instead, serious cardiovascular complications such as myocarditis, heart failure and acute myocardial infarction in patients previously compromised may follow influenza infections.

Reye syndrome is a rare complication observed mainly in children and is a rapidly progressive noninflammatory encephalopathy and fatty infiltration of the viscera, especially the liver, which results in severe hepatic dysfunction with elevated serum transaminase and ammonia levels. This syndrome is seen as a consequence of several viral infections, such as varicella, respiratory and gastrointestinal viral infections.

A wide spectrum of CNS disease can be observed during influenza A virus infections in humans,[123] ranging from irritability, drowsiness, boisterousness, and confusion to serious manifestations such as seizures, psychosis, delirium, and coma. Two specific CNS syndromes are associated with influenza infections: influenzal encephalopathy (may be fatal) and postinfluenzal encephalitis (extremely rare). In the latter, recovery is achieved in most cases.

6 GENESIS OF ANTIGENIC INFLUENZA VARIANTS AND PANDEMICS

The reason why influenza still remains a perennial clinical problem worldwide is its unparalleled ability to evade the host's immune system. The virus accomplishes this evasion by frequently altering the antigenicity of its surface proteins HA and NA. Antigenic differences in their NP and M1 allow influenza viruses to be classified as types A, B, or C influenza, while type A viruses further subtyping is based on the relatedness of their HA and NA glycoproteins. Currently, 16 HA (from H1 to H16) and 9 NA (from N1 to N9) subtypes have been identified in type A viruses.[85] Type A influenza viruses have been isolated from various animals, including humans, pigs, horses, sea mammals, and birds. Wild waterfowl, shorebirds, and gulls are the natural reservoirs of influenza A virus and can be infected with viruses harboring combinations of 16 different HA subtypes and nine different NA subtypes. Recently, two novel influenza A virus subtypes (H17N10 and H18N11) have been identified from two different bats, namely the little yellow-shouldered bat (*Sturnira lilium*) and the flat-faced fruit-eating bat (*Artibeus jamaicensis planirostris*).[124,125] Influenza viruses of these subtypes have not been isolated from any other animal species and it is actually unknown whether these viruses might be able to cross the species barrier. However, putative functional domains present on some viral proteins are conserved; the N-terminal domain of the H17N10 polymerase subunit PA which has preserved its functional and structural features. On the other hand, structural and functional analyses of the H18/N11 subtype indicate that sialic acids are not used for virus attachment and particle release, suggesting bat's derived influenza viruses employ a different mechanism of attachment and triggering of membrane fusion in order to gain access into the host cells. Taken together, these findings indicate that bats host a potentially important pool of influenza viruses that may likely represent a novel zoonotic reservoir for mammals.[126] The epidemiology of human influenza viruses is dictated by their continuous antigenic variation to escape the host immune response. Influenza viruses possess two different and complex mechanisms that allow them to reinfect humans and cause disease, namely antigenic drift (accumulation of mutations over time) and antigenic shift (rearrangement of viral RNA segments in cells infected with two or more different viruses, also known as "reassortment"). For human influenza A viruses, the evolutionary rates differ among the proteins, likely reflecting differences in the selective pressure of the host.[127] In fact, the HA and NA proteins evolve faster than the PB2, PB1, PA, NP, and M1 proteins, because residue substitutions in the antigenic domains of the surface glycoproteins HA and NA are likely to produce selective advantages by allowing mutated strains to evade preexisting immunity.[128] Antigenic drift, therefore, occurs as a result of point mutations in specific genes leading to minor, gradual, antigenic changes in the HA or NA proteins.[129] At the nucleotide level, the reported mutation rates range from $\sim 5 \times 10^{-4}$ to $\sim 8 \times 10^{-3}$ nucleotide substitutions per site per year.[130] Due to the relatively low fidelity of the RdRP, which also lacks a proofreading capacity (5′–3′ exonuclease activity), a common characteristic of all RNA polymerases, influenza constantly accumulates mutations within its genome during replication. It is important to note that these mutations may be silent or may alter the virulence and pathogenicity of the virus. For instance, if a highly pathogenic avian virus acquires the necessary mutations that facilitate its ability to efficiently enter and replicate in humans, then the virus can become a serious threat to humans. The same would apply for a virus of minor pathogenicity but is very efficacious in its human-to-human spread when it eventually accumulates enough mutations to become highly virulent. The selection of antigenic drift variants is a sequential event with the stepwise accumulation of mutations while an antigenic shift can occur all of a sudden, and at irregular and unpredictable intervals. This second mechanism of antigenic change is much less frequent since it only occurs when two different viruses coinfect the same individual. The genetic mutational mechanism is an exchange of whole segments between the two viruses. This is only possible because the influenza virus genome consists of eight separated RNA segments, therefore, coinfection of one host cell with two different viruses can result in progeny viruses containing gene segments of both parental viruses. In this way, by reassortment of genomic segments, a new virus is created, resulting in an unpredictable pathogenicity of the new virus, which may possess a surface antigenic pattern completely novel to the population. In fact, considering that this reassortment event usually occurs between viruses coming from different host-species (eg, a human virus and an avian virus), the new virus may immediately present HA and/or NA antigenic determinants from the avian strain against which the human population lacks any significant immunity. Once the pandemic strain is created (generally by antigenic shift), the new virus can further change its virulence determinants as it continues to replicate (generally by antigenic drift).

Some influenza virus strains are largely asymptomatic in chickens [and are considered low pathogenic avian influenza (LPAI) viruses], while other strains cause severe disease in chickens that is often fatal within 48 h [and are considered highly pathogenic avian influenza (HPAI) viruses]. Outbreaks of HPAI viruses can cause disaster for the poultry industry.[131] *Anseriformes* (mainly ducks, geese, and swans) and *Charadriiformes* (mainly gulls, terns, and waders) are considered the natural host reservoirs of LPAI viruses.[127] In wild birds, the LPAI viruses predominantly infect epithelial cells of the intestinal tract[132,133] and are subsequently excreted in the faeces. However, infection of wild birds with LPAI viruses is typically subclinical and occurs in the absence of obvious lesions.[134,135] LPAI viruses can cross the species barrier and infect wild marine mammals as well as domesticated mammals and birds. Swine (*Sus scrofa domesticus*) can be easily infected by avian influenza viruses. Disease manifestations can range from an acute respiratory tract disease to an inapparent infection.[136] Pigs have been traditionally perceived to act as a "mixing vessel" to facilitate the generation of novel reassortant influenza viruses. This is because swine respiratory epithelial cells express both $\alpha2,3$- and $\alpha2,6$-linked sialic acids[136] and the simultaneous presence of both SAα2,6Gal and SAα2,3Gal would allow swine to be potentially infected by both avian and human viruses. However, more recent data showed that SAα2,3Gal receptors on swine tracheal cells are in reality not that abundant as previously considered,[137] but the importance of swine in the reassortment mechanism is to be found in the presence of large numbers of pigs in close proximity to other animal species, especially domesticated water birds,[138] increasing the risk of an interspecies transmission event, at least in some world areas. However, it was also reported that the crucial determinant for influenza tropism is the structure of the underlying glycocalyx, not exclusively the terminal SA linkage.[139] Avian viruses prefer SA in a cone-like topology that is present in both SAα2,3Gal and SAα2,6Gal with short underlying glycans and allows HA to contact Neu5Ac and galactose sugars in a trisaccharide motif. Human viruses are reported to prefer an umbrella-like glycan topology, which is unique to SAα2,6Gal with long underlying glycans. Therefore, the $\alpha2,6$ alone is not sufficient for human transmission and the SA topology is of importance to conditionate influenza tropism, not just the SA linkage present.[139] A further event that may lead to pandemics is represented by the fact that terrestrial poultry can easily become infected with numerous different LPAI viruses (eg. H5, H6, H7, H9 viruses) which become endemic in the population.[140,141] However, the most relevant consequence of poultry infections with LPAI viruses is the high possibility of evolution in these species from LPAI viruses into HPAI viruses by the usual antigenic drift mechanism of point mutations. As earlier described, HA undergoes a posttranslational cleavage into HA1 and HA2 subunits by host proteases, with the generation of a fusogenic domain at the amino terminus of HA2 that induces fusion between the viral envelope and the endosomal membrane.[142] Proteolytic activation is fundamental for viral infectivity and organ dissemination,[143] and the HA glycoprotein has a key role in influenza virus pathogenicity.[144] HA0 of seasonal human influenza A viruses, as well as LPAI viruses have the consensus cleavage site motif XT/X-R and are selectively processed by a limited number of trypsin-type processing proteases, while in HPAI viruses, endoproteolytic processing of the HA0 occurs through ubiquitous cellular processing proteases, which recognize a multibasic consensus cleavage site motifs, such as, R-X-K/R-R and K-X-K/R-R.[145] Influenza viruses of subtypes H5 and H7 may become highly pathogenic after introduction into poultry and cause outbreaks of HPAI. The switch from an LPAI phenotype to the HPAI phenotype of these H5 and H7 influenza A viruses is achieved by the introduction of basic amino acid residues into the HA0 cleavage site by substitution or insertion, resulting in the so-called multibasic cleavage site (MBCS), which facilitates systemic virus replication.[146–148] The main processing proteases reported for human influenza viruses in animals and humans that recognize single basic motifs are pancreatic trypsin,[143] plasmin from calf and chicken,[149] blood clotting factor Xa from chick embryo,[150] tryptase Clara from rat lungs,[151] mini-plasmin from rat lungs,[152] ectopic anionic trypsin from rat lungs,[153] porcine mast cell tryptase,[154] tryptase TC30 from porcine lungs[155] and transmembrane protease serine (TMPRSS) 2, and type II membrane protein human airway trypsin-like protease (HAT).[156] In addition to the host cellular proteases, microbial proteases also proteolytically activate influenza virus HA0 in bacterial infection of the airways and may play a role in the spread of the virus.[157] The processing proteases for the single basic motif are predominantly distributed in the respiratory and intestinal tract, and most of the epidemic human influenza viruses known to date are pneumotropic in which the virus proliferates in these organs. On the other hand, the multiple basic residues motifs are readily cleaved by ubiquitously present intracellular processing proteases, such as furin and proprotein convertases (PCs)5/6.[158–160] In addition, type II membrane serine proteases (TTSP) of mosaic serine protease large form (MSPL) and its splice variant transmembrane protease serine 13 (TMPRSS13) have recently been identified.[161] The effect of the experimental introduction of an MBCS into a primary LPAI H6N1 virus, A/Mallard/Sweden/81/2002, has recently been described and such introduction resulted in trypsin-independent replication in vitro and enhanced pathogenesis in a chicken model.[162]

Influenza A viruses can, therefore, cause seasonal diseases as well pandemic episodes which are mainly characterized by the fact that, due to massive genetic variations, novel strains to which humans have little or no immunity, can be generated.

All of the known subtypes of influenza A viruses can infect birds, except subtypes H17N10 and H18N11, which have only been found in bats and only two influenza A virus subtypes (ie, H1N1 and H3N2) are currently in general circulation among humans. There are also some subtypes that infect other animals, such as, H7N7 which can cause flu disease in horses, and H3N8 virus infection that can also be found in dogs. Historical data suggests that there may have been at least 14 pandemics over the past 500 years (1509–2009), which is approximately one pandemic every 36 years. But our understanding of pandemic influenza viruses at the viral level is quite limited since pandemic viruses have only been isolated from 1957 onward, and the 1918 pandemic virus was reconstructed using an "archaeovirological" approach.[163] We do not yet know the subtypes or genetic makeup of the pandemics before 1918, therefore we will consider as "pandemic era" the events from 1918 onward. In the last 100 years we have experienced five pandemics which happened in the years: (1) 1918–1919; (2) 1957–1958; (3) 1968; (4) 1977–1978; and (5) 2009–2010.

The most important data on these pandemics are summarized in Table 1.2.

Table 1.2 Details on the Four Major Pandemics of the Last 100 Years

Years	1918–19	1957–58	1968–69	2009–10
Name	Spanish flu	Asian flu	Hong Kong flu	Swine flu
Area of emergence	Unclear	Southern China	Southern China	Mexico – Southern USA
Subtype	H1N1	H2N2	H3N2	H1N1
Estimated mortality worldwide	20–50 million	1–4 million	1–2 million	125–570 thousands
Estimated fatality rate	2–3%	0.2%	0.2%	0.03%
Age groups most affected	Young adults	Children	Across all age groups	Young people 5–40 years of age
Details	Since no pre-1918 influenza viruses are available, the origin and eventual shift from an animal host remain unknown. The high mortality was the result of bacterial pneumonia	This virus emerged as a linear descendent of the 1918 H1N1 with aquisition of three novel genes from an unknown avian virus (HA, NA, and PB1)	Reassortment event with an avian influenza virus with novel H3 and PB1 gene segments. The other six genes were retained from the 1957 H2N2	HA, polymerase genes, NP and NS derived from a triple-reassortant virus circulating in North American swine, while the NA and M gene derived from the Eurasian avian-like swine H1N1 lineage. The triple-reassortant itself comprised genes derived from avian (PB2 and PA), human H3N2 (PB1) and classical swine (HA, NP, and NS) lineages

The 1918–1919 Pandemic Virus: The Spanish influenza pandemic first struck in 1918 as an unusual respiratory disease that was associated with a disproportionate increase in deaths among young adults. The origin of the Spanish influenza is not clear, but reconstructions of the viral genomes from the tissues of several victims demonstrated that the causative agent was an avian-descended H1N1 virus. Epidemiological features of the pandemic were also unprecedented, including its appearance in up to three waves within the first year and the extremely high mortality registered. The pandemic killed an estimated 50 million people.[164]

The 1957–1958 Pandemic Virus: The "Asian flu" originated in the Yunan Province of China in the spring of 1957, spreading rapidly to South-East Asia and Japan, and subsequently to Australia, Indonesia and India, and after to Europe, Africa, North and South America, and the Caribbean. In just 6 months, the pandemic had covered the globe. A second wave occurred during the autumn of 1957. In addition, if the origin of the virus has been connected to the 1918 H1N1 pandemic virus, reassortment events took place to generate a new virus especially for its external antigenic makeup. The gene segments encoding HA and NA were replaced by an avian-like H2 subtype HA and an N2 subtype NA,[165] with the other five gene segments retained from the 1918-derived H1N1 lineage. Altogether, it affected about 40–50% of the global population, with 25–30% experiencing clinical disease.

The 1968 Pandemic Virus: This pandemic also originated in China, in 1968, spreading to Hong Kong and thereafter to India, Australia, Europe and USA. The 1968 H3N2, named the "Hong Kong", pandemic, was caused by a reassortment event between a circulating human H2N2 virus and an avian strain. The new configuration was derived by acquiring novel HA (H3 subtype) and PB1 gene segments.[166] Compared to the earlier pandemics of the century, the Hong Kong flu was relatively mild (reduction of the duration and severity of the clinical illness), possibly as a result of immunity in the population against the N2 subtype NA which the new pandemic H3N2 virus shared with the circulating H2N2 virus.

The 1977–1978 Pandemic Virus: Strictly speaking, there was a fourth pandemic, the Russian Flu, in the 20th century. It was sparked off by an H1N1 strain which appeared in 1977. This pandemic was considered a benign pandemic, or not a "true" pandemic, and primarily involved subjects born after 1950 (so people aged under 25 years), since the older population already presented a protective immunity as a consequence of prior experience with the H1N1 strain. Molecular characterization of this virus showed that both the HA and NA antigens were remarkably similar to those of the 1950s.[167] It is believed to not be probable that an influenza virus could have been maintained in nature for 20 years without accumulating mutations, therefore it has been suggested that the 1977 epidemic resulted from the accidental release of a laboratory strain from the 1950s.

The 2009–2010 Pandemic Virus: The 2009 H1N1 pandemic virus was derived by reassortment between two preexisting swine influenza viruses, a North American swine H1N2 "triple-reassortant" lineage virus and a Eurasian H1N1 swine lineage virus (Fig. 1.4). It is not currently known whether the novel virus emerged first in humans or swine.[168,169] The 2009 pandemic virus is also connected to the 1918 pandemic virus.[170] In fact, it derives from multiple reassortment events ultimately ascribed to the "classical" swine H1N1 lineage that circulated enzootically in swine in North America since 1918.

The novel H1N1 virus was first detected in a widespread outbreak in Mexico in March–April 2009,[171] but may have been circulating in people as early as late 2008.[172,173] While severe pneumonia has been described, especially associated with the initial Mexican outbreak,[98] most cases in the United States and in other countries have been self-limited, and appeared clinically similar to seasonal influenza.[174] Human infection with H1N1 have generally resulted in low mortality, although certain subgroups (including pregnant women, chronic medical conditions sufferers, immunosuppressed people) had significantly higher risk of severe disease. The 2009 H1N1 pandemic derived its NA and M gene segments from the European avian-like H1N1 lineage and its remaining six gene segments (PB2, PB1, PA, HA, NP, and NS) from the North American swine H1N2 "triple" reassortant lineage. The HA, NP, and NS gene segments of this lineage are derived from the classical swine H1N1 (1918 origin) lineage.[173]

Notwithstanding the evidence that the combination of one of the HA subtypes with one of the NA subtypes allows for a wide genetic diversity which could lead to a pandemic event, within the brief period of modern molecular virology, of the 16 HA subtypes known to exist, pandemics have only been caused by viruses of the H1, H2, and H3 subtypes. All of the other HA subtypes have, up to now, been demonstrated to produce novel influenza strains with a very low or nonexistent capacity for human-to-human transmission. Nevertheless, the possibility of further adaptation of the virus in humans and

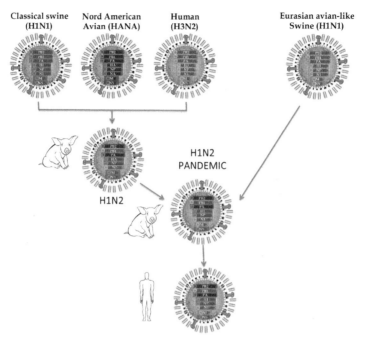

FIGURE 1.4 Antigenic Shift of the Generation of the Pandemic Virus H1N1 of 2009

the possibility to acquire a sustained human-to-human transmission must be taken into consideration assuming that the nature of the next pandemic virus cannot be predicted, but that it will arise from one of the 16 known HA subtypes in avian or mammalian species. In recent years, several novel reassorted influenza viruses (eg, H5N1, H7N9, H9N2, H10N8) have trespassed the host-species barrier and are under surveillance by the scientific community and public health systems. It is still unclear whether these viruses can actually cause pandemics or just isolated episodes. The ecological conditions in many Asian countries are such that new viruses with pandemic potential may arise relatively easily. These conditions include the year-round circulation of influenza viruses, along with the density of human, swine, and domestic poultry populations, often living in close proximity. These conditions may facilitate genetic reassortment or direct transfer of avian viruses to humans.

The prominent subtypes of avian influenza A viruses known to infect both birds and humans are described below.

Influenza A(H5N1) was first detected in 1996 in geese in China and reported in humans in 1997 during a poultry outbreak in Hong Kong.[175] It has since been detected in poultry and wild birds in more than 50 countries in Africa, Asia, Europe, and the Middle East. Since their reemergence in 2003,[112] highly pathogenic avian influenza (HPAI) A(H5N1) viruses of the A/goose/Guangdong/1/96 hemagglutinin (HA) lineage have become enzootic in six countries (Bangladesh, China, Egypt, India, Indonesia, and Vietnam).[176,177] More than 700 human infections with Asian HPAI H5N1 viruses have been reported since 2003 and Indonesia, Vietnam, and Egypt have reported the highest number of human HPAI Asian H5N1 cases to date. The first report of a human infection with Asian H5N1 in the Americas was in Canada in 2014 and occurred in a traveler who had recently visited China.[178] Most human infections with HPAI Asian H5N1 have occurred after prolonged and close contact with sick or dead poultry or wild infected birds. Human-to-human spread is rare and approximately 60% of the cases are fatal. One human infection in China was caused by an A(H5N6) virus in 2014.[179]

The influenza A(H7N9) virus is one subgroup among the larger group of H7 viruses, which normally circulate among birds. This particular A(H7N9) virus was not detected in humans until it was found in March 2013 in China. The patients were two men aged 87 and 27 years and a woman aged 35 years. All three patients already presented unfavorable medical conditions and died.[180] Two waves were identified, the first from Feb. to May 2013, with 133 cases, and the second from Oct. 2013 to Feb. 2014. By the end of May 2014, a total of 439 cases of H7N9 infection had been reported to the WHO, with a case fatality rate of 30%.[181] Illness in humans may include conjunctivitis and/or upper respiratory tract symptoms. Live poultry markets were indicated to likely be the principal source for human infections because most of the infected patients were reported to have had recent contacts with live poultry.[182] The novel H7N9 virus was of avian origin with HA derived from the duck H7N3 subtype, the NA gene from a migratory bird H7N9 virus in Korea and the six internal genes from poultry H9N2 viruses.[180] It seems that this novel reassortant could be transmitted from poultry to humans more easily than H5N1 and a recent epidemiological study showed that a considerable proportion of people involved in the poultry industry have been exposed to H7, suggesting a high diffusion of mild or asymptomatic infections.[183]

Influenza A H9 viruses have been identified worldwide in wild birds and poultry and they are all LPAI viruses. Rare, sporadic H9N2 virus infections of humans have been reported to cause generally mild upper respiratory tract illness. The first cases of infection with an H9N2 influenza strain were reported in China and affected two children.[184] The symptoms of infection with H9N2 are mild in comparison with those of H5N1 infection.[185] At a molecular level, it has been described that the internal genes of H9N2 are similar to those of the H5N1 that caused the infections in 1997,[186] therefore, H9N2 has been posed under control of health authorities as a potential pandemic strain.

Influenza A (H10N8) viruses have circulated mainly in wild birds since their first detection in quails in Italy in 1965.[187] The first H10N8 avian influenza virus was isolated from a duck in Guangdong in the beginning of 2012.[188] This novel avian influenza strain appeared for the first time in humans by the end of 2013. So far, three cases have been recorded, all in China, with two fatal outcomes. The virus isolated from the human cases was genetically distinguishable from bird H10N8 viruses, with long genetic distances between their HA and NA genes, and also contained the H9N2-derived internal genes.[189] It is quite difficult to assess the real pandemic potential posed by H10N8, in fact, despite the severe infections caused in humans, leading to death in two out of three cases, the reported human cases are too sporadic to allow any sensible epidemiological study. Furthermore, no other cases have been reported since February 2014.

Finally, we should note the appearance in Taiwan of influenza A H6N1 in 2013 when this novel virus was identified in a 20 year-old female patient.[190] As of September 2014, there has been no documentation of person-to-person transmission of the virus or further cases. Molecular characterization of H6N1 revealed that it is a typical avian influenza virus of low pathogenicity likely derived from reassortment events from different H6N1 lineages circulating in chickens in Taiwan, which might not replicate and propagate well in the upper airway of mammals.[191] On the other hand, a single HA mutation of the human isolate might have increased its mammalian receptor binding ability and, hence, its pathogenicity in humans. Based on these data, and considering the cocirculation and potential reassortment with several other influenza subtypes, this H6N1 virus merits consideration.

7 INFLUENZA TREATMENT

The most adopted strategy to fight influenza infections is through vaccination, however, vaccines need to be reformulated each year due to the genetic variability of the virus and they are not always completely protective. Moreover, in case of pandemics, it is hardly impossible to predict the newly emerging strain and, therefore, containment of pandemics by vaccination is restrained by the time-limiting factor (the period going from the understanding of the pandemic risk to the worldwide diffusion of the infection). The available drugs are somewhat limited in their use by the rapidly increasing drug resistances, therefore, additional antiinfluenza therapeutics are urgently needed. Thus, the rest of the chapter will discuss the current and past treatments that have been available for influenza in recent years together with some of the most interesting compounds under development. The first identified inhibitor against influenza virus was amantadine[192] (Fig. 1.5a). Amantadine is an amine derivative of adamantane consisting of an adamantane backbone that has an amino group substituted at one of the four methyne positions. Rimantadine is a closely related derivative of adamantane with similar biological properties. Both compounds are M2 ion channel inhibitors and were licensed as first generation antiinfluenza drugs and are still commercially available as Symmetrel (Endo Pharmaceuticals) and Flumadine (Forest Pharmaceuticals), respectively. The M2 transmembrane protein serves to selectively transport protons inside the viral particle for directing the uncoating of the particle and the release of RNPs into the cell cytosol. Both drugs exert an action on M2 ion channels by intercalating into the interior of the channel.[193] In fact, they both bind to the N-terminal channel lumen disabling its pump activity. Their efficacy is limited to influenza A only, since influenza B M2 channel pore lumen contains polar, and not hydrophobic, amino acids.[74] Furthermore, rapid emergence of drug-resistant virus strains (almost 100% for the circulating seasonal H3N2 and 2009 pandemic flu) restricts the use of both drugs

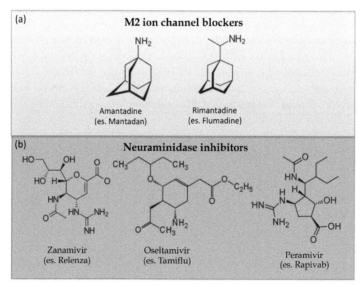

FIGURE 1.5 Compounds to Fight Influenza Virus

(a) M2 ion channel inhibitors; (b) neuraminidase inhibitors.

for prevention and treatment in favor of new-generation influenza drugs, the NA inhibitors (Fig. 1.5b). The second generation of antiinfluenza drugs act in the late stage of infection by inhibiting the release of virus particles from infected cells. Influenza virus NA is one of the two major surface glycoprotein antigens embedded in the viral envelope. Its main activity is to cleave off the terminal SA from the host cell receptors to release of progeny viruses from the membranes of infected cells, to which the virus would remain linked via the interaction with HA. Blocking the NA sialidase activity ends up in trapping the newly formed viruses on the cell surface, impairing their ability to infect neighboring cells and, therefore, impeding the spread of the infection in the organism.[194,195] NA inhibitors show a broad-spectrum of activity since both influenza A and B, as well as different NA subtypes keep the structural characteristics highly conserved.[196] NA inhibitors are structural analogs of sialic acid and were first synthesized in the 1970s.[197] Two NA inhibitors have been approved by regulatory authorities in the United States and Europe, Zanamivir (inhaled drug, 10 mg/dose; trademark Relenza, GlaxoSmith-Kline) and Oseltamivir (oral drug, 75 mg/dose; trademark Tamiflu, Roche).[198,199] Both drugs require twice-daily administration for treatment. Furthermore, Peramivir and Laninamivir have been approved by regulatory authorities in some parts of Asia.[200] Peramivir is a cyclopentane compound that potently and selectively inhibits influenza virus NA,[201] and an intravenous formulation has been approved for influenza treatment (single intravenous drip infusion, 300 mg/dose; Rapiacta) in Japan and as Peramiflu in South Korea. The US Food and Drug Administration (FDA) has only recently approved intravenous use of peramivir (Rapivab, manufactured by BioCryst Pharmaceuticals), as a drug available for adult influenza patients who have trouble taking an oral or inhaled antiviral, while in Europe ending of phase III clinical trials are approaching. The other compound, Laninamivir (R-125489), is structurally similar to zanamivir and its octanoyl prodrug, laninamivir octanoate (CS-8958) is used for influenza treatment in Japan where it was approved and is commercially available as Inavir (Daiichi Sankyo Co., Ltd, Tokyo).[202,203] In contrast to the currently available drugs, which require multiple dosing, for example, twice-daily for 5 days, it was confirmed that the single inhalation of CS-8958 is sufficient to treat influenza in humans.[204] Although the target site for the NA inhibitors is highly conserved, again a number of resistant mutations arise as a consequence of drug-induced selective pressure.[205] The mutations include substitutions in both NA and HA that reduce either the affinity of NA for the inhibitor or the affinity of HA for the sialylated glycoproteins.

Among the compounds under development, a nucleoside analog is currently in phase III clinical trials by the FDA, favipiravir. Favipiravir (T-705; 6-fluoro-3-hydroxy-2-pyrazinecar-boxamide) is a pyrazine derivative which was identified by Furuta el al. in 2002.[206] T-705 has been shown to inhibit influenza A, B, and C viruses in vitro and in an in vivo model.[206] T-705 is converted by cellular kinases to its active form, ribofuranosyl triphosphate, and the mechanism of action is through direct inhibition of viral replication and transcription, which represent an exclusive feature among antiinfluenza drugs. A combination of favipiravir with NA inhibitors has demonstrated a synergistic activity; therefore, broadening the therapeutic options for the management of infections by highly pathogenic avian strains and for severely ill patients. Since human cells do not present RdRP domains but are conserved among different families of RNA viruses, a favipivar mechanism targeting RNA viral polymerases makes this molecule an attractive drug candidate for other RNA viruses including agents belonging to the navirus, bunyavirus, and flavivirus families.[207]

Further compounds in actual development can be grouped for their mechanism of action and can be divided considering the phase of the viral cycle they interrupt. One of the viral targets of considerable interest is the viral RNA-dependent RNA polymerase (RdRP), which as earlier described, is a heterotrimer composed of subunits PB1, PB2, and PA. The three polymerase subunits interact with each other, in particular the N-terminus of PB1 interacts with the C-terminus of PA,[208] while the C-terminus of PB1 binds the N-terminus of PB2;[209] in addition, a weak transient interaction has been proposed for PA and PB2.[210] We already described favipiravir, the best characterized antiinfluenza nucleoside analog, but in view of the fact that interactions between the three polymerase subunits are essential for function, and therefore conserved along different viral strains, a synthetic peptide corresponding to residues 1–37 of PB2 was shown to inhibit the PB1/PB2 interaction in vitro, but also a peptide derived from amino acids 731–757 of PB1 can disrupt the interaction between the subunits.[211–213] A further antiviral target is the HA and the attachment or fusion mechanisms in the early stages of the viral infections. Several small molecules and peptides have been developed and of note among them are EB peptide and its derivatives, FluPep and arbidol (ARB; 1-metyl-2-phenyl-thiomethyl-3-carbotoxy-4-dimetylaminomethyl-5-hydroxy-6-bromoindolehydrochloridemonohydrate).

EB is a 20 amino acid peptide derived from signal sequence of fibroblast growth factor 4 showing a broad-spectrum activity against human, swine, and avian influenza A H1N1, H2N2, H3N2, H5N1, H5N9, and H7N3 strains and influenza B viruses. Several data have demonstrated the ability of the peptide to inhibit viral attachment.[214,215] FluPep is a mix of predominantly hydrophobic α-helical peptides capable of interaction with HA blocking the viral fusion. These peptides are derived from Tkip peptide, which is a mimetic for the suppressor of cytokine signaling protein, known to be active in modulating inflammatory cytokine responses and known as an effective antiviral drug against Poxviruses.[216] A variety of influenza subtypes were inhibited by FluPep in nanomolar concentrations in cell culture.[217] Finally, the small molecule arbidol exhibits a broad-spectrum of activity not only against influenza viruses A and B but also against other viruses such as RSV, parainfluenza virus, coxsackie virus, rhinovirus, hepatitis B virus (HBv), and HCV.[218,219] This drug has been approved in Russia and in China for treatment and prophylaxis of influenza viruses. Studies with viruses resistant to ARB bearing mutations which map in the HA2 subunit have confirmed that ARB interacts with HA and acts by stabilizing its structure, thus preventing the low pH-induced fusogenic change of HA2.[220] Recently, biochemical studies showed that ARB interacts both with cell membrane phospholipids and with aromatic residues of viral glycoproteins on the surface of enveloped viruses.[221] This mechanism of action of ARB can prevent the fusogenic change in viral glycoproteins required for membrane fusion and could explain the broad-spectrum of antiviral activity of this compound.

8 CONCLUSIONS

Influenza A viruses are zoonotic pathogens that continuously circulate and change in several animal hosts, including birds, pigs, horses, and humans. The emergence of novel virus strains that are capable of causing human epidemics or pandemics needs to be considered a serious possibility; therefore, surveillance and characterization (by high-throughput whole-genome sequencing) of naturally occurring influenza viruses is an imperative task of the world health system to establish drift variants and genetic changes that might be of epidemiological relevance. A great progress has been achieved in recent years in our understanding of the host tropism and virulence of influenza viruses which could lead to additional and enhanced antiviral interventions. Drugs directed against diverse viral or host targets and against distinct steps of influenza virus replication cycle would represent a terrific weapon to challenge infections by several strategies also able to minimize the development of resistant viruses. Therefore, influenza pathogenesis remains an indispensable field of study to elucidate the molecular requirements to understand both viral and host factors that play a crucial part in disease severity and outcome.

REFERENCES

1. Potter CW. A history of influenza. *J Appl Microbiol* 2001;**91**:572–9.
2. Shope RE, Lewis P. Swine influenza : I. Experimental transmission and pathology. *J Exp Med* 1931;**54**:349–59.
3. Smith W, Andrews CH, Laidlaw PP. A virus obtained from influenza patients. *Lancet* 1933;**225**:66–8.
4. Burnet FM. Influenza virus on the developing egg I. changes associated with the development of an egg-passage strain of virus. *Br J Exper Pathol* 1936;**17**:282–95.
5. Krossoy B, Hordvik I, Nilsen F, Nylund A, Endresen C. The putative polymerase sequence of infectious salmon anemia virus suggests a new genus within the Orthomyxoviridae. *J Virol* 1999;**73**:2136–42.
6. Kosoy OI, Lambert AJ, Hawkinson DJ, et al. Novel thogotovirus associated with febrile illness and death, United States 2014. *Emerg Infect Dis* 2015;**21**:760–4.
7. Allison AB, Ballard JR, Tesh RB, et al. Cyclic avian mass mortality in the northeastern United States is associated with a novel orthomyxovirus. *J Virol* 2015;**89**:1389–403.
8. Osterhaus AD, Rimmelzwaan GF, Martina BE, Bestebroer TM, Fouchier RA. Influenza B virus in seals. *Science* 2000;**288**:1051–3.
9. Shaw ML, Palese P. Orthomyxoviridae. In: Knipe DM, Howley PM, editors. *Fields Virology.* sixth ed. Philadelphia: Lippincott Williams & Wilkins; 2013. p. 1151–85.
10. Resa-Infante P, Jorba N, Coloma R, Ortin J. The influenza virus RNA synthesis machine: advances in its structure and function. *RNA Biol* 2011;**8**:207–15.
11. Gerber M, Isel C, Moules V, Marquet R. Selective packaging of the influenza A genome and consequences for genetic reassortment. *Trends Microbiol* 2014;**22**:446 55.
12. Perez JT, Varble A, Sachidanandam R, et al. Influenza A virus-generated small RNAs regulate the switch from transcription to replication. *PNAS* 2010;**107**:11525–30.
13. Perez JT, Zlatev I, Aggarwal S, et al. A small-RNA enhancer of viral polymerase activity. *J Virol* 2012;**86**:13475–85.
14. Rodriguez-Frandsen A, Alfonso R, Nieto A. Influenza virus polymerase: functions on host range, inhibition of cellular response to infection and pathogenicity. *Virus Res* 2015;**209**:23–38.
15. Area E, Martin-Benito J, Gastaminza P, et al. 3D structure of the influenza virus polymerase complex: localization of subunit domains. *PNAS* 2004;**101**:308–13.
16. Kobayashi M, Toyoda T, Ishihama A. Influenza virus PB1 protein is the minimal and essential subunit of RNA polymerase. *Arch Virol* 1996;**141**:525–39.
17. Blaas D, Patzelt E, Kuechler E. Identification of the cap binding protein of influenza virus. *Nucleic Acids Res* 1982;**10**:4803–12.
18. Braam J, Ulmanen I, Krug RM. Molecular model of a eucaryotic transcription complex: functions and movements of influenza P proteins during capped RNA-primed transcription. *Cell* 1983;**34**:609–18.
19. Dias A, Bouvier D, Crepin T, et al. The cap-snatching endonuclease of influenza virus polymerase resides in the PA subunit. *Nature* 2009;**458**:914–8.
20. Yuan P, Bartlam M, Lou Z, et al. Crystal structure of an avian influenza polymerase PA(N) reveals an endonuclease active site. *Nature* 2009;**458**:909–13.
21. Skehel JJ, Waterfield MD. Studies on the primary structure of the influenza virus hemagglutinin. *PNAS* 1975;**72**:93–7.
22. Deom CM, Schulze IT. Oligosaccharide composition of an influenza virus hemagglutinin with host-determined binding properties. *J Biol Chem* 1985;**260**:14771–4.

23. Schulze IT. Effects of glycosylation on the properties and functions of influenza virus hemagglutinin. *J Infect Dis* 1997;**176**(Suppl. 1):S24–8.
24. Veit M, Serebryakova MV, Kordyukova LV. Palmitoylation of influenza virus proteins. *Biochem Soc Trans* 2013;**41**:50–5.
25. Skehel JJ, Wiley DC. Receptor binding and membrane fusion in virus entry: the influenza hemagglutinin. *Annu Rev Biochem* 2000;**69**:531–69.
26. Ritchey MB, Palese P, Kilbourne ED. RNAs of influenza A, B, and C viruses. *J Viroly* 1976;**18**:738–44.
27. Ortega J, Martin-Benito J, Zurcher T, Valpuesta JM, Carrascosa JL, Ortin J. Ultrastructural and functional analyses of recombinant influenza virus ribonucleoproteins suggest dimerization of nucleoprotein during virus amplification. *J Virol* 2000;**74**:156–63.
28. Baudin F, Bach C, Cusack S, Ruigrok RW. Structure of influenza virus RNP. I. Influenza virus nucleoprotein melts secondary structure in panhandle RNA and exposes the bases to the solvent. *EMBO J* 1994;**13**:3158–65.
29. Martin K, Helenius A. Transport of incoming influenza virus nucleocapsids into the nucleus. *J Virol* 1991;**65**:232–44.
30. O'Neill RE, Jaskunas R, Blobel G, Palese P, Moroianu J. Nuclear import of influenza virus RNA can be mediated by viral nucleoprotein and transport factors required for protein import. *J Biol Chem* 1995;**270**:22701–4.
31. Paine PL, Moore LC, Horowitz SB. Nuclear envelope permeability. *Nature* 1975;**254**:109–14.
32. Husain M, Gupta CM. Interactions of viral matrix protein and nucleoprotein with the host cell cytoskeletal actin in influenza viral infection. *Curr Sci* 1997;**73**:40–7.
33. Digard P, Elton D, Bishop K, Medcalf E, Weeds A, Pope B. Modulation of nuclear localization of the influenza virus nucleoprotein through interaction with actin filaments. *J Virol* 1999;**73**:2222–31.
34. Gong J, Xu W, Zhang J. Structure and functions of influenza virus neuraminidase. *Curr Med Chem* 2007;**14**:113–22.
35. Vasin AV, Temkina OA, Egorov VV, Klotchenko SA, Plotnikova MA, Kiselev OI. Molecular mechanisms enhancing the proteome of influenza A viruses: an overview of recently discovered proteins. *Virus Res* 2014;**185**:53–63.
36. Chen W, Calvo PA, Malide D, et al. A novel influenza A virus mitochondrial protein that induces cell death. *Nature Med* 2001;**7**:1306–12.
37. Wise HM, Foeglein A, Sun J, et al. A complicated message: identification of a novel PB1-related protein translated from influenza A virus segment 2 mRNA. *J Virol* 2009;**83**:8021–31.
38. Yamayoshi S, Watanabe M, Goto H, Kawaoka Y. Identification of A novel viral protein expressed from the PB2 segment of influenza A virus. *J Virol* 2016;**90**:444–56.
39. Jagger BW, Wise HM, Kash JC, et al. An overlapping protein-coding region in influenza A virus segment 3 modulates the host response. *Science* 2012;**337**:199–204.
40. Schrauwen EJ, de Graaf M, Herfst S, Rimmelzwaan GF, Osterhaus AD, Fouchier RA. Determinants of virulence of influenza A virus. *Eur J Clin Microbiol Infect Dis* 2014;**33**:479–90.
41. Muramoto Y, Noda T, Kawakami E, Akkina R, Kawaoka Y. Identification of novel influenza A virus proteins translated from PA mRNA. *J Virol* 2013;**87**:2455–62.
42. Wise HM, Hutchinson EC, Jagger BW, et al. Identification of a novel splice variant form of the influenza A virus M2 ion channel with an antigenically distinct ectodomain. *PLoS Pathog* 2012;**8** e1002998.
43. Selman M, Dankar SK, Forbes NE, Jia JJ, Brown EG. Adaptive mutation in influenza A virus non-structural gene is linked to host switching and induces a novel protein by alternative splicing. *Emerg Microbes Infect* 2012;**1** e42.
44. Matrosovich MN, Gambaryan AS, Teneberg S, et al. Avian influenza A viruses differ from human viruses by recognition of sialyloligosaccharides and gangliosides and by a higher conservation of the HA receptor-binding site. *Virology* 1997;**233**:224–34.

45. Wilks S, de Graaf M, Smith DJ, Burke DF. A review of influenza haemagglutinin receptor binding as it relates to pandemic properties. *Vaccine* 2012;**30**:4369–76.

46. Stevens J, Blixt O, Glaser L, et al. Glycan microarray analysis of the hemagglutinins from modern and pandemic influenza viruses reveals different receptor specificities. *J Mol Biol* 2006;**355**:1143–55.

47. Wilson IA, Skehel JJ, Wiley DC. Structure of the haemagglutinin membrane glycoprotein of influenza virus at 3 A resolution. *Nature* 1981;**289**:366–73.

48. Sauter NK, Bednarski MD, Wurzburg BA, et al. Hemagglutinins from two influenza virus variants bind to sialic acid derivatives with millimolar dissociation constants: a 500-MHz proton nuclear magnetic resonance study. *Biochemistry* 1989;**28**:8388–96.

49. Patterson S, Oxford JS, Dourmashkin RR. Studies on the mechanism of influenza virus entry into cells. *J Gen Virol* 1979;**43**:223–9.

50. Matlin KS, Reggio H, Helenius A, Simons K. Infectious entry pathway of influenza virus in a canine kidney cell line. *J Cell Biol* 1981;**91**:601–13.

51. Yoshimura A, Kuroda K, Kawasaki K, Yamashina S, Maeda T, Ohnishi S. Infectious cell entry mechanism of influenza virus. *J Virol* 1982;**43**:284–93.

52. Lakadamyali M, Rust MJ, Zhuang X. Endocytosis of influenza viruses. *Microbes and infection/Institut Pasteur* 2004;**6**:929–36.

53. Grove J, Marsh M. The cell biology of receptor-mediated virus entry. *J Cell Biol* 2011;**195**:1071–82.

54. Galloway CJ, Dean GE, Marsh M, Rudnick G, Mellman I. Acidification of macrophage and fibroblast endocytic vesicles in vitro. *PNAS* 1983;**80**:3334–8.

55. Perez L, Carrasco L. Involvement of the vacuolar H(+)-ATPase in animal virus entry. *J Gen Virol* 1994;**75**(Pt 10):2595–606.

56. Carr CM, Kim PS. A spring-loaded mechanism for the conformational change of influenza hemagglutinin. *Cell* 1993;**73**:823–32.

57. Bullough PA, Hughson FM, Skehel JJ, Wiley DC. Structure of influenza haemagglutinin at the pH of membrane fusion. *Nature* 1994;**371**:37–43.

58. Tsurudome M, Gluck R, Graf R, Falchetto R, Schaller U, Brunner J. Lipid interactions of the hemagglutinin HA2 NH2-terminal segment during influenza virus-induced membrane fusion. *J Biol Chem* 1992;**267**:20225–32.

59. Weber T, Paesold G, Galli C, Mischler R, Semenza G, Brunner J. Evidence for H(+)-induced insertion of influenza hemagglutinin HA2 N-terminal segment into viral membrane. *J Biol Chem* 1994;**269**:18353–8.

60. Durrer P, Galli C, Hoenke S, et al. H + -induced membrane insertion of influenza virus hemagglutinin involves the HA2 amino-terminal fusion peptide but not the coiled coil region. *J Biol Chem* 1996;**271**:13417–21.

61. Markovic I, Leikina E, Zhukovsky M, Zimmerberg J, Chernomordik LV. Synchronized activation and refolding of influenza hemagglutinin in multimeric fusion machines. *J Cell Biol* 2001;**155**:833–44.

62. Tatulian SA, Hinterdorfer P, Baber G, Tamm LK. Influenza hemagglutinin assumes a tilted conformation during membrane fusion as determined by attenuated total reflection FTIR spectroscopy. *EMBO J* 1995;**14**:5514–23.

63. Chernomordik LV, Frolov VA, Leikina E, Bronk P, Zimmerberg J. The pathway of membrane fusion catalyzed by influenza hemagglutinin: restriction of lipids, hemifusion, and lipidic fusion pore formation. *J Cell Biol* 1998;**140**:1369–82.

64. Spruce AE, Iwata A, White JM, Almers W. Patch clamp studies of single cell-fusion events mediated by a viral fusion protein. *Nature* 1989;(**342**):555–8.

65. Melikyan GB, Niles WD, Cohen FS. Influenza virus hemagglutinin-induced cell-planar bilayer fusion: quantitative dissection of fusion pore kinetics into stages. *J Gen Virol* 1993;**102**:1151–70.

66. Melikyan GB, Niles WD, Peeples ME, Cohen FS. Influenza hemagglutinin-mediated fusion pores connecting cells to planar membranes: flickering to final expansion. *J Gen Virol* 1993;**102**:1131–49.

67. Stegmann T. Membrane fusion mechanisms: the influenza hemagglutinin paradigm and its implications for intracellular fusion. *Traffic* 2000;**1**:598–604.
68. Ruigrok RW, Barge A, Durrer P, Brunner J, Ma K, Whittaker GR. Membrane interaction of influenza virus M1 protein. *Virology* 2000;**267**:289–98.
69. Fontana J, Steven AC. At low pH, influenza virus matrix protein M1 undergoes a conformational change prior to dissociating from the membrane. *J Virol* 2013;**87**:5621–8.
70. Noton SL, Medcalf E, Fisher D, Mullin AE, Elton D, Digard P. Identification of the domains of the influenza A virus M1 matrix protein required for NP binding, oligomerization and incorporation into virions. *J Gen Virol* 2007;**88**:2280–90.
71. Pinto LH, Holsinger LJ, Lamb RA. Influenza virus M2 protein has ion channel activity. *Cell* 1992;**69**:517–28.
72. Wharton SA, Belshe RB, Skehel JJ, Hay AJ. Role of virion M2 protein in influenza virus uncoating: specific reduction in the rate of membrane fusion between virus and liposomes by amantadine. *J Gen Virol* 1994;**75**(Pt 4):945–8.
73. Zhirnov OP. Solubilization of matrix protein M1/M from virions occurs at different pH for orthomyxo- and paramyxoviruses. *Virology* 1990;**176**:274–9.
74. Pinto LH, Lamb RA. The M2 proton channels of influenza A and B viruses. *J Biol Chem* 2006;**281**:8997–9000.
75. Chou YY, Heaton NS, Gao Q, Palese P, Singer RH, Lionnet T. Colocalization of different influenza viral RNA segments in the cytoplasm before viral budding as shown by single-molecule sensitivity FISH analysis. *PLOS Pathog* 2013;**9** e1003358.
76. Cros JF, Palese P. Trafficking of viral genomic RNA into and out of the nucleus: influenza Thogoto and Borna disease viruses. *Virus Res* 2003;**95**:3–12.
77. Engelhardt OG, Smith M, Fodor E. Association of the influenza A virus RNA-dependent RNA polymerase with cellular RNA polymerase II. *J Virol* 2005;**79**:5812–8.
78. Plotch SJ, Bouloy M, Krug RM. Transfer of 5′-terminal cap of globin mRNA to influenza viral complementary RNA during transcription in vitro. *PNAS* 1979;**76**:1618–22.
79. Bouloy M, Morgan MA, Shatkin AJ, Krug RM. Cap and internal nucleotides of reovirus mRNA primers are incorporated into influenza viral complementary RNA during transcription in vitro. *J Virol* 1979;**32**:895–904.
80. Bouloy M, Plotch SJ, Krug RM. Both the 7-methyl and the 2'-O-methyl groups in the cap of mRNA strongly influence its ability to act as primer for influenza virus RNA transcription. *PNAS* 1980;**77**:3952–6.
81. Krug RM. Priming of influenza viral RNA transcription by capped heterologous RNAs. *Curr Top Microbiol Immunol* 1981;**93**:125–49.
82. Smith GL, Hay AJ. Replication of the influenza virus genome. *Virology* 1982;**118**:96–108.
83. Enami M, Sharma G, Benham C, Palese P. An influenza virus containing nine different RNA segments. *Virology* 1991;**185**:291–8.
84. Bancroft CT, Parslow TG. Evidence for segment-nonspecific packaging of the influenza a virus genome. *J Virol* 2002;**76**:7133–9.
85. Wright PF, Neumann G, Kawaoka Y. Orthomyxoviruses. In: Knipe DM, Howley PM, editors. *Fields virology*. 6th ed. Philadelphia: Lippincott Williams & Wilkins; 2013. p. 1186–243.
86. Tamerius J. Influenza is generally regarded as a seasonal disease for temperate climate regions. *Environ Health Perspect* 2011;**119**:439–55.
87. Azziz Baumgartner E, Dao CN, Nasreen S, et al. Seasonality, timing, and climate drivers of influenza activity worldwide. *J Infect Dis* 2012;**206**:838–46.
88. Noble GR. Epidemiological and clinical aspects of influenza. In: Beare AS, editor. *Basic and applied influenza research*. Boca Raton: CRC Press; 1982. p. 11–50.
89. Hall CB. The spread of influenza and other respiratory viruses: complexities and conjectures. *Clin Infect Dis* 2007;**45**:353–9.

90. Patrozou E, Mermel LA. Does influenza transmission occur from asymptomatic infection or prior to symptom onset? *Public Health Rep* 2009;**124**:193–6.
91. Glezen WP. Serious morbidity and mortality associated with influenza epidemics. *Epidemiol Rev* 1982;**4**:25–44.
92. Wilkinson TM, Li CK, Chui CS, et al. Pre-existing influenza-specific CD4+ T cells correlate with disease protection against influenza challenge in humans. *Nature Med* 2012;**18**:274–80.
93. Couch RB, Kasel JA. Immunity to influenza in man. *Annu Rev Microbiol* 1983;**37**:529–49.
94. Hayward AC, Fragaszy EB, Bermingham A, et al. Comparative community burden and severity of seasonal and pandemic influenza: results of the Flu Watch cohort study. *Lancet Respir Med* 2014;**2**:445–54.
95. Sonoguchi T, Sakoh M, Kunita N, Satsuta K, Noriki H, Fukumi H. Reinfection with influenza A (H2N2, H3N2, and H1N1) viruses in soldiers and students in Japan. *J Infect Dis* 1986;**153**:33–40.
96. de Jong JC, Claas EC, Osterhaus AD, Webster RG, Lim WL. A pandemic warning? *Nature* 1997;**389**:554.
97. Centers for Disease, C., and Prevention. Serum cross-reactive antibody response to a novel influenza A (H1N1) virus after vaccination with seasonal influenza vaccine. *MMWR Morb Mortal Wkly Rep* 2009;**58**:521–4.
98. Chowell G, Bertozzi SM, Colchero MA, et al. Severe respiratory disease concurrent with the circulation of H1N1 influenza. *N Engl J Med* 2009;**361**:674–9.
99. Hancock K, Veguilla V, Lu X, et al. Cross-reactive antibody responses to the 2009 pandemic H1N1 influenza virus. *N Engl J Med* 2009;**361**:1945–52.
100. Itoh Y, Shinya K, Kiso M, et al. In vitro and in vivo characterization of new swine-origin H1N1 influenza viruses. *Nature* 2009;**460**:1021–5.
101. Thompson WW, Shay DK, Weintraub E, et al. Mortality associated with influenza and respiratory syncytial virus in the United States. *JAMA* 2003;**289**:179–86.
102. Elliot AJ, Fleming DM. Influenza and respiratory syncytial virus in the elderly. *Expert Rev Vaccines* 2008;**7**:249–58.
103. Neuzil KM, Reed GW, Mitchel EF, Simonsen L, Griffin MR. Impact of influenza on acute cardiopulmonary hospitalizations in pregnant women. *Am J Epidemiol* 1998;**148**:1094–102.
104. Nicholls JM, Chan MC, Chan WY, et al. Tropism of avian influenza A (H5N1) in the upper and lower respiratory tract. *Nature Med* 2007;**13**:147–9.
105. Shinya K, Ebina M, Yamada S, Ono M, Kasai N, Kawaoka Y. Avian flu: influenza virus receptors in the human airway. *Nature* 2006;**440**:435–6.
106. Frank AL, Taber LH, Wells CR, Wells JM, Glezen WP, Paredes A. Patterns of shedding of myxoviruses and paramyxoviruses in children. *J Infect Dis* 1981;**144**:433–41.
107. Ip DK, Lau LL, Chan KH, et al. The dynamic relationship between clinical symptomatology and viral shedding in naturally acquired seasonal and pandemic influenza virus infections. *Clin Infect Dis* 2015.
108. Zambon MC. The pathogenesis of influenza in humans. *Rev Med Virol* 2001;**11**:227–41.
109. Hers JF, Mulder J. Broad aspects of the pathology and pathogenesis of human influenza. *Am Rev Respir Dis* 1961;**83**(2):84–97 Pt 2.
110. Guarner J, Paddock CD, Shieh WJ, et al. Histopathologic and immunohistochemical features of fatal influenza virus infection in children during the 2003–2004 season. *Clin Infect Dis* 2006;**43**:132–40.
111. Gill JR, Sheng ZM, Ely SF, et al. Pulmonary pathologic findings of fatal 2009 pandemic influenza A/H1N1 viral infections. *Arch Pathol Lab Med* 2010;**134**:235–43.
112. Peiris JS, Yu WC, Leung CW, et al. Re-emergence of fatal human influenza A subtype H5N1 disease. *Lancet* 2004;**363**:617–9.
113. Subbarao K, Klimov A, Katz J, et al. Characterization of an avian influenza A (H5N1) virus isolated from a child with a fatal respiratory illness. *Science* 1998;**279**:393–6.
114. To KF, Chan PK, Chan KF, et al. Pathology of fatal human infection associated with avian influenza A H5N1 virus. *J Med Virol* 2001;**63**:242–6.

115. Ungchusak K, Auewarakul P, Dowell SF, et al. Probable person-to-person transmission of avian influenza A (H5N1). *N Engl J Med* 2005;**352**:333–40.
116. Hinshaw VS, Olsen CW, Dybdahl-Sissoko N, Evans D. Apoptosis: a mechanism of cell killing by influenza A and B viruses. *J Virol* 1994;**68**:3667–73.
117. Kuiken T, Taubenberger JK. Pathology of human influenza revisited. *Vaccine* 2008;**26**(Suppl 4):D59–66.
118. Rynda-Apple A, Robinson KM, Alcorn JF. Influenza and bacterial superinfection: illuminating the immunologic mechanisms of disease. *Infect Immun* 2015;**83**:3764–70.
119. Rothberg MB, Haessler SD, Brown RB. Complications of viral influenza. *Am J Med* 2008;**121**:258–64.
120. Walsh EE, Cox C, Falsey AR. Clinical features of influenza A virus infection in older hospitalized persons. *J Am Geriatr Soc* 2002;**50**:1498–503.
121. Smith AM, McCullers JA. Secondary bacterial infections in influenza virus infection pathogenesis. *Curr Top Microbiol Immunol* 2014;**385**:327–56.
122. Agyeman P, Duppenthaler A, Heininger U, Aebi C. Influenza-associated myositis in children. *Infection* 2004;**32**:199–203.
123. Studahl M. Influenza virus and CNS manifestations. *J Clin Virol* 2003;**28**:225–32.
124. Tong S, Zhu X, Li Y, et al. New world bats harbor diverse influenza A viruses. *PLOS Pathog* 2013;**9** e1003657.
125. Tong S, Li Y, Rivailler P, et al. A distinct lineage of influenza A virus from bats. *PNAS* 2012;**109**:4269–74.
126. Mehle A. Unusual influenza A viruses in bats. *Viruses* 2014;**6**:3438–49.
127. Webster RG, Bean WJ, Gorman OT, Chambers TM, Kawaoka Y. Evolution and ecology of influenza A viruses. *Microbiol Rev* 1992;**56**:152–79.
128. Murphy BR, Clements ML. The systemic and mucosal immune response of humans to influenza A virus. *Curr Top Microbiol Immunol* 1989;**146**:107–16.
129. Wilson IA, Cox NJ. Structural basis of immune recognition of influenza virus hemagglutinin. *Annu Rev Immunol* 1990;**8**:737–71.
130. Chen R, Holmes EC. Avian influenza virus exhibits rapid evolutionary dynamics. *Mol Biol Evol* 2006;**23**:2336–41.
131. Short KR, Richard M, Verhagen JH, et al. One health, multiple challenges: the inter-species transmission of influenza A virus. *One Health* 2015;**1**:1–13.
132. Daoust PY, Kibenge FS, Fouchier RA, van de Bildt MW, van Riel D, Kuiken T. Replication of low pathogenic avian influenza virus in naturally infected Mallard ducks (*Anas platyrhynchos*) causes no morphologic lesions. *J Wildl Dis* 2011;**47**:401–9.
133. Hofle U, Van de Bildt MW, Leijten LM, et al. Tissue tropism and pathology of natural influenza virus infection in black-headed gulls (*Chroicocephalus ridibundus*). *Avian Pathol* 2012;**41**:547–53.
134. Franca MS, Brown JD. Influenza pathobiology and pathogenesis in avian species. *Curr Top Microbiol Immunol* 2014;**385**:221–42.
135. Pantin-Jackwood MJ, Swayne DE. Pathogenesis and pathobiology of avian influenza virus infection in birds. *Rev Sci Tech* 2009;**28**:113–36.
136. Janke BH, Influenza. A virus infections in swine: pathogenesis and diagnosis. *Vet Pathol* 2014;**51**:410–26.
137. Trebbien R, Larsen LE, Viuff BM. Distribution of sialic acid receptors and influenza A virus of avian and swine origin in experimentally infected pigs. *Virol J* 2011;**8**:434.
138. Gray GC, Trampel DW, Roth JA. Pandemic influenza planning: shouldn't swine and poultry workers be included? *Vaccine* 2007;**25**:4376–81.
139. Chandrasekaran A, Srinivasan A, Raman R, et al. Glycan topology determines human adaptation of avian H5N1 virus hemagglutinin. *Nat Biotechnol* 2008;**26**:107–13.
140. Huang K, Zhu H, Fan X, et al. Establishment and lineage replacement of H6 influenza viruses in domestic ducks in southern China. *J Virol* 2012;**86**:6075–83.

141. Chu YC, Cheung CL, Hung Leung CY, et al. Continuing evolution of H9N2 influenza viruses endemic in poultry in southern China. *Influenza Other Respir Viruses* 2011;**5**(Suppl. 1):68–71.

142. White J, Kartenbeck J, Helenius A. Membrane fusion activity of influenza virus. *EMBO J* 1982;**1**:217–22.

143. Klenk HD, Rott R, Orlich M, Blodorn J. Activation of influenza A viruses by trypsin treatment. *Virology* 1975;**68**:426–39.

144. Webster RG, Rott R. Influenza virus A pathogenicity: the pivotal role of hemagglutinin. *Cell* 1987;**50**:665–6.

145. Horimoto T, Kawaoka Y. Influenza: lessons from past pandemics, warnings from current incidents. *Nat Rev Microbiol* 2005;**3**:591–600.

146. Banks J, Speidel ES, Moore E, et al. Changes in the haemagglutinin and the neuraminidase genes prior to the emergence of highly pathogenic H7N1 avian influenza viruses in Italy. *Arch Virol* 2001;**146**:963–73.

147. Bosch FX, Garten W, Klenk HD, Rott R. Proteolytic cleavage of influenza virus hemagglutinins: primary structure of the connecting peptide between HA1 and HA2 determines proteolytic cleavability and pathogenicity of Avian influenza viruses. *Virology* 1981;**113**:725–35.

148. Horimoto T, Kawaoka Y. Molecular changes in virulent mutants arising from avirulent avian influenza viruses during replication in 14-day-old embryonated eggs. *Virology* 1995;**206**:755–9.

149. Lazarowitz SG, Goldberg AR, Choppin PW. Proteolytic cleavage by plasmin of the HA polypeptide of influenza virus: host cell activation of serum plasminogen. *Virology* 1973;**56**:172–80.

150. Gotoh B, Ogasawara T, Toyoda T, Inocencio NM, Hamaguchi M, Nagai Y. An endoprotease homologous to the blood clotting factor X as a determinant of viral tropism in chick embryo. *EMBO J* 1990;**9**:4189–95.

151. Kido H, Yokogoshi Y, Sakai K, et al. Isolation and characterization of a novel trypsin-like protease found in rat bronchiolar epithelial Clara cells. *A possible activator of the viral fusion glycoprotein. J Biol Chem.* 1992;**267**:13573–9.

152. Murakami M, Towatari T, Ohuchi M, et al. Mini-plasmin found in the epithelial cells of bronchioles triggers infection by broad-spectrum influenza A viruses and Sendai virus. *Eur J Biochem* 2001;**268**:2847–55.

153. Towatari T, Ide M, Ohba K, et al. Identification of ectopic anionic trypsin I in rat lungs potentiating pneumotropic virus infectivity and increased enzyme level after virus infection. *Eur J Biochem* 2002;**269**:2613–21.

154. Chen Y, Shiota M, Ohuchi M, et al. Mast cell tryptase from pig lungs triggers infection by pneumotropic Sendai and influenza A viruses. Purification and characterization. *Eur J Biochem* 2000;**267**:3189–97.

155. Sato M, Yoshida S, Iida K, Tomozawa T, Kido H, Yamashita M. A novel influenza A virus activating enzyme from porcine lung: purification and characterization. *Biol Chem* 2003;**384**:219–27.

156. Bottcher E, Matrosovich T, Beyerle M, Klenk HD, Garten W, Matrosovich M. Proteolytic activation of influenza viruses by serine proteases TMPRSS2 and HAT from human airway epithelium. *J Virol* 2006;**80**:9896–8.

157. Tashiro M, Ciborowski P, Klenk HD, Pulverer G, Rott R. Role of *Staphylococcus protease* in the development of influenza pneumonia. *Nature* 1987;**325**:536–7.

158. Klenk HD, Garten W. Host cell proteases controlling virus pathogenicity. *TIM* 1994;**2**:39–43.

159. Horimoto T, Nakayama K, Smeekens SP, Kawaoka Y. Proprotein-processing endoproteases PC6 and furin both activate hemagglutinin of virulent avian influenza viruses. *J Virol* 1994;**68**:6074–8.

160. Remacle AG, Shiryaev SA, Oh ES, et al. Substrate cleavage analysis of furin and related proprotein convertases. A comparative study. *J Biol Chem* 2008;**283**:20897–906.

161. Kido H, Okumura Y, Takahashi E, et al. Host envelope glycoprotein processing proteases are indispensable for entry into human cells by seasonal and highly pathogenic avian influenza viruses. *J Mol Genet Med* 2008;**3**:167–75.

162. Munster VJ, Schrauwen EJ, de Wit E, et al. Insertion of a multibasic cleavage motif into the hemagglutinin of a low-pathogenic avian influenza H6N1 virus induces a highly pathogenic phenotype. *J Virol* 2010;**84**:7953–60.

163. Taubenberger JK, Hultin JV, Morens DM. Discovery and characterization of the 1918 pandemic influenza virus in historical context. *Antivir Ther* 2007;**12**:581–91.
164. Taubenberger JK, Morens DM. Influenza: the once and future pandemic. *Public Health Rep* 2010;**125**(Suppl 3):16–26.
165. Scholtissek C, von Hoyningen V, Rott R. Genetic relatedness between the new 1977 epidemic strains (H1N1) of influenza and human influenza strains isolated between 1947 and 1957 (H1N1). *Virology* 1978;**89**:613–7.
166. Kawaoka Y, Krauss S, Webster RG. Avian-to-human transmission of the PB1 gene of influenza A viruses in the 1957 and 1968 pandemics. *J Virol* 1989;**63**:4603–8.
167. Nakajima K, Desselberger U, Palese P. Recent human influenza A (H1N1) viruses are closely related genetically to strains isolated in 1950. *Nature* 1978;**274**:334–9.
168. Garten RJ, Davis CT, Russell CA, et al. Antigenic and genetic characteristics of swine-origin 2009 A(H1N1) influenza viruses circulating in humans. *Science* 2009;**325**:197–201.
169. Dunham EJ, Dugan VG, Kaser EK, et al. Different evolutionary trajectories of European avian-like and classical swine H1N1 influenza A viruses. *J Virol* 2009;**83**:5485–94.
170. Morens DM, Taubenberger JK, Fauci AS. The persistent legacy of the 1918 influenza virus. *N Engl J Med* 2009;**361**:225–9.
171. Perez-Padilla R, de la Rosa-Zamboni D, Ponce de Leon S, et al. Pneumonia and respiratory failure from swine-origin influenza A (H1N1) in Mexico. *N Engl J Med* 2009;**361**:680–9.
172. Fraser C, Donnelly CA, Cauchemez S, et al. Pandemic potential of a strain of influenza A (H1N1): early findings. *Science* 2009;**324**:1557–61.
173. Smith GJ, Vijaykrishna D, Bahl J, et al. Origins and evolutionary genomics of the 2009 swine-origin H1N1 influenza A epidemic. *Nature* 2009;**459**:1122–5.
174. Novel Swine-Origin Influenza AVIT, Dawood FS, Jain S, et al. Emergence of a novel swine-origin influenza A (H1N1) virus in humans. *N Engl J Med* 2009;**360**:2605–15.
175. Centers for Disease C and Prevention. Isolation of avian influenza A(H5N1) viruses from humans--Hong Kong, May-December 1997. *MMWR Morb Mortal Wkly Rep* 1997;**46**:1204–7.
176. Tran TH, Nguyen TL, Nguyen TD, et al. Avian influenza A (H5N1) in 10 patients in Vietnam. *N Engl J Med* 2004;**350**:1179–88.
177. Dudley JP. Age-specific infection and death rates for human A(H5N1) avian influenza in Egypt. *Euro Surveill* 2009;14.
178. Maurer-Stroh S, Li Y, Bastien N, et al. Potential human adaptation mutation of influenza A(H5N1) virus, Canada. *Emerg Infect Dis* 2014;**20**:1580–2.
179. Yang ZF, Mok CK, Peiris JS, Zhong NS. Human Infection with a Novel Avian Influenza A(H5N6) Virus. *N Engl J Med* 2015;**373**:487–9.
180. Gao R, Cao B, Hu Y, et al. Human infection with a novel avian-origin influenza A (H7N9) virus. *N Engl J Med* 2013;**368**:1888–97.
181. WHO. *Human infections with avian influenza A(H7N9) virus—summary of surveillance and investigation findings.* Geneva, Switzerland: WHO; 2014.
182. Liu J, Xiao H, Wu Y, et al. H7N9: a low pathogenic avian influenza A virus infecting humans. *Curr Opin Virol* 2014;**5**:91–7.
183. Wang D, Yang L, Gao R, et al. Genetic tuning of the novel avian influenza A(H7N9) virus during interspecies transmission, China, 2013. *Euro Surveill* 2014;**19**(25):pii: 20836.
184. Peiris M, Yuen KY, Leung CW, et al. Human infection with influenza H9N2. *Lancet* 1999;**354**:916–7.
185. Butt KM, Smith GJ, Chen H, et al. Human infection with an avian H9N2 influenza A virus in Hong Kong in 2003. *J Clin Microbiol* 2005;**43**:5760–7.
186. Guan Y, Shortridge KF, Krauss S, Webster RG. Molecular characterization of H9N2 influenza viruses: were they the donors of the "internal" genes of H5N1 viruses in Hong Kong? *PNAS.* 1999;**96**:9363–7.

187. De Marco MA, Campitelli L, Foni E, et al. Influenza surveillance in birds in Italian wetlands (1992–1998): is there a host restricted circulation of influenza viruses in sympatric ducks and coots? *Vet Microbiol* 2004;**98**:197–208.

188. Jiao P, Cao L, Yuan R, et al. Complete genome sequence of an H10N8 avian influenza virus isolated from a live bird market in Southern China. *J Virol* 2012;**86**:7716.

189. Chen H, Yuan H, Gao R, et al. Clinical and epidemiological characteristics of a fatal case of avian influenza A H10N8 virus infection: a descriptive study. *Lancet* 2014;**383**:714–21.

190. Wei SH, Yang JR, Wu HS, et al. Human infection with avian influenza A H6N1 virus: an epidemiological analysis. *Lancet Respir Med* 2013;**1**:771–8.

191. Shi W, Shi Y, Wu Y, Liu D, Gao GF. Origin and molecular characterization of the human-infecting H6N1 influenza virus in Taiwan. *Protein Cell* 2013;**4**:846–53.

192. Davies WL, Grunert RR, Haff RF, et al. Antiviral activity of 1-adamantanamine (amantadine). *Science* 1964;**144**:862–3.

193. Leonov H, Astrahan P, Krugliak M, Arkin IT. How do aminoadamantanes block the influenza M2 channel, and how does resistance develop? *J Am Chem Soc* 2011;**133**:9903–11.

194. Matrosovich MN, Matrosovich TY, Gray T, Roberts NA, Klenk HD. Neuraminidase is important for the initiation of influenza virus infection in human airway epithelium. *J Virol* 2004;**78**:12665–7.

195. Suzuki T, Takahashi T, Guo CT, et al. Sialidase activity of influenza A virus in an endocytic pathway enhances viral replication. *J Virol.* 2005;**79**:11705–15.

196. Yen HL, Hoffmann E, Taylor G, et al. Importance of neuraminidase active-site residues to the neuraminidase inhibitor resistance of influenza viruses. *J Virol* 2006;**80**:8787–95.

197. Schulman JL, Palese P. Susceptibility of different strains of influenza A virus to the inhibitory effects of 2-deoxy-2,3-dehydro-n-trifluoroacetylneuraminic acid (FANA). *Virology* 1975;**63**:98–104.

198. von Itzstein M, Wu WY, Kok GB, et al. Rational design of potent sialidase-based inhibitors of influenza virus replication. *Nature* 1993;**363**:418–23.

199. von Itzstein M, Thomson R. Anti-influenza drugs: the development of sialidase inhibitors. *Handb Exp Pharmacol* 2009;**189**:111–54.

200. Burnham AJ, Baranovich T, Govorkova EA. Neuraminidase inhibitors for influenza B virus infection: efficacy and resistance. *Antiviral Res* 2013;**100**:520–34.

201. Sidwell RW, Smee DF. Peramivir (BCX-1812, RWJ-270201): potential new therapy for influenza. *Expert Opin Investig Drugs* 2002;**11**:859–69.

202. Koyama K, Takahashi M, Oitate M, et al. CS-8958, a prodrug of the novel neuraminidase inhibitor R-125489, demonstrates a favorable long-retention profile in the mouse respiratory tract. *Antimicrob Agents Chemother* 2009;**53**:4845–51.

203. Kiso M, Kubo S, Ozawa M, et al. Efficacy of the new neuraminidase inhibitor CS-8958 against H5N1 influenza viruses. *PLOS Pathog* 2010;**6** e1000786.

204. Yamashita M. Laninamivir and its prodrug, CS-8958: long-acting neuraminidase inhibitors for the treatment of influenza. *Antiviral Chem Chemother* 2010;**21**:71–84.

205. Samson M, Pizzorno A, Abed Y, Boivin G. Influenza virus resistance to neuraminidase inhibitors. *Antiviral Res* 2013;**98**:174–85.

206. Furuta Y, Takahashi K, Fukuda Y, et al. In vitro and in vivo activities of anti-influenza virus compound T-705. *Antimicrob Agents Chemother* 2002;**46**:977–81.

207. Furuta Y, Gowen BB, Takahashi K, Shiraki K, Smee DF, Barnard DL. Favipiravir (T-705), a novel viral RNA polymerase inhibitor. *Antiviral Res* 2013;**100**:446–54.

208. Ohtsu Y, Honda Y, Sakata Y, Kato H, Toyoda T. Fine mapping of the subunit binding sites of influenza virus RNA polymerase. *Microbiol Immunol* 2002;**46**:167–75.

209. Poole EL, Medcalf L, Elton D, Digard P. Evidence that the C-terminal PB2-binding region of the influenza A virus PB1 protein is a discrete alpha-helical domain. *FEBS Lett* 2007;**581**:5300–6.

210. Hemerka JN, Wang D, Weng Y, et al. Detection and characterization of influenza A virus PA-PB2 interaction through a bimolecular fluorescence complementation assay. *J Virol* 2009;**83**:3944–55.

211. Ghanem A, Mayer D, Chase G, et al. Peptide-mediated interference with influenza A virus polymerase. *J Virol* 2007;**81**:7801–4.

212. Wunderlich K, Mayer D, Ranadheera C, et al. Identification of a PA-binding peptide with inhibitory activity against influenza A and B virus replication. *PloS One* 2009;**4** e7517.

213. Li C, Ba Q, Wu A, Zhang H, Deng T, Jiang T. A peptide derived from the C-terminus of PB1 inhibits influenza virus replication by interfering with viral polymerase assembly. *FEBS J* 2013;**280**:1139–49.

214. Jones JC, Turpin EA, Bultmann H, Brandt CR, Schultz-Cherry S. Inhibition of influenza virus infection by a novel antiviral peptide that targets viral attachment to cells. *J Virol* 2006;**80**:11960–7.

215. Jones JC, Settles EW, Brandt CR, Schultz-Cherry S. Identification of the minimal active sequence of an anti-influenza virus peptide. *Antimicrob Agents Chemother* 2011;**55**:1810–3.

216. Ahmed CM, Dabelic R, Waiboci LW, Jager LD, Heron LL, Johnson HM. SOCS-1 mimetics protect mice against lethal poxvirus infection: identification of a novel endogenous antiviral system. *J Virol* 2009;**83**:1402–15.

217. Nicol MQ, Ligertwood Y, Bacon MN, Dutia BM, Nash AA. A novel family of peptides with potent activity against influenza A viruses. *J Gen Virol* 2012;**93**:980–6.

218. Boriskin YS, Leneva IA, Pecheur EI, Polyak SJ. Arbidol: a broad-spectrum antiviral compound that blocks viral fusion. *Curr Med Chem* 2008;**15**:997–1005.

219. Pecheur EI, Lavillette D, Alcaras F, et al. Biochemical mechanism of hepatitis C virus inhibition by the broad-spectrum antiviral arbidol. *Biochemistry* 2007;**46**:6050–9.

220. Leneva IA, Russell RJ, Boriskin YS, Hay AJ. Characteristics of arbidol-resistant mutants of influenza virus: implications for the mechanism of anti-influenza action of arbidol. *Antiviral Res* 2009;**81**:132–40.

221. Teissier E, Zandomeneghi G, Loquet A, et al. Mechanism of inhibition of enveloped virus membrane fusion by the antiviral drug arbidol. *PloS One* 2011;**6**:e15874.

INFLUENZA VIRAL INFECTION IN THE RESPIRATORY SYSTEM—POTENTIAL WAYS OF MONITORING

2

S.C.B. Gopinath*,**, T. Lakshmipriya*, U. Hashim*, M.K. Md Arshad*, R.M. Ayub*, T. Adam*,†

*Universiti Malaysia Perlis, Institute of Nano Electronic Engineering (INEE), Kangar, Perlis, Malaysia; **Universiti Malaysia Perlis, School of Bioprocess Engineering, Arau, Perlis, Malaysia; †Universiti Malaysia Perlis, School of Electronic Engineering Technology, Faculty of Engineering Technology, Kangar, Perlis, Malaysia

1 INTRODUCTION

Respiratory diseases caused by different viruses usually affect the lung, followed by the trachea and upper part of the respiratory track. Influenza is one of the seasonal RNA viruses that leads to respiratory disease and which spreads easily via airborne droplets. Influenza has three types: Influenza A, B, and C. Among these, A and B are common in humans and C is not common. Influenza virus with ~100 nm in diameter is formed by different proteins and RNA[1,2] (Fig. 2.1). Influenza A and B have eight RNA strands and C has seven. Influenza has been classified by its two major proteins [hemagglutinin (HA) and neuraminidase (NA)] on its surface, which are mainly involved in the multiplication and infection. Until now, 17 HAs (H1–H17) and nine NAs (N1–N9) have been found in different influenza strains. Influenza virus is named according to the number of HA and NA, such as H1N1, H2N2, H3N2, H5N1 and so on.[3–5] Structural studies on these surface proteins elucidate further understanding and help precise classification (Fig. 2.2a,b). New strains of influenza emerge due to seasonal variations, genetic reassortments, the involvement of intermediate hosts, and other environmental influences[6,7] (Fig. 2.3). Currently available vaccines cannot prevent the appearance of new strains and, therefore, it is vital to develop novel detection systems to diagnose influenza viruses, which paves the way to generate vaccinations to prevent new influenza viruses.

Currently, in most cases the antibody-based probes that are predominantly used for influenza virus detection[5,8] cannot differentiate among influenza subtypes. Other probes such as aptamer and glycan have been developed to take a step to distinguish influenza subtypes.[3,9–16] These probes have allowed several sensing strategies to be generated from laboratory to industrial scale, and which connects the gap between the different scales.[16] On the other hand, in nanotechnology top-down and bottom-up approaches have been used to develop the current plaforms.[17–23] Currently, the primary aim of sensors is to bring the bench-top sensors to the front-line of medical practitioners. In this chapter, we will discuss the different strategies involved in the detection of influenza disease with novel sensing methods.

The Microbiology of Respiratory System Infections. http://dx.doi.org/10.1016/B978-0-12-804543-5.00002-6

33

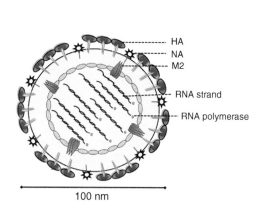

FIGURE 2.1 Image of an Intact Influenza Virus

It has a size of about 100 nm in diameter. The major surface proteins are hemagglutinin (HA) and neuraminidase (NA).

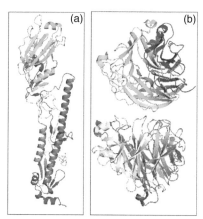

FIGURE 2.2 Crystal Structures of Influenza Surface Proteins

3D structures of Hemagglutinin (HA: PDB Accession: 2VIU) and Neuraminidase (NA: PDB Accession: 3SAL) are displayed.

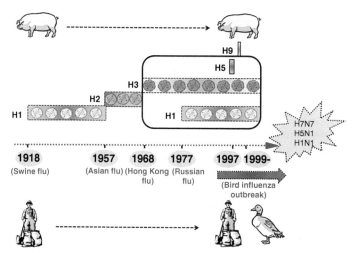

FIGURE 2.3 Emergence of Influenza Viruses

Different influenza pandemics are indicated. Influenza hosts are displayed.

2 PROBES INVOLVED IN INFLUENZA DETECTION

As stated earlier, the choice of probe (antibody, aptamer, and glycan) is the important factor in the development of sensors for the diagnosis of influenza (Fig. 2.4a–c). These probes have been generated for influenza viruses having higher affinities to the surface proteins (Hemagglutinin-HA; neuraminidase-NA), and they facilitate the whole virus detection. These probes can also be generated for proteins

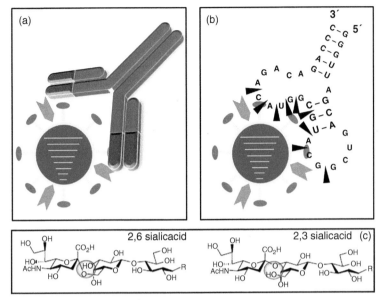

FIGURE 2.4 Probes used for influenza detection

(a) Antibody as probe, (b) aptamer as probe, and (c) glycan as probe. Sialic acids are shown for glycan-based probes α-2,6 and α-2,3.

residing inside the virus, such as structural proteins, and in this case the virus has to lyse before being diagnosed. Both antibody and aptamer bind, based on affinities to the target, whereas glycan has both affinity and functional interaction to the target/host cell. Usually, in most viruses glycan mediates the infection, which helps to attach to the host cells. During infection or interaction, the HA of influenza will interact with the glycan present on the host cell.[24] After multiplication of viral particles, NA cleaves these glycans (sialic acids) to infect neighboring cells.

2.1 INFLUENZA DETECTION BY ANTIBODY

Antibodies can be generated by injecting the antigen of interest into the animal or a recombinant antibody can be prepared using the DNA sequence of the antigen. Several sensing systems have been demonstrated for influenza using suitable antiinfluenza antibodies. The binding of the antibody and the target (antigen) from the influenza virus generates a signal that can be used to identify the influenza viruses. As stated above, to detect a whole influenza virus, the antigen should be either hemagglutinin or neuraminidase (Fig. 2.2). Gopinath et al.[4,5] used HA antibody to detect the influenza viruses and discriminated between HA of human and bird influenza viruses. Again in the case of anti-HA antibody, this can be either anti-HA1 or anti-HA2 antibodies. Researchers have used enzyme linked immunosorbent assay (ELISA) to detect influenza viruses with the HA1 antibodies. In most cases, using antibodies means that we can only differentiate types of influenza viruses. Because of its larger size, in several cases we cannot differentiate subtypes. To substitute antibodies, researchers are currently looking at using aptamers as a probe to detect and differentiate influenza viruses.

2.2 INFLUENZA DETECTION BY APTAMER

Aptamer is an artificial chemical antibody that is generated from the randomized nucleic acid library by three simple steps: binding, separation, and amplification. The selected aptamer has a high binding affinity with the target molecule.[25–30] In most cases, aptamer was found to be better than antibodies. Additionally, aptamers have more advantages, such as being easy to prepare, having no variation with different preparations, higher sensitive, they are easy to modify, and they are nonimmunogenic. In the past, several generated aptamers against influenza viruses have been reported to have high affinity. Moreover, since aptamer binds with only a few bases on the target molecule, it has a higher chance to differentiate subtypes of influenza viruses. Among the different target/antigens from influenza viruses, HA is a suitable target for designing the detection strategy because of its high abundance (Fig. 2.4b). HA accounts for 80% of the influenza surface proteins; next to HA, NA molecules are higher in number on influenza surface. Gopinath et al.[10,11] have selected the aptamers against both influenza A (H3N2) and B (Johannesburg) viruses, using either HA or whole virus. The same team also generated the aptamer against H1N1; the selected aptamer has a high binding affinity with the HA protein.[14] Lakshmipriya et al.[31] have selected the aptamers that can be used against intact influenza B virus (B/Tokio) and HA protein of influenza B virus (B/Jilin).

2.3 INFLUENZA DETECTION BY GLYCAN

The initiation of viral infection is highly dependent on the presence of receptor molecules on the host cells and it is mediated predominantly by either α-2,6 or α-2,3 linked sialic acid (glycan) chains in human and bird (avian) influenza, respectively[32] (Fig. 2.4c). However, following the emergence of different strains in the past several decades, the specificities of these glycans have become varied. Some exceptional cases have been reported, irrespective of glycan specificities, due to special viral adaptations. Furthermore, the avian influenza virus could also infect humans when a close physical touch between human and bird occurred. Different amino acids from the HA protein have been found to be specifically involved in the host interaction. Different sensing strategies have previously been formulated based on glycan to diagnose and discriminate influenza viruses.[33,34] The main advantage of sialic acid based sensing is to discriminate avian and human influenza viruses.

3 DETECTION OF INFLUENZA VIRUS

3.1 IMMUNOCHROMATOGRAPHIC TEST

Using the above probes, several sensing methods have been reported and have achieved different levels by both qualitative and quantitative detection. With qualitative detection, immunochromatographic test (ICT) has been demonstrated to be reliable. ICT operates based on the lateral flow of solutions, with different set-ups of pad. This system immobilized with gold nanoparticle (GNP)-conjugated antiinfluenza antibody. Upon the interaction of the GNP-conjugated antibody and influenza virus, there will be an accumulation of this complex, which causes the appearance of red colored lines that indicate an influenza positive sample. At the same time, it displays a control red-line and no color formation with a negative line (influenza A or B), depending on the samples to be analyzed (Fig. 2.5). However, this system does not predominantly demonstrate for subtyping (within influenza A or B) purposes.

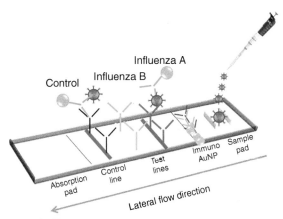

FIGURE 2.5 Immunochromatography Test

This is a lateral flow system operating based on different pads. It uses antiinfluenza antibodies conjugated with gold nanoparticles.

Furthermore, physicians have claimed that ICT can only detect influenza when the patient has attained a high fever. This means that it could not detect an influenza virus in the earlier stages and it needs the higher viral count that happens with the on-set of high fever. Even though ICT has been successful in several instances, for quantitative analyses it is necessary to look for established sensors. The following section will discuss sensors that can detect the target from influenza virus in a quantitative manner.

3.2 SURFACE PLASMON RESONANCE

Surface Plasmon Resonance (SPR) is the favorite sensor system to analyze biomolecular interactions on the sensing surface in the lower femtomolar range. In an SPR system, light hits a prism and the substrate is then reflected. The angle of the reflected light will be changed depending on the molecules binding on the substrate. Changes in the reflection are considered as the real binding of the ligand and analyte. Using the SPR based principle, various kinds of sensor have been proposed to detect the target against its appropriate partner, such as antibody or aptamer. Among SPR based sensors, BIACORE is the most well established sensing system and it is used to detect biomolecules with higher sensitivity. In the BIACORE system, various sensing surfaces are readily available. Basically, the entire gold surface in this system was modified with different linkers, which are suited for different biomolecular interactive analyses. Using the BIACORE system, Gopinath and colleagues have analyzed the interactions of influenza and aptamers in several different studies with human influenza A viruses.[10,11,14,33] With these aptamers, they could reach a picomolar level sensitivity during analysis of influenza belonging to H3N2 and influenza B. With a further step, they attained a femtomolar level sensitivity during analysis of aptamer and H1N1 (swine flu) virus. Similarly, Suenaga et al. have made different studies with avian influenza viruses on SPR.[34–37]

3.3 SURFACE PLASMON FLUORESCENCE SPECTROSCOPY

Surface plasmon resonance fluorescence spectroscopy (SPFS) works on a similar principle to SPR, the only difference is that SPFS uses fluorescence to detect the analyte. Lakshmipriya et al.[31] have detected intact influenza B virus and HA protein of influenza B using SPFS by their appropriate aptamer or antibody. They compared the interaction between antibody and aptamer against influenza viruses and showed an improved detection level. Furthermore, they found that aptamer behaves better than antibody and that it could discriminate influenza types and subtypes. Comparative studies between SPFS and radio isotope labeling revealed the better performance of SPFS. The authors attested to the higher success rate of the operation of SPR, SPFS, and other similar systems, such as waveguide mode sensors, which are is based on Kretschmann configuration.

3.4 WAVEGUIDE MODE SENSOR

A waveguide mode sensor is used to detect analyte molecules in a solution and has been used to diagnose influenza virus. Even though the principles of SPR and a waveguide mode sensor are similar, in the waveguide mode system the light that hits the prism passes through the surface as a guiding mode and not as a surface mode.[38,39] Depending on the molecular binding on the substrate environment, the reflected light will be changed, and changes in the reflection are considered as the real binding. Gopinath et al.[10,11] have demonstrated the use of the binding events for the detection of an influenza virus against antiinfluenza antibody. In their study, they used an antibody raised in the laboratory against H3N2 influenza and had better discriminating ability than commercial antiinfluenza antibody. This anti-H3N2 antibody is able to discriminate types and subtypes of influenza viruses among H3N2, H1N1, and influenza B. To improve the sensitivity of their waveguide mode sensor, they conjugated GNP on the surface of influenza viruses. They were able to attain a sensitivity level of 8×10^3 PFU/mL. They also demonstrated waveguide mode detection of influenza in the absence of GNP and in the presence of dye materials.[40]

3.5 GOLD NANOPARTICLE BASED COLORIMETRIC ASSAY

GNP-based colorimetric assay has been formulated to detect different kinds of molecule with the naked eye. This assay is more suitable for controlled assembly and disassembly of aptamers or antibodies on the GNP in the presence (or absence) of target molecules. Gold has naturally no charge, but GNP will get a surface charge, which may be either positive or negative charge depending on the reagents that are used. Prepared GNP is usually in a dispersed condition (red colored solution) and when we add charged ions they induce aggregation (blue/purple-colored solution).[19] Usually monovalent or divalent ions can be used to induce the aggregation. Sodium chloride (NaCl) has predominantly been used to induce the aggregation. When aptamer or antibody is used with dispersed GNP, it keeps the same status (dispersed red solution), even in the presence of NaCl, and the aptamer/antibody will bind to GNP by electrostatic interaction or chemical reaction. However, in the presence of appropriate target molecules to the aptamer/antibody, GNP will be aggregated (blue solution) in the presence of NaCl. Under this condition, a complex of aptamer/antibody target will be formed and this will induce the aggregation. Lee et al.[41] have developed a colorimetric assay for influenza detection, they used glycan (sialic acid) as the probe and reached a sensitivity of up to 512 HA titer (for influenza B/Victoria and B/Yamagata). In this case, the assay is slightly modified, but this is not like controlled assembly and disassembly. They formulated that in the presence of influenza virus, the complex will be formed with sialic acid-conjugated GNP and this will induce virus mediated aggregation. Gopinath et al.[19] have revealed the nonfouling effect of HA protein on the GNP and they compared their results with other molecules.

3.6 DISC PLATFORM—INTERFEROMETRY

As mentioned above, sensing platforms have been reported to be appropriate systems for detection of influenza virus and other molecules. However, most of the sensing designs are suitable for analyzing a limited number of samples. A sensing platform/system suitable for high-throughput screening with the facility to analyze higher number of samples is highly appreciated. The disc platform is suitable for high-throughput sampling. The disc platform is basically designed based on a commercial compact disc (CD) or digital versatile disc (DVD). Varma et al.[42] have shown CD-based detections of biomolecules with antigen–antibody interactions. Later, Gopinath et al.[43–45] have shown the progression with the DVD platform and analyzed different samples, including influenza viruses. These two platforms have spiral tracks on the surface that can accommodate several thousands of samples. The cost of the sensing surface is cheaper and easier for surface functionalization. In most cases, gold was used as the top substrate layer and it can be easily linked to biomolecules through thiol groups. Gopinath et al.[44] have used five layer structures, with a top gold layer, for biomolecular interactive analyses. For the interactive analysis of influenza virus-A against its antibody, they attained a dissociation constant to the lower nanomolar level. This level is comparable to the detection level of BIACORE with the same molecules.

3.7 FLUORESCENT CAPTURING

In most cases, these platforms are defined as label-free, except for SPFS. However, other fluorescent labelling techniques are also used for the diagnosis of influenza. For example, fluorescent labelling has the advantage of higher sensitivity, although it has been argued that it is difficult to label the molecule due to the unavailability of the site for labelling or because the labelling site may be hidden. Several different labelling materials, such as rhodamine, cyanine, Alexa Flour, and fluorescein, are routinely used for different bio-labelling strategies (Fig. 2.6a,b). In the SPFS method for influenza virus detection, Lakshmipriya et al.[31] used cyanine as the fluorescent material to label the aptamer. Nomura et al.[8] used Alexa Flour 700 for the detection of influenza virus. They used a new kind of SPR system without involving the use of a prism (prism-free). A CCD camera system has been used to capture the

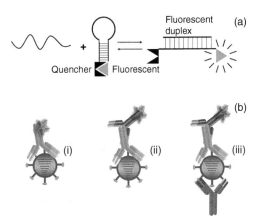

FIGURE 2.6 Fluorescent-Based Detection Strategies

(a) Aptamer-conjugated fluorescent; (b) antibody-conjugated fluorescent; (i–iii) different ways of detection.

fluorescent signaling upon interaction of influenza virus and antiinfluenza H3N2 antibody. With this system, the authors have shown a clear discrimination between influenza subtypes using antibody.

3.8 ENZYME LINKED IMMUNOSORBENT ASSAY

ELISA is the gold standard method to detect a wide range of target molecules assisted with appropriate partner molecules. ELISA is used not only for the detection but also for the basic screening of many important diseases, such as HIV, influenza, and so on. Gopinath et al.[14] have demonstrated ELISA based detection of influenza virus using anti-H3N2 antibody and discriminated against other influenza viruses. Due to its high sensitivity and selectivity, ELISA helps to identify the target molecules, even in the human crude samples (serum, urine, and saliva). Until now, antibodies have commonly been used as a probe to detect biomolecules in ELISA. In general, the binding of the target (antigen) and the probe (antibody) on the ELISA plate was detected by the enzyme (eg, horseradish peroxidase (HRP), alkaline phosphatase) conjugated with a secondary antibody and detected by chromogenic substrates. Biotin-streptavidin conjugation has also been used to improve the ELISA methods. In this case, biotinylated secondary antibody was detected by streptavidin-conjugated enzyme. Various kinds of pattern are used in the ELISA method to improve the detection methods, such as direct, indirect, sandwich, and competitive ELISA. In most of the cases, antibody was used as the probe due its strong binding, stability, and selectivity. The sandwich ELISA is usually conducted by the monoclonal or polyclonal antibody of the specific target (Fig. 2.7). In some cases, only the Fc region of the antibody has been used as the capture molecule on the ELISA plate. After aptamer generation, some researchers have used aptamer as the probe instead of antibody. This method is called aptamer linked immunosorbent assay (ALISA).[46] Since aptamer has a higher sensitivity than the antibody, it is possible to increase the

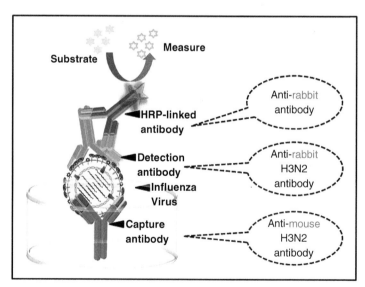

FIGURE 2.7 Enzyme Linked Immunosorbent Assay (ELISA)

There are basically three types of ELISAs (direct, indirect, and sandwich). Here a sandwich ELISA is shown as an example.

limit of detection when aptamer is used as the probe. In addition, there is a possibility to do a sandwich assay with two different aptamers for the same target. Since both aptamer and antibody are suitable as detection molecules, there is a higher possibility of increasing the limit of detection with the sandwich patterns using aptamer and antibody.

4 CONCLUSIONS

Currently, the treatment for influenza is focused against the neuraminidase protein, which is one of the major proteins on the surface of the influenza virus, by a so called drug "Tamiflu." Current vaccines cannot prevent all types of influenza viruses because current vaccination is only suitable for particular influenza viruses, while every year different subtypes of influenza viruses are emerging. Therefore, it is necessary to generate a new vaccine, and a good starting point for this purpose is to diagnose new influenza strains and generate an efficient detection system that is suitable for early diagnosis. Moreover, the use of currently available antibodies does not allow us to create a detection system which can distinguish influenza subtypes. Development of a complement detection system which uses different molecules, such as antibody with aptamer, aptamer with glycan or glycan with antibody, will increase the possibility of developing novel detection systems. Further optimization of studies performed with viruses/molecules other than influenza viruses can be coupled for influenza detection strategies.[47-52]

REFERENCES

1. Noda T, Sagara H, Yen A, Takada A, Kida H, Cheng RH, Kawaoka Y. Architecture of ribonucleo-protein complexes in influenza a virus particles. *Nature* 2006;**439**:490–2.
2. Shima T, Fujimaki M, Yoshida A, Gopinath SCB, Kuwahara M, Ohki Y, Awazu K. Detection of influenza viruses attached to an optical disk. *J Biomat Nanobiotechnol* 2013;**4**:145–50.
3. Gopinath SCB. Anti-viral aptamers. *Arch Virol* 2007;**152**:2137–57.
4. Gopinath SCB, Awazu K, Fujimaki M, Shimizu K, Shima T. Observations of immuno-gold conjugates on influenza viruses using waveguide-mode sensor. *PLoS One* 2013;**8**:e69121.
5. Gopinath SCB, Awazu K, Fujimaki M, Shimizu K. Evaluation of Anti-A/Udorn/307/1972 antibody specificity to Influenza viruses using a waveguide mode sensor. *PLoS One* 2013;**8**:e81396.
6. Webster RG, Bean WJ, Gorman OT, Chambers TM, Kawaoka Y. Evolution and ecology of influenza A viruses. *Microbiol Rev* 1992;**56**:152–79.
7. Neumann G, Noda T, Kawaoka Y. Emergence and pandemic potential of swine- origin H1N1 influenza virus. *Nature* 2009;**459**:931–9.
8. Nomura K, Gopinath SCB, Lakshmipriya T, Fukuda N, Wang X, Fujimaki M. An angular fluidic channel for prism-free surface-plasmon-assisted fluorescence capturing. *Nat Commun* 2013;**4**:2855.
9. Jeon SH, Kayhan B, Ben-Yedidia T, Arnon R. A DNA aptamer prevents influenza infection by blocking the receptor binding region of the viral hemagglutinin. *J Biol Chem* 2004;**279**:48410–9.
10. Gopinath SCB, Misono TS, Kawasaki K, Mizuno M, Imai T, Odagiri T, Kumar PKR. An RNA aptamer that distinguishes between closely related human influenza viruses and inhibits hemagglutinin-mediated membrane fusion. *J Gen Virol* 2006;**87**:479–87.
11. Gopinath SCB, Sakamaki Y, Kawasaki K, Kumar PKR. An efficient RNA aptamer against human influenza B virus hemagglutinin. *J Biochem* 2006;**139**:837–46.
12. Gopinath SCB, Awazu K, Fujimaki M. Waveguide-mode sensors as aptasensors. *Sensors* 2012;**12**:2136–51.

13. Gopinath SCB, Tang TH, Yeng C, Citartan M, Tominaga J, Lakshmipriya T. Sensing strategies for influenza surveillance. *Biosens Bioelectron* 2014;**61**:357–69.

14. Gopinath SCB, Kumar PKR. Aptamers that bind to the hemagglutinin of recent pandemic H1N1 and inhibits efficiently the agglutination. *Acta Biomaterialia* 2013;**9**:8932–41.

15. Hidari KIPJ, Suzuki T. Glycan receptor for influenza virus. *Open Antimicrob Agents J* 2010;**2**:26–33.

16. Fathil MF, Md Arshad MK, Gopinath SC, Hashim U, Adzhri R, Ayub RM, Ruslinda AR, Nuzaihan MNM, Azman AH, Zaki M, Tang TH. Diagnostics on acute myocardial infarction: cardiac troponin biomarkers. *Biosens Bioelectron* 2015;**70**:209–20.

17. Yuan Y, Gopinath SCB, Kumar PKR. Regeneration of commercial Biacore chips to analyze biomolecular interactions. *Opt Engg* 2011;**50** 034402-1-6.

18. Perumal V, Hashim U, Gopinath SCB, Haarindraprasad R, Ravichandran M, Balakrishnan SR, Poopalan P. A new nano-worm structure from gold-nanoparticle mediated random curving of zinc oxide nanorod. *Biosens Bioelectron* 2016;**78**:14–22.

19. Gopinath SCB, Lakshmipriya T, Awazu K. Colorimetric detection of controlled assembly and disassembly of aptamers on unmodified gold nanoparticles. *Biosen Bioelectron* 2014;**51**:115–23.

20. Haarindraprasad R, Hashim U, Gopinath SCB, Kashif M, Veeradasan P, Balakrishnan SR, Foo KL. Low temperature annealed zinc oxide nanostructured thin film-based transducer: characterization for sensing applications. *PloS One* 2015;**10**:e0132755.

21. Perumal V, Hashim U, Gopinath SCB, Haarindraprasad R, Foo KL, Balakrishnan SR, Poopalan P. 'Spotted nanoflowers'-gold-seeded zinc oxide nanohybrid for selective bio-capture. *Sci Rep* 2015;**5**:12231.

22. Balakrishnan SR, Hashim U, Gopinath SCB, Poobalan P, Hariram R, Omar I, Dhahi TS, Haarindraprasad R, Veeradasan P. Polysilicon nanogap lab-on-a-chip: a point-of-care immunosensing human chorionic gonadotropin in clinical human urine. *Plos One* 2015;**10**:e0137891.

23. Nadzirah Sh, Azizah N, Hashim U, Gopinath SCB, Kashif M. Titanium dioxide nanoparticle-based interdigitated electrodes: a novel current to voltage DNA biosensor recognizes *E. coli* O157:H7. *Plos One* 2015;**10**:e0139766.

24. Skehel JJ, Wiley DC. Receptor binding and membrane fusion in virus entry: the influenza hemagglutinin. *Ann Rev Biochem* 2000;**69**:531–69.

25. Ellington AD, Szostak JW. In vitro selection of RNA molecules that bind specific ligands. *Nature* 1990;**346**:818–22.

26. Kumarevel T, Gopinath SCB, Nishikawa S, Mizuno H, Kumar PKR. Identification of important chemical groups of the hut mRNA for HutP interactions that regulates the hut operon in *Bacillus subtilis*. *Nucleic Acids Res* 2004;**32**:3904–12.

27. Gopinath SCB, Balasundaresan D, Akitomi J, Mizuno H. An RNA aptamer that discriminates bovine factor IX from human factor IX. *J Biochem* 2006;**140**:667–76.

28. Gopinath SCB, Awazu K, Fujimaki M, Sugimoto K, Ohki Y, Komatsubara T, Tominaga J, Gupta KC, Kumar PKR. Influence of nonometric holes on the sensitivity of waveguide plasmon: a label-free nano-sensor to analyze RNA-aptamer ligand interactions. *Anal Chem* 2008;**80**:6602–9.

29. Gopinath SCB, Hayashi K, Kumar PKR. Aptamer that binds to the gD protein of herpes simplex virus-1 and efficiently inhibits viral entry. *J Virol* 2012;**86**:6732–44.

30. Gopinath SCB. Aptamers. In: Meyers RA, editor. *Encyclopedia of analytical chemistry*. Chichester: John Wiley; 2011. p. 93–120.

31. Lakshmipriya T, Fujimaki M, Gopinath SCB, Awazu K. Generation of anti-influenza Aptamers using systematic evolution of ligands by exponential enrichment for sensing applications. *Langmuir* 2013;**29**:15107–15.

32. Suzuki Y. Sialobiology of influenza: molecular mechanism of host range variation of influenza viruses. *Biol Pharm Bulletin* 2005;**28**:399–408.

33. Gopinath SCB, Awazu K, Fujimaki M. Detection of influenza viruses by a waveguide-mode sensor. *Anal Method* 2010;**2**:1880–4.

34. Suenaga E, Kumar PKR. An aptamer that binds efficiently to the hemagglutinins of highly pathogenic avian influenza viruses (H5N1 and H7N7) and inhibits hemagglutinin–glycan interactions. *Acta Biomaterialia* 2014;**10**:1314–23.

35. Suenaga E, Mizuno H, Kumar PKR. Monitoring influenza hemagglutinin and glycan interactions using surface plasmon resonance. *Biosens Bioelectron* 2012;**32**:195–201.

36. Gopinath SCB, Awazu K, Fujimaki M, Shimizu K. Neu5Acα2,6Gal and Neu5Acα2,3Gal receptor specificities on influenza viruses determined by a waveguide-mode sensor. *Acta Biomaterialia* 2013;**9**:5080–7.

37. Suenaga E, Mizuno H, Kumar PKR. Influenza virus surveillance using surface plasmon resonance. *Virulence* 2012;**3**:464–70.

38. Gopinath SCB, Misono T, Kumar PKR. Prospects of ligand-induced aptamers. *Crit Rev Anal Chem* 2008;**38**:34–47.

39. Fujimaki M, Nomura K, Sato K, Kato T, Gopinath SCB, Wang X, Awazu K, Ohki Y. Detection of coloured nanomaterials using evanescent field-based waveguide sensors. *Opt Exp* 2010;**18**:15732–40.

40. Gopinath SCB, Awazu K, Fujimaki M, Sugimoto K, Ohki Y, Komatsubara T, Tominaga J, Kumar PKR. Monitoring surface-assisted biomolecular assembly by means of evanescent-field-coupled waveguide-mode nanobiosensors. *Anal Bioanal Chem* 2009;**394**:481–8.

41. Lee C, Gaston MA, Weiss AA, Zhang P. Colorimetric viral detection based on sialic acid stabilized gold nanoparticles. *Biosens Bioelectron* 2013;**42**:236–41.

42. Varma MM, Nolte DD, Inerowicz HD, Regnier FE. Spinning-disk self-referencing interferometry of antigen–antibody recognition. *Opt Lett* 2004;**29**:950–2.

43. Gopinath SCB, Awazu K, Kumar PKR, Tominaga J. Monitoring biomolecular interactions on a digital versatile disc: a BioDVD platform technology. *ACS Nano* 2008;**2**:1885–95.

44. Gopinath SCB, Awazu K, Fons P, Tominaga J, Kumar PKR. A sensitive multilayered structure suitable for biosensing on the BioDVD platform. *Anal Chem* 2009;**81**:4963–70.

45. Gopinath SCB, Kumar PKR, Tominaga J. A BioDVD media with multilayered structure is suitable for analyzing biomolecular interactions. *J Nanosci Nanotechnol* 2011;**11**:5682–8.

46. Toh SY, Citartan M, Gopinath SCB, Tang TH. Aptamer as a modular replacement of antibodies in enzyme-linked immunosorbent assay. *Biosens Bioelectron* 2015;**64**:392–403.

47. Anbu P, Gopinath SCB, Hilda A, Mathivanan N, Annadurai G. Secretion of keratinolytic enzymes and keratinolysis by *Scopulariopsis brevicaulis* and *Trichophyton mentagrophytes*: regression analysis. *Canadian J Microbiol* 2006;**52**:1060–9.

48. Gopinath SCB, Shikamoto Y, Mizuno H, Kumar PKR. Snake venom-derived factor IX binding protein specifically blocks the Gla domain-mediated-membrane binding of human factors IX and X. *Biocheml J* 2007;**405**:351–7.

49. Gopinath SCB, Anbu P, Lakshmipriya T, Hilda A. Strategies to characterize fungal lipases. *BioMed Res Int* 2013;**2013** ID 154549.

50. Poongodi GL, Suresh N, Gopinath SCB, Chang T, Inoue S, Inoue Y. Dynamic change of NCAM polysialylation on human neuroblastoma (IMR-32) and rat pheochromocytoma (PC-12) cells during growth and differentiation. *J Biol Chem* 2002;**277**:28200–11.

51. Gopinath SCB. Anti-coagulant aptamers. *Thromb Res* 2008;**122**:838–47.

52. Gopinath SCB, Hilda A, Lakshmipriya T, Annadurai G, Anbu P. Purification of lipase from *Geotrichum candidum*: conditions optimized for enzyme production using Box-Behnken design. *World Microbiol Biotechnol* 2003;**19**:681–9.

SARS CORONAVIRUS INFECTIONS OF THE LOWER RESPIRATORY TRACT AND THEIR PREVENTION

3

N. Petrovsky

Vaxine Pty Ltd, Department of Endocrinology, Flinders Medical Centre, Adelaide, Australia;
Flinders University, Faculty of Medicine, Adelaide, Australia

1 INTRODUCTION

The severe acute respiratory syndrome (SARS) coronavirus is a positive-stranded RNA virus, 29.7 kb in length with approximately 14 open reading frames.[1] It was identified as a human pathogen for the first time in 2003 as part of an intensive investigation into the cause of a series of fatal pneumonia cases that started in Hong Kong but then rapidly spread to over 30 different countries.[2,3] The outbreak was eventually brought under control by quarantine measures but not before more than 8000 people worldwide were infected and there were over 800 confirmed deaths. This translated into an overall case fatality rate of ∼10% but with mortality rates approaching 50% in the elderly.[4] Those most likely to get serious complications or die from SARS virus infection were individuals over 65 years of age or who had a chronic illness, such as, diabetes or hepatitis.

SARS virus is spread by aerosol in the form of respiratory droplets or through close personal contact, being absorbed through mucous membranes. It has a typical incubation period between initial infection and development of symptoms of 3–7 days. Symptoms include a high fever, dry cough, shortness of breath, headache, muscle aches, sore throat, fatigue, and diarrhea. Lung histology in fatal cases revealed a marked diffuse alveolar damage, inflammatory cell infiltrate, bronchial epithelial denudation, loss of cilia, and squamous metaplasia.[5] Deaths resulted from respiratory, heart and/or liver failure. Infected individuals recovering from coronavirus infections may become susceptible to reinfection due to rapidly waning immunity.[6,7] In fact, those with waning immunity may be at risk of even more severe disease upon coronavirus reinfection.[8] SARS homologous reinfection studies showed that although immune animals cleared lung virus much faster than naive animals, the incidence and severity of lung inflammation was not reduced.[9] This suggests that illness severity is not just dictated by SARS viral load but is also influenced by host factors. Given the possibility of future human outbreaks, development of a safe and effective coronavirus vaccine platform would be beneficial. In particular, lessons learned from development of SARS vaccines may also be applicable to other coronavirus infections, such as, the recently emerged Middle East Respiratory Syndrome (MERS) coronavirus.[10]

2 INACTIVATED WHOLE VIRUS VACCINES

Initial attempts to produce SARS vaccine candidates were based on traditional methods where SARS virus was grown in cell culture under biosafety level (BSL)-3 conditions and then inactivated with formaldehyde, beta-propriolactone, ultraviolet irradiation, or a combination of these. These initial inactivated whole virus vaccines provided modest protection in animal models, inducing low titers of neutralizing antibody and earlier lung virus clearance but did not completely prevent infection.[11] One such inactivated vaccine was administered to a small number of human subjects in a phase 1 clinical trial and was shown to be able to induce neutralizing antibodies.[12] Although no immediate safety issues were identified, the subjects were not exposed to SARS virus and hence any risk of vaccine exacerbation of lung immune-pathology by this vaccine remains unknown.

Of note, animals immunized with similar inactivated vaccines developed a severe lung eosinophilic immunopathology when challenged with SARS virus.[13–16] This lung pathology was exacerbated further when SARS vaccines were formulated with alum adjuvant in an attempt to increase their immunogenicity.[9] Despite inactivated vaccines combined with alum adjuvant being able to reduce virus replication in young animals, they failed to reduce virus replication in older animals in which their use was associated with severe lung eosinophilic pathology postchallenge.[13] This finding is particularly concerning, as the elderly are a major target population for SARS vaccines, being most at risk of complications and mortality. If the animal data translates to humans, this suggests that inactivated vaccines might not only fail to protect the elderly but may even increase their risk of serious illness if exposed to SARS virus.

The problem of lung eosinophilic pathology was also seen in mice immunized with Venezuelan equine encephalitis virus replicon particles expressing the SARS nucleocapsid protein.[17] Similarly, mice immunized with vaccinia virus encoding the SARS nucleocapsid protein developed a severe pneumonia characterized by increased Th2 and Th1 cytokines, reduced IL-10 and TGF-beta, thickening of the alveolar epithelium and lung infiltration by eosinophils, lymphocytes, and neutrophils.[15] Hence, given these problems of poor immunogenicity, difficulty of BSL-3 manufacture, poor efficacy in the elderly, and risk of exacerbated disease due to eosinophilic lung immunopathology, inactivated whole virus formulations are poor candidates for safe and effective SARS vaccines.

3 RECOMBINANT SPIKE PROTEIN VACCINES

A major advance in vaccine development was the identification that the SARS virus spike (S) protein mediates cell entry via its ability to bind angiotensin-converting enzyme 2 and CD209L, thereby triggering virus endocytosis into target cells.[18,19] A human monoclonal antibody binding the S protein N-terminal domain was shown to be able to block infection, thereby identifying S protein as a major target of SARS virus neutralizing antibodies.[20] Consistent with this, monkeys could be protected against SARS infection by intranasal immunization with a S protein-encoding live parainfluenza vector.[21] S protein was also shown to be the target of CD4 and CD8 T cell responses suggesting these may also be important to SARS protection.[22] A recombinant S protein vaccine was manufactured using an insect cell expression system but was found to be considerably less immunogenic that inactivated whole virus vaccine, requiring ~100 times more antigen to achieve the same level of immunogenicity.[23] Attempts to improve the immunogenicity of S protein vaccine by formulation with alum adjuvant again resulted in severe lung eosinophilic immunopathology in response to SARS virus infection, marking this as another potentially unsafe approach.[13,16] This confirmed that the problem of lung eosinophilic immunopathology was not just confined to inactivated or nucleocapsid protein vaccines but was a more general problem of vaccines made from any SARS virus antigen.

4 SARS-ASSOCIATED EOSINOPHILIC LUNG IMMUNOPATHOLOGY

What is the mechanistic basis of lung eosinophilic immunopathology? This question is not just of academic interest as SARS-associated lung immunopathology bears a striking resemblance to the lung eosinophilic pathology previously seen with an alum-adjuvanted formalin-inactivated respiratory syncytial virus (RSV) vaccine. In clinical trials this vaccine caused increased mortality when immunized children became infected with RSV.[24] Hence, any SARS vaccine that induces lung eosinophilic immunopathology could be equally unsafe in human subjects.

Lung eosinophilic immunopathology did not arise in naive mice pretreated with anti-S protein antibody before challenge suggesting it is not due to an antibody-dependent enhancement mechanism.[17] Lung immunopathology might instead be due to an aberrant host immune response to SARS virus, exacerbated by vaccine priming. As the ability of inactivated or recombinant vaccines to induce eosinophilic immunopathology is dramatically exacerbated by formulation with alum adjuvant, a known Th2-polarising adjuvant, this implies that the problem might be vaccines that prime an excessive Th-2-response and/or that fail to prime for a sufficient Th1 response. SARS virus itself mediates broad ranging immune modulatory effects that might be expected to lead to strong Th2 immune bias. S protein binds lung surfactant protein D, a collectin found in the lung that activates macrophages but not dendritic cells (DC).[25] Autopsy lung samples of patients dying after SARS infection revealed down regulation of type 1 interferon and CXCL10, a chemokine involved in T cell recruitment and inhibition of the STAT1 pathway.[26] Conversely, SARS virus enhances production of interleukin-6 (IL-6) and IL-8, inflammatory cytokines that inhibit the ability of DC to prime T cells.[27] GU-rich ssRNAs from SARS virus activate TLR7 and TLR8 in mice and thereby induce high levels of pro-inflammatory cytokines including TNF-α, IL-6, and IL-12 leading to lethal acute lung injury.[28] Thus SARS virus has the capacity to exacerbate lung inflammation while at the same time inhibiting antiviral interferon responses.

SARS virus uses multiple mechanisms to inhibit the host type 1 interferon response with open reading frame (ORF) 3b, ORF 6, and nucleocapsid proteins all playing a major role; nucleocapsid protein inhibits the synthesis of interferon while ORF 3b and ORF 6 proteins inhibit both interferon synthesis and signaling.[29] In addition, ORF 6 protein inhibits nuclear translocation of STAT1. The importance of STAT1 to protection is demonstrated by STAT1 knockout (STAT1$^{-/-}$) mice being unable to clear SARS virus infection, resulting in increased virus lethality.[30] Furthermore, STAT1$^{-/-}$ mice infected with SARS virus had evidence of T cell and macrophage dysregulation with increased alternatively activated macrophages and a Th2-biased immune response. As STAT6 is essential for development of alternatively activated macrophage, the importance of alternatively activated macrophages to SARS pathology was able to be demonstrated using STAT1/STAT6 double-knockout mice, which exhibited reduced lung disease in response to SARS virus challenge.[31] On the host side, CD8 T cells are responsible for virus clearance and adoptive transfer of immune splenocytes or SARS-specific T cells reduced lung virus titers and enhanced survival.[32] The importance of Th1-cellular immunity to viral control may help explain the many mechanisms that SARS virus has developed to subvert Th1-cellular responses. By suppressing Th1 responses, the SARS virus may thereby impart an unbalanced Th2 bias to the antiviral lung immune response, which could amplify and exaggerate any preexisting Th2 bias already imparted by immunization with a Th-2 biased SARS vaccine. This thereby provides a plausible explanation for the vaccine-exacerbated lung eosinophilic pathology observed in recipients of alum-adjuvanted or unadjuvanted inactivated whole virus and recombinant S protein vaccines. This is supported by data from a ferret SARS virus reinfection model where reinfected animals or ferrets previously immunized with an alum-adjuvanted inactivated vaccine candidate failed to mount an effective type 1 interferon response.[33] In mice, SARS-associated lung pathology was shown to be prevented by pretreatment of alveolar macrophages with poly(I:C), a TLR3 agonist, which may reflect its ability to prime for a strong type 1 interferon response.[34] Hence the aberrant host immune response causing SARS-associated lung immunopathology is complex, with roles played by alternatively activated macrophages, neutrophils, NFkB activation, lack of a type 1 interferon response, and excess IL-Iβ, IL-6, and TNF production due to inflammasome activation by the SARS envelope protein-encoded ion channel.[31,35–39]

5 SARS PATHOLOGY IN THE ELDERLY

Elderly human subjects infected with SARS virus experienced an extremely high mortality rate, approaching 50%.[4] SARS-infected aged macaques similarly developed more severe pathology than young adult animals, even though viral replication levels were similar. This suggests that increased SARS mortality in the elderly may reflect as much an aberrant host response as toxic effects of the virus. In keeping with this, aged macaques showed greater NFkB activation in response to SARS infection and this was associated with increased inflammatory gene expression but lower expression of type I interferon.[40] Treatment of aged animals with type 1 interferon reduced expression of inflammatory genes, such as IL-8, and reduced lung pathology, despite not changing lung virus levels.[40] It has been shown that aged mice exhibit increased lung prostaglandin D2 in response to lung infection that correlates with impairment of DC migration and lower T cell responses.[41] More severe clinical disease in aged animals could be attenuated by blocking prostaglandin D2 with small-molecule antagonists.[41] Thus, multiple factors likely contribute to the increased mortality in elderly subjects suffering from SARS infection, including intrinsic DC defects, increased production of DC-inhibitory factors, such as prostaglandin D2, predisposition to an increased inflammatory response, and reduced Th1 immune responses. An ongoing challenge is to develop a SARS virus vaccine strategy that is able to overcome all of these obstacles and is thereby safe and effective in the elderly.

6 PREVENTION OF VACCINE-EXACERBATED SARS LUNG IMMUNOPATHOLOGY

There is a critical need to identify suitable SARS vaccines that do not run the risk of exacerbating coronavirus-associated lung immunopathology. Advax™ is a polysaccharide adjuvant based on microcrystalline particles of β-D-[2-1]poly(fructo-furanosyl)α-D-glucose (delta inulin).[42–44] It has been shown to enhance vaccine immunity against a wide variety of pathogens, including influenza,[45,46] Japanese encephalitis,[47] West Nile virus,[48] and hepatitis B,[49] enhancing both humoral and cellular immunity, and inducing a balanced Th1-Th2 response.[46,49] It has also been shown to be safe and nonreactogenic in human vaccine trials.[50,51] When combined with either recombinant S protein or inactivated SARS, Advax™ adjuvant alone or combined with a TLR9 agonist enhanced neutralizing antibody and cellular immune responses, reduced lung viral load and completely protected against SARS lethality.[52] Notably, animals immunized with Advax™ adjuvant formulations had minimal or no lung eosinophilic pathology postchallenge in stark contrast to the severe immunopathology observed in animals immunized with antigen alone or combined with alum adjuvant (Fig. 3.1) thereby highlighting the critical importance of adjuvant selection for SARS vaccine development. Protection against lung immunopathology correlated with a strong memory Th1 response in the Advax™-immunized animals.[52]

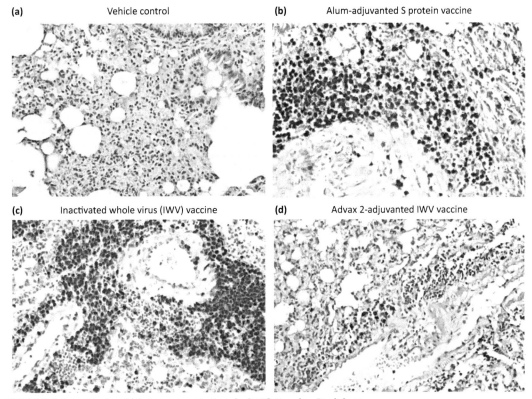

| (a) Vehicle control | (b) Alum-adjuvanted S protein vaccine |
| (c) Inactivated whole virus (IWV) vaccine | (d) Advax 2-adjuvanted IWV vaccine |

FIGURE 3.1 Lung Eosinophilic Immunopathology in SARS Vaccine Recipients

Mice were immunized with various SARS antigen formulations then challenged with SARS virus. Lungs were harvested day 6 postchallenge for staining with H&E plus a specific eosinophil stain (*brown*).[52] After SARS challenge, mice that received only vehicle control had marked lung inflammation but minimal eosinophils (a) By contrast, mice immunized with either alum-adjuvanted S protein (b) or unadjuvanted inactivated whole virus (c) exhibited florid lung eosinophil infiltration. However, minimal eosinophils were seen in mice immunized with the same inactivated whole virus vaccine formulated with Advax-2 adjuvant (d).

Pictures courtesy of Bi-Hung Peng, University of Texas Medical Branch, Galveston, Texas.

TLR agonists including lipopolysaccharide, poly(U) or poly(I:C) when formulated with inactivated SARS virus, have similarly been shown to help prevent lung immunopathology.[53] What all of these successful strategies have in common is that they prime for a Th1-immune response, suggesting that this is the key to avoiding lung eosinophilic immunopathology.

7 CONCLUSIONS AND FUTURE PROSPECTS

Protection against SARS virus involves coordinated action of memory CD4 and CD8 T cells and neutralizing antibody. Vaccines that fail to induce Th1 immunity against the SARS virus run the risk of exacerbating lung disease.[54] Hence, the key to any successful SARS vaccine will be the inclusion of an adjuvant able to induce robust cellular immunity together with long-lived neutralizing antibodies. Important questions remain on how best to protect elderly subjects, the most vulnerable population for lethal human coronaviruses. Another important issue is how best to provide protection against heterologous SARS virus strains. Unfortunately, with the passage of time since the SARS epidemic, funding for SARS research has dried up and some of these questions may never be answered. While the SARS epidemic was halted by quarantine measures, there is an ongoing need for a safe and effective vaccine platform that could be used in the event of future human coronavirus threats. Notably, the MERS coronavirus has emerged to become a major human threat. Unlike SARS, MERS has not been halted by quarantine procedures, creating an urgent need for a safe and effective MERS vaccine.[55] It is not yet known whether MERS vaccines will suffer from the same problems as SARS vaccines, such as, low immunogenicity and lung eosinophilic immunopathology, but this seems likely. Hence, similar strategies as described above will be required to make MERS vaccines safe and effective. MERS will almost certainly not be the last new coronavirus to cause human disease, identifying the need for ongoing research into the unique host-pathogen interactions contributing to coronavirus pathology and infection outcomes, and to develop an effective coronavirus vaccine platform.

ACKNOWLEDGMENTS

NP is supported by Federal funds from the National Institute of Allergy and Infectious Diseases, National Institute of Health, Department of Health and Human Services under contracts HHSN272200800039C and U01AI061142. The content is solely the responsibility of the author and the funders played no role in the writing of this manuscript.

REFERENCES

1. Marra MA, Jones SJ, Astell CR, Holt RA, Brooks-Wilson A, Butterfield YS, et al. The genome sequence of the sars-associated coronavirus. *Science* 2003;**300**:1399–404.
2. Drosten C, Gunther S, Preiser W, van der Werf S, Brodt HR, Becker S, et al. Identification of a novel coronavirus in patients with severe acute respiratory syndrome. *N Engl J Med* 2003;**348**:1967–76.
3. Rota PA, Oberste MS, Monroe SS, Nix WA, Campagnoli R, Icenogle JP, et al. Characterization of a novel coronavirus associated with severe acute respiratory syndrome. *Science* 2003;**300**:1394–9.

4. Graham RL, Donaldson EF, Baric RS. A decade after sars: strategies for controlling emerging coronaviruses. *Nat Rev Microbiol* 2013;**11**:836–48.

5. Nicholls J, Dong XP, Jiang G, Peiris M. Sars: clinical virology and pathogenesis. *Respirology* 2003;**8**(Suppl):S6–8.

6. Callow KA, Parry HF, Sergeant M, Tyrrell DA. The time course of the immune response to experimental coronavirus infection of man. *Epidemiol Infect* 1990;**105**:435–46.

7. Tang F, Quan Y, Xin ZT, Wrammert J, Ma MJ, Lv H, et al. Lack of peripheral memory B cell responses in recovered patients with severe acute respiratory syndrome: a six-year follow-up study. *J Immunol* 2011;**186**:7264–8.

8. Wang SF, Tseng SP, Yen CH, Yang JY, Tsao CH, Shen CW, et al. Antibody-dependent sars coronavirus infection is mediated by antibodies against spike proteins. *Biochem Biophys Res Commun* 2014;**451**:208–14.

9. Clay C, Donart N, Fomukong N, Knight JB, Lei W, Price L, et al. Primary severe acute respiratory syndrome coronavirus infection limits replication but not lung inflammation upon homologous rechallenge. *J Virol* 2012;**86**:4234–44.

10. Assiri A, McGeer A, Perl TM, Price CS, Al Rabeeah AA, Cummings DA, et al. Hospital outbreak of middle east respiratory syndrome coronavirus. *N Engl J Med* 2013;**369**:407–16.

11. Darnell ME, Plant EP, Watanabe H, Byrum R, St Claire M, Ward JM, et al. Severe acute respiratory syndrome coronavirus infection in vaccinated ferrets. *J Infect Dis* 2007;**196**:1329–38.

12. Lin JT, Zhang JS, Su N, Xu JG, Wang N, Chen JT, et al. Safety and immunogenicity from a phase I trial of inactivated severe acute respiratory syndrome coronavirus vaccine. *Antivir Ther* 2007;**12**:1107–13.

13. Bolles M, Deming D, Long K, Agnihothram S, Whitmore A, Ferris M, et al. A double-inactivated severe acute respiratory syndrome coronavirus vaccine provides incomplete protection in mice and induces increased eosinophilic proinflammatory pulmonary response upon challenge. *J Virol* 2011;**85**:12201–15.

14. See RH, Zakhartchouk AN, Petric M, Lawrence DJ, Mok CP, Hogan RJ, et al. Comparative evaluation of two severe acute respiratory syndrome (SARS) vaccine candidates in mice challenged with SARS coronavirus. *J Gen Virol* 2006;**87**:641–50.

15. Yasui F, Kai C, Kitabatake M, Inoue S, Yoneda M, Yokochi S, et al. Prior immunization with severe acute respiratory syndrome (SARS)-associated coronavirus (Sars-Cov) nucleocapsid protein causes severe pneumonia in mice infected with SARS-Cov. *J Immunol* 2008;**181**:6337–48.

16. Tseng CT, Sbrana E, Iwata-Yoshikawa N, Newman PC, Garron T, Atmar RL, et al. Immunization with Sars Coronavirus vaccines leads to pulmonary immunopathology on challenge with the Sars Virus. *PLoS ONE* 2012;**7**:e35421.

17. Deming D, Sheahan T, Heise M, Yount B, Davis N, Sims A, et al. Vaccine efficacy in senescent mice challenged with recombinant Sars-Cov bearing epidemic and zoonotic spike variants. *PLoS Med* 2006;**3**:e525.

18. Li W, Moore MJ, Vasilieva N, Sui J, Wong SK, Berne MA, et al. Angiotensin-converting enzyme 2 is a functional receptor for the Sars Coronavirus. *Nature* 2003;**426**:450–4.

19. Jeffers SA, Tusell SM, Gillim-Ross L, Hemmila EM, Achenbach JE, Babcock GJ, et al. Cd209l (L-Sign) is a receptor for severe acute respiratory syndrome coronavirus. *Proc Natl Acad Sci USA* 2004;**101**:15748–53.

20. Berger A, Drosten C, Doerr HW, Sturmer M, Preiser W. Severe acute respiratory syndrome (Sars)—paradigm of an emerging viral infection. *J Clin Virol* 2004;**29**:13–22.

21. Bukreyev A, Lamirande EW, Buchholz UJ, Vogel LN, Elkins WR, St Claire M, et al. Mucosal immunisation of African Green Monkeys (*Cercopithecus aethiops*) with an attenuated parainfluenza virus expressing the Sars Coronavirus spike protein for the prevention of Sars. *Lancet* 2004;**363**:2122–7.

22. Huang J, Cao Y, Du J, Bu X, Ma R, Wu C. Priming with Sars Cov S DNA and boosting with Sars Cov S epitopes specific for Cd4+ and Cd8+ T cells promote cellular immune responses. *Vaccine* 2007;**25**:6981–91.

23. Zhou Z, Post P, Chubet R, Holtz K, McPherson C, Petric M, et al. A recombinant baculovirus-expressed S Glycoprotein vaccine elicits high titers of Sars-associated coronavirus (Sars-Cov) neutralizing antibodies in mice. *Vaccine* 2006;**24**:3624–31.

24. Openshaw PJ, Culley FJ, Olszewska W. Immunopathogenesis of vaccine-enhanced Rsv disease. *Vaccine* 2001;**20**(Suppl. 1):S27–31.
25. Leth-Larsen R, Zhong F, Chow VT, Holmskov U, Lu J. The Sars coronavirus spike glycoprotein is selectively recognized by lung surfactant protein D and activates macrophages. *Immunobiology* 2007;**212**:201–11.
26. Kong SL, Chui P, Lim B, Salto-Tellez M. Elucidating the molecular physiopathology of acute respiratory distress syndrome in severe acute respiratory syndrome patients. *Virus Res* 2009;**145**:260–9.
27. Yoshikawa T, Hill T, Li K, Peters CJ, Tseng CT. Severe acute respiratory syndrome (Sars) Coronavirus-induced lung epithelial cytokines exacerbate Sars pathogenesis by modulating intrinsic functions of monocyte-derived macrophages and dendritic cells. *J Virol* 2009;**83**:3039–48.
28. Li Y, Chen M, Cao H, Zhu Y, Zheng J, Zhou H. Extraordinary Gu-rich single-strand Rna identified from Sars coronavirus contributes an excessive innate immune response. *Microbes Infect* 2013;**15**:88–95.
29. Kopecky-Bromberg SA, Martinez-Sobrido L, Frieman M, Baric RA, Palese P. Severe acute respiratory syndrome coronavirus open reading frame (Orf) 3b, Orf 6, and nucleocapsid proteins function as interferon antagonists. *J Virol* 2007;**81**:548–57.
30. Zornetzer GA, Frieman MB, Rosenzweig E, Korth MJ, Page C, Baric RS, et al. Transcriptomic analysis reveals a mechanism for a prefibrotic phenotype in Stat1 knockout mice during severe acute respiratory syndrome coronavirus infection. *J Virol* 2010;**84**:11297–309.
31. Page C, Goicochea L, Matthews K, Zhang Y, Klover P, Holtzman MJ, et al. Induction of alternatively activated macrophages enhances pathogenesis during severe acute respiratory syndrome coronavirus infection. *J Virol* 2012;**86**:13334–49.
32. Zhao J, Zhao J, Perlman S. T cell responses are required for protection from clinical disease and for virus clearance in severe acute respiratory syndrome coronavirus-infected mice. *J Virol* 2010;**84**:9318–25.
33. Cameron MJ, Kelvin AA, Leon AJ, Cameron CM, Ran L, Xu L, et al. Lack of innate interferon responses during Sars coronavirus infection in a vaccination and reinfection Ferret model. *PLoS ONE* 2012;**7**:e45842.
34. Zhao J, Wohlford-Lenane C, Zhao J, Fleming E, Lane TE, McCray Jr PB, et al. Intranasal treatment with poly(I*C) protects aged mice from lethal respiratory virus infections. *J Virol* 2012;**86**:11416–24.
35. Fett C, DeDiego ML, Regla-Nava JA, Enjuanes L, Perlman S. Complete protection against severe acute respiratory syndrome coronavirus-mediated lethal respiratory disease in aged mice by immunization with a mouse-adapted virus lacking E protein. *J Virol* 2013;**87**:6551–9.
36. Nieto-Torres JL, DeDiego ML, Verdia-Baguena C, Jimenez-Guardeno JM, Regla-Nava JA, Fernandez-Delgado R, et al. Severe acute respiratory syndrome coronavirus envelope protein ion channel activity promotes virus fitness and pathogenesis. *PLoS Pathog* 2014;**10**:e1004077.
37. Smits SL, van den Brand JM, de Lang A, Leijten LM, van Ijcken WF, van Amerongen G, et al. Distinct severe acute respiratory syndrome coronavirus-induced acute lung injury pathways in two different nonhuman primate species. *J Virol* 2011;**85**:4234–45.
38. Rockx B, Baas T, Zornetzer GA, Haagmans B, Sheahan T, Frieman M, et al. Early upregulation of acute respiratory distress syndrome-associated cytokines promotes lethal disease in an aged-mouse model of severe acute respiratory syndrome coronavirus infection. *J Virol* 2009;**83**:7062–74.
39. DeDiego ML, Nieto-Torres JL, Regla-Nava JA, Jimenez-Guardeno JM, Fernandez-Delgado R, Fett C, et al. Inhibition of Nf-Kappab-mediated inflammation in severe acute respiratory syndrome coronavirus-infected mice increases survival. *J Virol* 2014;**88**:913–24.
40. Smits SL, de Lang A, van den Brand JM, Leijten LM, van IWF, Eijkemans MJ, et al. Exacerbated innate host response to Sars-Cov in aged non-human primates. *PLoS Pathog* 2010;**6**:e1000756.
41. Zhao J, Zhao J, Legge K, Perlman S. Age-related increases in Pgd(2) expression impair respiratory Dc migration, resulting in diminished T cell responses upon respiratory virus infection in mice. *J Clin Invest* 2011;**121**:4921–30.
42. Petrovsky N, Aguilar JC. Vaccine adjuvants: current state and future trends. *Immunol Cell Biol* 2004;**82**:488–96.

43. Petrovsky N, Cooper PD. Carbohydrate-based immune adjuvants. *Expert Rev Vaccines* 2011;**10**:523–37.

44. Cooper PD, Petrovsky N. Delta inulin: a novel, immunologically active, stable packing structure comprising beta-ᴅ-[2 -> 1] poly(fructo-furanosyl) alpha-ᴅ-glucose polymers. *Glycobiology* 2011;**21**:595–606.

45. Layton RC, Petrovsky N, Gigliotti AP, Pollock Z, Knight J, Donart N, et al. Delta inulin polysaccharide adjuvant enhances the ability of split-virion H5n1 vaccine to protect against lethal challenge in ferrets. *Vaccine* 2011;**29**:6242–51.

46. Honda-Okubo Y, Saade F, Petrovsky N. Advax, a polysaccharide adjuvant derived from delta inulin, provides improved influenza vaccine protection through broad-based enhancement of adaptive immune responses. *Vaccine* 2012;**30**:5373–81.

47. Larena M, Prow NA, Hall RA, Petrovsky N, Lobigs M. Je-Advax vaccine protection against Japanese encephalitis virus mediated by memory B cells in the absence of Cd8+ T cells and pre-exposure neutralizing antibody. *J Virol* 2013;**87**:4395–402.

48. Petrovsky N, Larena M, Siddharthan V, Prow NA, Hall RA, Lobigs M, et al. An inactivated cell culture Japanese encephalitis vaccine (Je-Advax) formulated with delta inulin adjuvant provides robust heterologous protection against West Nile encephalitis via cross-protective memory B cells and neutralizing antibody. *J Virol* 2013;**87**:10324–33.

49. Saade F, Honda-Okubo Y, Trec S, Petrovsky N. A novel hepatitis B vaccine containing a, a polysaccharide adjuvant derived from delta inulin, induces robust humoral and cellular immunity with minimal reactogenicity in preclinical testing. *Vaccine* 2013;**31**:1999–2007.

50. Gordon DL, Sajkov D, Woodman RJ, Honda-Okubo Y, Cox MM, Heinzel S, et al. Randomized clinical trial of immunogenicity and safety of a recombinant H1n1/2009 pandemic influenza vaccine containing advax polysaccharide adjuvant. *Vaccine* 2012;**30**:5407–16.

51. Gordon D, Kelley P, Heinzel S, Cooper P, Petrovsky N. Immunogenicity and safety of advax, a novel polysaccharide adjuvant based on delta inulin, when formulated with hepatitis B surface antigen: a randomized controlled phase 1 study. *Vaccine* 2014;**32**:6469–77.

52. Honda-Okubo Y, Barnard D, Ong CH, Peng BH, Tseng CT, Petrovsky N. Severe acute respiratory syndrome-associated coronavirus vaccines formulated with delta inulin adjuvants provide enhanced protection while ameliorating lung eosinophilic immunopathology. *J Virol* 2015;**89**:2995–3007.

53. Iwata-Yoshikawa N, Uda A, Suzuki T, Tsunetsugu-Yokota Y, Sato Y, Morikawa S, et al. Effects of toll-like receptor stimulation on eosinophilic infiltration in lungs of Balb/C mice immunized with Uv-inactivated severe acute respiratory syndrome-related coronavirus vaccine. *J Virol* 2014;**88**:8597–614.

54. Channappanavar R, Fett C, Zhao J, Meyerholz DK, Perlman S. Virus-specific memory Cd8 T cells provide substantial protection from lethal severe acute respiratory syndrome coronavirus infection. *J Virol* 2014;**88**:11034–44.

55. Ma C, Wang L, Tao X, Zhang N, Yang Y, Tseng CT, et al. Searching for an ideal vaccine candidate among different mers coronavirus receptor-binding fragments—the importance of immunofocusing in subunit vaccine design. *Vaccine* 2014;**32**:6170–6.

THE MIDDLE EAST RESPIRATORY SYNDROME CORONAVIRUS RESPIRATORY INFECTION: AN EMERGING INFECTION FROM THE ARABIAN PENINSULA

4

J.A. Al-Tawfiq*,**, Z.A. Memish[†,‡]

*Speciality Internal Medicine Department, Johns Hopkins Aramco Healthcare, Dhahran, Kingdom of Saudi Arabia; **Department of Medicine, Indiana University School of Medicine, Indianapolis, IN, United States; [†]Ministry of Health, Riyadh, Kingdom of Saudi Arabia; [‡]Alfaisal University, College of Medicine, Riyadh, Kingdom of Saudi Arabia

1 INTRODUCTION

Coronaviruses (CoV) are a group of viruses known to cause mild to severe diseases in humans. Known human coronaviruses causing disease belong to the genera alpha-coronavirus and beta-coronavirus. These viruses usually cause mild upper respiratory tract disease in humans. The Middle East Respiratory Syndrome Coronavirus (MERS-CoV) belongs to the beta-coronaviruses and was first identified in the Kingdom of Saudi Arabia in 2012.[1] The virus was isolated from the sputum of a 60-year-old man who presented with community acquired pneumonia and subsequently developed a fatal disease associated with acute renal failure and respiratory failure.[1] Since Apr. 2012 to date, the virus has caused a total of 1611 cases including 575 deaths that were reported by the World Health Organization in 26 countries.[2] The majority of these cases occurred in the Arabian Peninsula and the other cases were linked to this geographic area, usually through travel. The disease has a wide range of clinical presentation and epidemiology.[3–7] The clinical spectrum ranges from mild disease to a rapidly fatal disease. The presence of asymptomatic cases was also described. Three main factors contribute to the transmission of MERS-CoV, these are the virus, the host, and the environment. Cases occurred as sporadic patients, limited intrafamilial transmission, and clusters of healthcare associated transmissions. The sporadic cases may result from camel to human transmission with subsequent cases being secondary cases among human contacts. The virus seems to have a peculiar tendency to cause healthcare-associated transmissions as exemplified by multiple hospital outbreaks, as will be discussed later. The emergence of MERS-CoV caused great attention to the emergent respiratory pathogens and the potential for global spread of the disease with the current spread of globalization. Understanding the pathogen, the mode of transmission, and the spectrum of the diseases allows the development of preventive measures and the application of effective infection control practices. The prospect for the development of a novel therapy or the use of previous therapy for the treatment of MERS-CoV would further enhance our abilities to combat the disease. Here, we review the epidemiology of the disease, clinical presentations, and the outcome.

2 THE ORGANISM

Coronaviruses are parts of the Nidovirales order. The name stems from the presence of crown-like spikes on their surfaces. Coronaviruses were first identified as human pathogens in the mid-1960s. Coronaviruses are enveloped RNA viruses and there are four virus clusters within the Coronavirinae subfamily: alpha, beta, gamma, and delta coronaviruses. Pathogenic human coronaviruses are classified into the genera alpha-coronavirus (HCoV-229E and HCoV-NL63) and beta-coronavirus (HCoV-OC43, HCoV-HKU1, and SARS-CoV).[1] MERS-CoV emerged as a significant pathogen after the initial identification in 2012 from a patient with rapidly fatal community acquired pneumonia and is the first human coronavirus in lineage C of the beta-coronavirus genus.[1,8] The MERS-CoV virus is known to have multiple clades circulating in humans. In one study, four different phylogenetic MERS-CoV clades were circulating in Saudi Arabia in Sep. 2012 to May 2013.[9] Only one clade persisted at the end of the observation period.[9] The length of each clade was different: Al-Hasa clade from Apr. 21, 2013 to Jun. 22, 2013 (62 days), Riyadh_3 clade from Feb. 5, 2013 to Jul. 2, 2013 (147 days), Buraidah_1 clade from May 3, 2013 to Aug. 5, 2013 (84 days), and Hafr-Al-Batin_1 clade from Jun. 4, 2013 to Oct. 1, 2013 (119 days).[9] Most of the cases in the 2014 Jeddah outbreak belong to a single clade indicating human-to-human transmission.[10] The imported case into South Korea showed that the MERS-CoV is a recombinant of groups 3 and 5 elements and that the recombination event occurred in the second half of 2014.[11]

3 MERS-COV EPIDEMIOLOGY

Since Apr. 2012 to Oct. 2015, a total of 1611 cases including 575 deaths have been reported by the World Health Organization in 26 countries.[2] Most of these cases were reported from Saudi Arabia (Table 4.1). Multiple healthcare associated infections occurred within Saudi Arabia and contributed to the significant increase in the number of the cases. The most studied outbreaks occurred in Al-Hasa,[7] Jeddah,[12–16] and Riyadh.[12–16] The Al-Hasa outbreak occurred in Apr. 2013 and involved 23 confirmed cases and 11 probable cases of MERS-CoV in 4 hospitals.[7] In Mar.–Apr. 2014, a large number of cases were reported in Saudi Arabia and the United Arab Emirates.[12–16] During the 2014 Jeddah outbreak, a total of 14 hospitals were involved and they had a total of 128 cases.[10,13] The largest outbreak outside the Arabian Peninsula occurred in the Republic of Korea and was initiated by an index patient after returning from a trip to multiple countries in the Middle East (Bahrain/Saudi Arabia/UAE/Qatar).[16] In about 2 weeks, the outbreak involved 5 health care facilities and there were 63 cases.[17] Subsequently, the outbreak in the Republic of Korea involved 72 health care facilities and 6 health care facilities had nosocomial transmission.[18] The total number of cases as of Jun. 26, 2015 were 182 cases with 31 deaths.[19,20]

Table 4.1 Number of Cases and Deaths of MERS-CoV Among Most Frequent Countries

Country	Number of Cases	Number of Deaths (% Case Fatality Rate)
Saudi Arabia	1255	539 (43)
South Korea	185	36 (19.5)
United Arab Emirates	81	11 (13.6)
Jordan	35	14 (40)
Qatar	13	5 (38.5)
All countries	1611	275 (35.7)

4 CLINICAL PRESENTATIONS

The clinical presentation of MERS-CoV varies from asymptomatic or mildly symptomatic cases to severe and often fatal disease. A large number of the patients had underlying medical comorbidities.[3–7] These comorbidities include: diabetes mellitus (44%), cardiac disease (21%), renal failure (26%), hemodialysis (6.2%), and hypertension (24%) (Table 4.2).[6,7,21–25]

According to the Saudi Ministry of Health, 38% of the cases were primary, 45% were healthcare-associated infection, and 14% were household infections.[26] These numbers summarize three epidemiological pattern of the disease: sporadic cases occurring in the communities, probably from an animal contact, and human to human transmission as a result of healthcare-associated infection and intrafamilial transmission of MERS-CoV.[3,5,7,27–29]

Most of the affected patients were adults with a mean age of 56 years (range: 14–94) years[4,21] and a number of pediatric cases were described.[30–32] A study of 1898 combined nasal and throat swabs yielded no MERS-CoV by PCR in children <2 years of age in Jordan.[33] The relative low number of MERS-CoV in children is not readily explained.

Although, initially MERS-CoV cases were severe requiring intensive care unit services, subsequent cases included less severe disease.[34] The proportion of asymptomatic cases varied from 0% to 30%.[34] The initial phase of the clinical illness is nonspecific and includes fever and mild nonproductive cough lasting several days.[4,7] Progressive pneumonia then follows with multiorgan failure and this may result in death with a case fatality rate of 30% to 60%.[4,21] Most of the patients present with fever (87%), cough (87%), and shortness of breath (48%) (Table 4.2).[4,7] About 35% of patients may have gastrointestinal symptoms such as: diarrhea (22%) and vomiting (17%). Of the total cases, 50% had 2 medical comorbidities, diabetes, and chronic renal disease.[4] Acute renal failure developed in a proportion of patients, and three patients developed neurological signs: altered level of consciousness, confusion or coma, ataxia, and focal motor deficit.[22]

Many nonspecific laboratory abnormalities exist in patients with MERS-CoV and include: leucopenia (14%), lymphopenia (34%), thrombocytopenia (36%), increased lactate dehydrogenase (LDH) (49%), and increased hepatic transaminases (11–15%).[4,7,21,22–25,35] Chest radiographic abnormalities include: increased bronchovascular markings (17%), unilateral infiltrate (43%), bilateral infiltrates (22%), and diffuse reticulonodular pattern (4%).[7] Other studies showed ground-glass opacity in 66% and consolidation in 18%.[36–37] In one study utilizing CT-scan imaging, the lower lobes were more commonly

Table 4.2 Most Common Underlying Comorbidities, Clinical Signs and Symptoms, and Laboratory Findings in Patients With MERS-CoV From Various Studies

	%
Comorbidities	
Diabetes Mellitus	44
Cardiac disease	20.7
Renal failure	25.9
Hemodialysis	6.2
Malignancy	1.6
Hypertension	23.8
Clinical signs and symptoms	
Fever	75.6
Dyspnea	61.7
Chest pain	15
Cough	62.2
Hemoptysis	8.3
Sore throat	6.7
Headache	9.8
Myalgia	15.5
Vomiting	20.7
Diarrhea	22.8
Weakness	18.7
Abdominal pain	14
Rhinorrhea	4.7
Lymphopenia	31.6
Thrombocytopenia	11.9

involved than the upper and middle lobes combined.[37] In fatal cases, the mean number of lung segments involved was 12.3 segments compared to 3.4 segments in those who survived.[37]

Laboratory diagnosis relies on respiratory tract samples for the detection of MERS-CoV using real-time reverse transcriptase polymerase-chain-reaction (RT-PCR). The virus may be detected in the lower and upper respiratory tract samples. Lower respiratory tract samples yielded better diagnostic results,[38] and had higher viral loads.[39] Lower respiratory tract samples had the highest viral loads (mean 5.01×10^6 copies/mL), compared with upper respiratory tract samples (2×10^4 copies/mL), urine (1.26×10^2 copies/mL), stool (1.58×10^4 copies/mL), and serum (2.51×10^3 copies/mL).[39] Serologic tests had been used for the diagnosis of MERS-CoV.[40,41] Data on the sensitivity and specificity of antibody tests for MERS-CoV are limited. In one study, the use of plaque reduction neutralization tests

(PRNT), microneutralisation (MN), MERS-spike pseudoparticle neutralization (ppNT) and MERS S1-enzyme-linked immunosorbent assay (ELISA) were found to be sensitive and specific.[41]

5 TREATMENT OF MERS-COV

The main therapeutic options for MERS-CoV infection are not known. In vitro, MERS-CoV is sensitive to alpha interferon (IFN-α).[42] No randomized controlled trials exists to establish the efficacy and side effects of any therapeutic modalities. Learning from the SARS experience, interferon and ribavirin was suggested as a therapy for MERS-CoV.[43] The combination of interferon-α2b and ribavirin prevented pneumonia in animals.[44] The first report of the use of ribavirin and interferon showed no survival advantage[45] because the combination was started late in the course of the disease.[45] A 14-day survival advantage was documented with this combination but there was no survival advantage at 28 days.[24] There was no difference in therapy between interferon-α2a with ribavirin and interferon-β1a with ribavirin in treating MERS-CoV.[25] In a case report from Greece, pegylated interferon, ribavirin, and lopinavir/ritonavir was initiated on day 13 of illness.[46] MERS-CoV was detectable in the respiratory tract secretions of the patient for 4 weeks after onset illness and viraemia lasted 2 days after initiation of therapy.[46]

6 PREVENTIVE AND CONTROL OF MERS-COV

The prospect for the control and prevention of MERS-CoV relies on the identification of the definite host, the interruption of the animal to human transmission, and the application of the proper infection control measures in the healthcare settings. The available data links dromedary camels with human cases of MERS-CoV.[47] A high prevalence of MERS-CoV antibodies was detected in dromedary camels from across the Arabian Peninsula, North Africa, and Eastern Africa.[48-54] In addition, viral MERS-CoV was detected in samples from dromedary camels in multiple locations in the Arabian Peninsula using RT-PCR.[52,54-61] The main infection control measures in healthcare settings include: contact isolation, droplet isolation, and airborne infection isolation precautions especially when during aerosol generating procedures.[62] The centers for disease control and preventions (CDC) recommends placing patients with suspected or confirmed MERS-CoV infection in an airborne infection isolation rooms (AIIR).[63]

7 SUMMARY

MERS-CoV infection is an emerging infectious disease with a high mortality rate. The exact incidence and prevalence of the disease was evaluated in a large population based survey using serology in the Kingdom of Saudi Arabia. The study showed that anti-MERS-CoV antibodies were present in 0·15% of 10,009 people.[40] The mean age of seropositive individuals was significantly younger than that of patients with reported, laboratory-confirmed, primary MERS (43·5 years vs 53·8 years), and that men had a higher antibody prevalence than did women [11 (0·25%) of 4341 vs two (0·05%) of 4378] and antibody prevalence was significantly higher in central versus coastal provinces [14 (0·26%) of 5479 vs one (0·02%) of 4529].[40] The diagnosis of MERS-CoV infection relies on detection of the virus using real-time RT-PCR. Currently, the best therapeutic options for MERS-CoV are not known and there are no available vaccines.

REFERENCES

1. Zaki AM, van Boheemen S, Bestebroer TM, Osterhaus AD, Fouchier RA. Isolation of a novel coronavirus from a man with pneumonia in Saudi Arabia. *N Engl J Med* 2012;**367**:1814–20.
2. World Health Organization. www.who.int/emergencies/mers-cov/en/.
3. AlBarrak AM, Stephens GM, Hewson R, Memish ZA. Recovery from severe novel coronavirus infection. *Saudi Med J* 2012;**33**:1265–9.
4. Assiri A, Al-Tawfiq JA, Al-Rabeeah AA, Al-Rabiah FA, Al-Hajjar S, Al-Barrak A, Flemban H, Al-Nassir WN, Balkhy HH, Al-Hakeem RF, Makhdoom HQ, Zumla AI, Memish ZA. Epidemiological, demographic, and clinical characteristics of 47 cases of Middle East respiratory syndrome coronavirus disease from Saudi Arabia: a descriptive study. *Lancet Infect Dis* 2013;**13**(9):752–61.
5. Bermingham A, Chand MA, Brown CS, Aarons E, Tong C, Langrish C, Hoschler K, Brown K, Galiano M, Myers R, Pebody RG, Green HK, Boddington NL, Gopal R, Price N, Newsholme W, Drosten C, Fouchier RA, Zambon M. Severe respiratory illness caused by a novel coronavirus, in a patient transferred to the United Kingdom from the Middle East, September 2012. *Euro Surveill* 2012;**17**(40):20290.
6. Buchholz U, Müller MA, Nitsche A, Sanewski A, Wevering N, Bauer-Balci T, Bonin F, Drosten C, Schweiger B, Wolff T, Muth D, Meyer B, Buda S, Krause G, Schaade L, Haas W. Contact investigation of a case of human novel coronavirus infection treated in a German hospital, October–November 2012. *Euro Surveill* 2013;**18**(8).
7. Assiri A, McGeer A, Perl TM, et al. Hospital outbreak of Middle East respiratory syndrome coronavirus. *N Engl J Med* 2013;**369**(5):407–16.
8. de Groot RJ, Baker SC, Baric RS, et al. Middle East respiratory syndrome coronavirus (MERS-CoV): announcement of the Coronavirus Study Group. *J Virol* 2013;**87**:7790–2.
9. Cotten M, Watson SJ, Zumla AI, et al. Spread, circulation, and evolution of the Middle East respiratory syndrome coronavirus. *MBio* 2014;**5**:e01013–62.
10. Drosten C, Muth D, Corman VM, Hussain R, Al Masri M, HajOmar W, Landt O, Assiri A, Eckerle I, Al Shangiti A, Al-Tawfiq JA, Albarrak A, Zumla A, Rambaut A, Memish ZA. An observational, laboratory-based study of outbreaks of Middle East respiratory syndrome coronavirus in Jeddah and Riyadh, kingdom of Saudi Arabia, 2014. *Clin Infect Dis* 2015;**60**(3):369.
11. Wang Y, Liu D, Shi W, Lu R, Wang W, Zhao Y, Deng Y, Zhou W, Ren H, Wu J, Wang Y, Wu G, Gao GF, Tan W. Origin and possible genetic recombination of the Middle East respiratory syndrome coronavirus from the first imported case in China: phylogenetics and coalescence analysis. *MBio* 2015;**6**(5) pii: e01280-15.
12. World Health Organization. Middle East respiratory syndrome coronavirus (MERS-CoV): summary of current situation, literature update and risk assessment–as of 5 February 2015. www.who.int/csr/disease/coronavirus_infections/mers-5-february-2015.pdf?ua=1; 2015.
13. Oboho IK, Tomczyk SM, Al-Asmari AM, et al. 2014 MERS-CoV outbreak in Jeddah—a link to health care facilities. *N Engl J Med* 2015;**372**:846.
14. World Health Organization. Middle East respiratory syndrome coronavirus (MERS-CoV) summary and literature update – as of 9 May 2014. www.who.int/csr/disease/coronavirus_infections/MERS_CoV_Update_09_May_2014.pdf?ua=1; 2015.
15. Fagbo SF, Skakni L, Chu DKW, et al. Molecular epidemiology of hospital outbreak of Middle East respiratory syndrome, Riyadh, Saudi Arabia. *Emerg Infect Dis* 2015 Nov;**21**(11):1981–8.
16. WHO. Middle East respiratory syndrome coronavirus (MERS-CoV)—Republic of Korea. Available at: www.who.int/csr/don/30-may-2015-mers-korea/en/; 2015.

17. FluTrackers. South Korea Coronavirus MERS Case List—including imported and exported cases. Available at: https://flutrackers.com/forum/forum/novel-coronavirus-ncov-mers-2012-2014/novel-coronavirus-who-chp-wpro-ecdc-oie-fao-moa-reports-and-updates/south-korea-coronavirus/732065-south-korea-coronavirus-mers-case-list-including-imported-and-exported-cases; 2015.

18. World Health Organization. Middle East respiratory syndrome coronavirus (MERS-CoV): summary and risk assessment of current situation in the Republic of Korea and China—as of 19 June 2015. Available at: www.who.int/emergencies/mers-cov/mers-cov-republic-of-korea-and-china-risk-assessment-19-june-2015.pdf?ua=1; 2015.

19. World health organization. Middle East respiratory syndrome coronavirus (MERS-CoV). MERS-CoV in Republic of Korea at a glance. Available at: www.wpro.who.int/outbreaks_emergencies/wpro_coronavirus/en/; 2015.

20. Cowling BJ, Park M, Fang VJ, Wu P, Leung GM, Wu JT. Preliminary epidemiological assessment of MERS-CoV outbreak in South Korea, May to June 2015. *Euro Surveill* 2015;**20**(25) pii=21163.

21. Al-Tawfiq JA, Memish ZA. Managing MERS-CoV in the healthcare setting. *Hosp Pract* 2015;**43**(3): 158–63.

22. Arabi YM, Harthi A, Hussein J, Bouchama A, Johani S, Hajeer AH, Saeed BT, Wahbi A, Saedy A, AlDabbagh T, Okaili R, Sadat M, Balkhy H. Severe neurologic syndrome associated with Middle East respiratory syndrome corona virus (MERS-CoV). *Infection* 2015;**43**(4):495–501.

23. Al-Tawfiq JA, Hinedi K, Ghandour J, Khairalla H, Musleh S, Ujayli A, Memish ZA. Middle East respiratory syndrome coronavirus: a case-control study of hospitalized patients. *Clin Infect Dis* 2014;**59**(2):160–5.

24. Omrani AS, Saad MM, Baig K, Bahloul A, Abdul-Matin M, Alaidaroos AY, Almakhlafi GA, Albarrak MM, Memish ZA, Albarrak AM. Ribavirin and interferon alfa-2a for severe Middle East respiratory syndrome coronavirus infection: a retrospective cohort study. *Lancet Infect Dis* 2014;**14**(11):1090–5.

25. Shalhoub S, Farahat F, Al-Jiffri A, Simhairi R, Shamma O, Siddiqi N, Mushtaq A. IFN-α2a or IFN-β1a in combination with ribavirin to treat Middle East respiratory syndrome coronavirus pneumonia: a retrospective study. *J Antimicrob Chemother* 2015;**70**(7):2129–32.

26. Saudi Ministry of Health. www.moh.gov.sa/en/CCC/PressReleases/Pages/Statistics-2015-11-07-001.aspx; 2015.

27. Memish ZA, Zumla AI, Al-Hakeem RF, Al-Rabeeah AA, Stephens GM. Family cluster of Middle East respiratory syndrome coronavirus infections. *N Engl J Med* 2013;**368**:2487–94.

28. Hijawi B, Abdallat M, Sayaydeh A, et al. Novel coronavirus infections in Jordan, April 2012: epidemiological findings from a retrospective investigation. *East Mediterr Health J* 2013;**19**(Suppl. 1):S12–8.

29. Health Protection Agency (HPA) UK Novel Coronavirus Investigation team. Evidence of person-to-person transmission within a family cluster of novel coronavirus infections, United Kingdom, February 2013. *Euro Surveill* 2013;**18**:20427.

30. WHO. Global alert and response (GAR): Middle East respiratory syndrome coronavirus (MERS-CoV)—update. July 7, 2013. www.who.int/csr/don/2013_07_07/en/index.html; 2015.

31. Memish ZA, Al-Tawfiq JA, Assiri A, AlRabiah FA, Al Hajjar S, Albarrak A, Flemban H, Alhakeem RF, Makhdoom HQ, Alsubaie S, Al-Rabeeah AA. Middle East respiratory syndrome coronavirus disease in children. *Pediatr Infect Dis J* 2014;**33**(9):904–6.

32. Thabet F, Chehab M, Bafaqih H, Al Mohaimeed S. Middle East respiratory syndrome coronavirus in children. *Saudi Med J* 2015;**36**(4):484–6.

33. Khuri-Bulos N, Payne DC, Lu X, Erdman D, Wang L, Faouri S, Shehabi A, Johnson M, Becker MM, Denison MR, Williams JV, Halasa NB. Middle East respiratory syndrome coronavirus not detected in children hospitalized with acute respiratory illness in Amman, Jordan, March 2010 to September 2012. *Clin Microbiol Infect* 2014;**20**(7):678–82.

34. Al-Tawfiq JA, Memish ZA. Middle East respiratory syndrome coronavirus: epidemiology and disease control measures. *Infect Drug Resist* 2014;**7**:281–7.

35. Eckerle I, Müller MA, Kallies S, Gotthardt DN, Drosten C. In-vitro renal epithelial cell infection reveals a viral kidney tropism as a potential mechanism for acute renal failure during Middle East Respiratory Syndrome (MERS) Coronavirus infection. *Virol J* 2013;**10**:359.

36. Das KM, Lee EY, Al Jawder SE, Enani MA, Singh R, Skakni L, Al-Nakshabandi N, AlDossari K, Larsson SG. Acute Middle East respiratory syndrome coronavirus: temporal lung changes observed on the chest radiographs of 55 patients. *AJR Am J Roentgenol* 2015;**205**(3):W267–74.

37. Das KM, Lee EY, Enani MA, AlJawder SE, Singh R, Bashir S, Al-Nakshbandi N, AlDossari K, Larsson SG. CT correlation with outcomes in 15 patients with acute Middle East respiratory syndrome coronavirus. *AJR Am J Roentgenol* 2015;**204**(4):736–42.

38. Memish ZA, Al-Tawfiq JA, Makhdoom HQ, Assiri A, Alhakeem RF, Albarrak A, Alsubaie S, Al-Rabeeah AA, Hajomar WH, Hussain R, Kheyami AM, Almutairi A, Azhar EI, Drosten C, Watson SJ, Kellam P, Cotten M, Zumla A. Respiratory tract samples, viral load, and genome fraction yield in patients with Middle East respiratory syndrome. *J Infect Dis* 2014;**210**(10):1590–4.

39. Corman VM, Albarrak AM, Omrani AS, Albarrak MM, Farah ME, Almasri M, Muth D, Sieberg A, Meyer B, Assiri AM, Binger T, Steinhagen K, Lattwein E, Al-Tawfiq J, Müller MA, Drosten C, Memish ZA. Viral shedding and antibody response in 37 patients with MERS-coronavirus infection. *Clin Infect Dis* 2016 Feb 15;**62**(4):477–83.

40. Müller MA, Meyer B, Corman VM, Al-Masri M, Turkestani A, Ritz D, Sieberg A, Aldabbagh S, Bosch BJ, Lattwein E, Alhakeem RF, Assiri AM, Albarrak AM, Al-Shangiti AM, Al-Tawfiq JA, Wikramaratna P, Alrabeeah AA, Drosten C, Memish ZA. Presence of Middle East respiratory syndrome coronavirus antibodies in Saudi Arabia: a nationwide, cross-sectional, serological study. *Lancet Infect Dis* 2015;**15**(5):559–64.

41. Park SW, Perera RA, Choe PG, Lau EH, Choi SJ, Chun JY, Oh HS, Song KH, Bang JH, Kim ES, Kim HB, Park WB, Kim NJ, Poon LL, Peiris M, Oh MD. Comparison of serological assays in human Middle East respiratory syndrome (MERS)-coronavirus infection. *Euro Surveill* 2015;**20**(41).

42. de Wilde AH, Raj VS, Oudshoorn D, Bestebroer TM, van Nieuwkoop S, Limpens RW, Posthuma CC, van der Meer Y, Bárcena M, Haagmans BL, Snijder EJ, van den Hoogen BG. MERS-coronavirus replication induces severe in vitro cytopathology and is strongly inhibited by cyclosporin A or interferon-α treatment. *J Gen Virol* 2013;**94**(Pt 8):1749–60.

43. Momattin H, Mohammed K, Zumla A, Memish ZA, Al-Tawfiq JA. Therapeutic options for Middle East respiratory syndrome coronavirus (MERS-CoV)--possible lessons from a systematic review of SARS-CoV therapy. *Int J Infect Dis* 2013;**17**(10):e792–8.

44. Falzarano D, de Wit E, Martellaro C, Callison J, Munster VJ, Feldmann H. Inhibition of novel βcoronavirus replication by a combination of interferon-α2b and ribavirin. *Sci Rep* 2013;**3**:1686.

45. Al-Tawfiq JA, Momattin H, Dib J, Memish ZA. Ribavirin and interferon therapy in patients infected with the Middle East respiratory syndrome coronavirus: an observational study. *Int J Infect Dis* 2014;**20**:42–6.

46. Spanakis N, Tsiodras S, Haagmans BL, Raj VS, Pontikis K, Koutsoukou A, Koulouris NG, Osterhaus AD, Koopmans MP, Tsakris A. Virological and serological analysis of a recent Middle East respiratory syndrome coronavirus infection case on a triple combination antiviral regimen. *Int J Antimicrob Agents* 2014;**44**(6):528–32.

47. Omrani S, Al-Tawfiq JA, Memsih ZA. Middle East respiratory syndrome coronavirus (MERS-CoV): animal to human interaction. *Pathog Glob Health* 2015;**109**(8):354–62.

48. Reusken CB, Haagmans BL, Muller MA, et al. Middle East respiratory syndrome coronavirus neutralizing serum antibodies in dromedary camels: a comparative serological study. *Lancet Infect Dis* 2013;**13**:859–66.

49. Reusken CB, Ababneh M, Raj VS, et al. Middle East Respiratory Syndrome coronavirus (MERS-CoV) serology in major livestock species in an affected region in Jordan, June to September 2013. *Euro Surveill* 2013;**18**:20662.

50. Hemida MG, Perera RA, Wang P, et al. Middle East Respiratory Syndrome (MERS) coronavirus seroprevalence in domestic livestock in Saudi Arabia, 2010 to 2013. *Euro Surveill* 2013;**18**:20659.

51. Alexandersen S, Kobinger GP, Soule G, Wernery U. Middle East respiratory syndrome coronavirus antibody reactors among camels in Dubai, United Arab Emirates, in 2005. *Transbound Emerg Dis* 2014;**61**:105–8.

52. Reusken CB, Messadi L, Feyisa A, et al. Geographic distribution of MERS coronavirus among dromedary camels, Africa. *Emerg Infect Dis* 2014;**20**:1370–4.

53. Nowotny N, Kolodziejek J. Middle East respiratory syndrome coronavirus (MERS-CoV) in dromedary camels, Oman, 2013. *Euro Surveill* 2014;**19**:20781.

54. Corman VM, Jores J, Meyer B, et al. Antibodies against MERS coronavirus in dromedary camels, Kenya, 1992-2013. *Emerg Infect Dis* 2014;**20**:1319–22.

55. Alagaili AN, Briese T, Mishra N, et al. Middle East respiratory syndrome coronavirus infection in dromedary camels in Saudi Arabia. *MBio* 2014;**5**:e00814–84.

56. Hemida MG, Chu DK, Poon LL, et al. MERS coronavirus in dromedary camel herd, Saudi Arabia. *Emerg Infect Dis* 2014;**20**:1231–4.

57. Wernery U, Corman VM, Wong EY, et al. Acute middle East respiratory syndrome coronavirus infection in livestock Dromedaries, Dubai, 2014. *Emerg Infect Dis* 2015;**21**:1019–22.

58. Meyer B, Muller MA, Corman VM, et al. Antibodies against MERS Coronavirus in Dromedary Camels, United Arab Emirates, 2003 and 2013. *Emerg Infect Dis* 2014;**20**:552–9.

59. Shirato K, Azumano A, Nakao T, et al. Middle East respiratory syndrome coronavirus infection not found in camels in Japan. *Jpn J Infect Dis* 2015;**68**:256–8.

60. Chan SM, Damdinjav B, Perera RA, et al. Absence of MERS-Coronavirus in Bactrian Camels, Southern Mongolia, November 2014. *Emerg Infect Dis* 2015;**21**:1269–71.

61. Khalafalla AI, Lu X, Al-Mubarak AI, Dalab AH, Al-Busadah KA, Erdman DD. MERS-CoV in upper Respiratory tract and lungs of dromedary camels, Saudi Arabia, 2013–2014. *Emerg Infect Dis* 2015;**21**:1153–8.

62. World Health Organization. Infection prevention and control during health care for probable or confirmed cases of novel coronavirus (nCoV) infection Interim guidance: 6 May 2013. www.who.int/csr/disease/coronavirus_infections/IPCnCoVguidance_06May13.pdf; 2013.

63. CDC. Interim Infection Prevention and Control Recommendations for Hospitalized Patients with Middle East Respiratory Syndrome Coronavirus (MERS-CoV). www.cdc.gov/coronavirus/mers/infection-prevention-control.html; 2013.

RESPIRATORY INFECTIONS OF THE HUMAN BOCAVIRUS

5

O. Schildgen, V. Schildgen

Witten/Herdecke University, Department of Pathology, gGmbH clinics of Cologne, Cologne, Germany

1 INTRODUCTION

The variant 1 of the human bocavirus (HBoV-1) that causes respiratory infections in primates and humans belongs to the family of Parvoviridae, subfamily Parvovirinae, genus *Bocaparvovirus*, was discovered originally in 2005 by Tobias Allander[1] and his team and represents together with the strains HBoV-3 and the gorilla bocavirus the species *Primate bocaparvovirus 1*.[2]

The discovery of HBoV-1 was one among a series of virus discoveries in the first decade of our millennium that was based on novel virus discovery systems developed in order to reduce the considerable number of cases in which a clinical diagnosis of a respiratory infection could not be confirmed by detection of a pathogen. Following the initial description of the virus, a huge number of clinical studies and case reports have been published which were supplemented by some basic research reports. Unfortunately, HBoV research still relies on clinical studies and case reports with accompanying cell culture studies as the major source of information on HBoV biology, as so far no animal model has been identified.

2 HBoV BIOLOGY

The human bocavirus (HBoV) was initially discovered in clinical samples from the respiratory tract of children suffering from respiratory infections of unknown etiology.[1] To date, HBoV is the fourth most detected respiratory virus, but as there is still no animal model or a broadly convertible cell culture available, Koch's modified postulates have not yet been completely fulfilled.[3]

Nevertheless, HBoV is the second parvovirus that is capable of infecting humans with the potential to cause clinical disease. Until HBoV was discovered, parvovirus B19 was the sole human parvovirus, which can hardly be cultured in in vitro cell cultures, likely because it strongly depends on the optimal cell cycle phase.[4–13] This latter fact hampered the development of potent and specific antivirals, tenacity studies, and the development of disinfectants active against human parvoviruses as surrogate pathogens with animal pathogenicity were used. The discovery of HBoV led to a couple of molecular findings that are of major general interest for the human parvovirus biology and clinics. Within a primary cell culture that productively replicated the human bocavirus, it was possible

The Microbiology of Respiratory System Infections. http://dx.doi.org/10.1016/B978-0-12-804543-5.00005-1

to identify the HBoV transcriptome, including the splicing variant of viral RNA.[14] This cell culture displayed for the first time a tool for the investigation of a human parvovirus in its natural infectious surrounding enabling follow up studies on the molecular biology of human parvoviruses in general, and HBoV in particular.

Unfortunately, the primary cell culture that enables HBoV growth in vitro is very expensive, requires a highly specialized laboratory, and is an error-prone cell culture, thus the availability of this technology is rather limited to a couple of laboratories worldwide, which in turn is a stumbling block for further research. In search for a broadly convertible replication system the group of Jianming Qiu from Kansas made a significant step forward: by establishing a plasmid based replicon-like system the group identified additional RNA species that are transcribed during the HBoV replication cycle.[15] The system is based on plasmids that contain the complete published HBoV sequence but are flanked by ITR-regions of the adeno-associated virus (AAV); the ITR-regions are terminal repeats containing palindromic sequences that form hairpin-like structures, which in turn are required for the replication of parvoviruses according to the so-called rolling hairpin mechanisms of replication.[16]

Although the hairpin-like structures of HBoV had yet not been described at that time, it was postulated that also the HBoV genome is flanked by such structures and that HBoV replicates its genome by the rolling hairpin mechanism, although this assumption is exclusively based on phylogenetic analogous conclusion rather than on experimental evidence. In theory, the rolling hairpin replication results in progeny genomes that occur in equal amounts of both polarities, while packaging of viral genomes is dependent on additional factors.[17–23] For almost four decades, it was postulated that all parvoviruses replicate according to this mechanism, although this replication model is solely based on experimental data obtained by the research on rodent parvoviruses. The model is characterized by a terminal hairpin dependent self-priming initiation of the viral genome replication and concatemeric replication intermediates of head-to-head or tail-to-tail replication intermediates. Based on an early publication of the postulated model in 1976 in *Nature*, this replication model became a dogma in the field of parvovirology and was deemed to be true for all parvoviruses. Interestingly, it was impossible to identify both genome polarities in clinical samples containing HBoV infected cells.[24] Thereby, NASBA analyses revealed that all HBoV strains package negative strand genome while only a minority also package the plus strand; this observation is compatible with another replication mechanism, known as rolling circle replication. A couple of systematic PCR-based analyses were performed to test the hypothesis if rolling circle replication may occur in HBoV infection and to decipher the unknown terminal hairpins.[25]

This approach identified DNA sequences that contained head-to-tail genome fragments linked by a newly identified linker stretch that had a partial by high homology to the minute virus of canine (MVC) ITR and to the ITR of bovine parvovirus. Most recently, it was shown that these sequences most likely represent the missing terminal hairpin like structures,[25,26] and it is likely that the virus was originally transmitted as a zoonosis (Fig. 5.1). Despite identifying the terminal sequences in clinical samples and also in cell cultures, not only a lack of self-priming activity of HBoV-genomes but also the lack of intermediates typical for rolling hairpin replication were observed. Instead, the samples contained head-to-tail structures. Several groups published similar observations, all tackling the dogma of parvovirus replication.[27–30] It is therefore important to know that the head-to-tail episomal form of HBoV differs from formerly described circular parvoviral episomes that have been shown to consist of circular closed genome dimers of head-to-head and tail-to-tail orientation.[31]

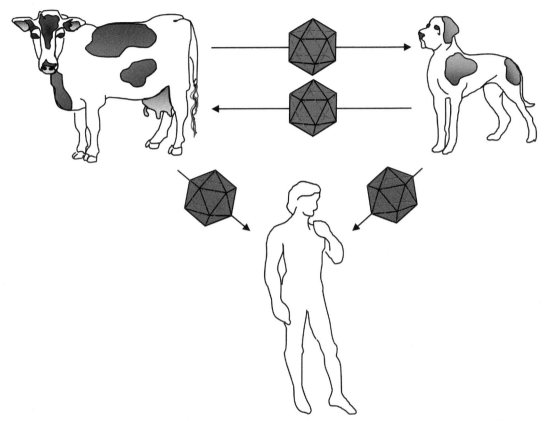

FIGURE 5.1 Overview of the Putative Zoonotic Transmission of Animal Bocaviruses to the Human Population

Based on sequence analyses, especially of the terminal sequences, a zoonotic event is likely, as HBoV-1 contains genome structures highly conserved from the canine minute virus and the bovine parvovirus.

Although the role of the linker sequence and the head-to-tail junction remains unclear, these findings were surprising as they support the hypothesis that HBoV replicates differently from nonhuman parvoviruses by possibly initiating a rolling circle mechanism, at least as an alternative route of replication.

Based on the newly identified sequences, the structure of the putative terminal repeats of the HBoV genome were predicted in silico.[26] Beyond that, the Kansas group developed a true full-length vector clone of HBoV which can be transfected to HEK-293 cells and produces a "recombinant wild type" human bocavirus that in turn is infectious for differentiated CuFi-8 cells.[32] CuFi-8 cells are derived from a patient with cystic fibrosis and can be grown as monolayer cultures that by change of the culturing media can be differentiated into a polarized respiratory epithelial structure that in turn supports HBoV replication.[32] This novel cell culture gives raise to the hypothesis that HBoV is a true serious pathogen because it induced a remarkable cytopathic effect in the polarized CuFi-8 cell line, which in turn is compatible with the assumption that clinical symptoms of an HBoV infection are caused by tissue damages due to the viral replication. Thereby, this infection model harbors a surprising feature that is a further hint for an alternative replication of the human bocavirus—if the full-length HBoV plasmid containing the hairpin sequences is transfected into HEK293 cells, then infectious progeny virions are produced, although based on the rolling hairpin model this process must be impossible because the free (!) hairpin sequences are believed to be essential for the replication. In contrast, in the plasmid they are flanked by the vector's backbone sequence, replication is possible although no helper plasmids are required as known for the dependoviruses. This simple observation strongly contradicts the model of rolling hairpin replication but in turn favours other replication models known for circular DNA, for example the rolling circle replication, which in the natural infection would produce head-to-tail concatemers.

Furthermore, clinical observations give raise to the hypothesis that the HBoV replication can be triggered or influenced by human herpesviruses, such as HHV-6, CMV, and Herpes simplex Virus. In this context, it is noteworthy that herpesviruses, especially HSV, are capable of initiating a rolling circle replication mechanism of replication in trans, as shown for SV40, which has a circular double stranded genome.[33]

Thereby, herpesviruses may either act as a trigger that arrests the host cell at transition from G1- to S-phase of the cell cycle or they could directly interact with the HBoV DNA supporting the replication by the herpesviral replication enzymes. The latter appears likely because head-to-tail intermediates are a feature of the rolling circle replication that may be initiated by a couple of viruses including the human herpesviruses type 1 and type 6.[33-40] These viruses (eg, AAV) in turn are able to act as helper viruses for the parvoviral subclass dependoviruses, that require those helper viruses for their replication.[36-40] Recently, a clinical case was observed in which the HBoV infection appeared to depend on coinfection and coreplication of human herpesvirus type 6. In this case, the HBoV infection persisted because of an immune disease but was terminated by antiviral therapy with cidofovir which is directed against HHV6.[41] This was the key observation leading to the assumption that HBoV is either sensitive to cidofovir or that a possible rolling circle HBoV replication is triggered by HHV6, which in turn would explain the high frequency of coinfections observed in case of HBoV.[40,42,43]

In 2011, two severe cases of respiratory failure in adults associated with HBoV infection, herpesvirus coinfection, and a history of lung fibrosis likely dedicated to the presence of chronic HBoV infection[44] indicate that the head-to-tail structures could have been episomal reservoirs enabling the virus' persistence as postulated by Kapoor and coworkers.[27] It may be speculated whether the persistence of HBoV episomes in the lung of the patients in analogy to a HBV infection, in which episomal cccDNA persists in the infected cell until the cell is targeted by the immune response or subjected to apoptosis, and in which this chronic state frequently goes ahead with a mild inflammation that is subclinical but finally could induce fibrosis, could have led to mild chronic inflammation eventually resulting in fibrosis of the lung, which could not be easily compensated as in the liver. In the context of a putative chronic HBoV infection or a persistence of HBoV at a subclinical level, it thus appears possible that HBoV could directly or indirectly, by interactions with the immune system, contribute to chronic lung disease such as idiopathic lung fibrosis.

Another recently detected novel feature of HBoV is the expression of more nonstructural proteins than concluded from our previous knowledge on parvovirus replication studies. Shen et al. have shown that besides NS1 three novel proteins named NS2, NS3, and NS4, are expressed during the viral replication, of which NS2 is believed to have a crucial role during the viral life cycle.[45]

3 EPIDEMIOLOGY

Like all respiratory pathogens (except SARS- and MERS-coronavirus) that cause respiratory infections, HBoV-1 is distributed worldwide and has been detected in patients from several regions of each continent.[46–98] However, unlike most other viruses that are known to peak seasonally in autumn and winter, HBoV infection peaks do not seem to be restricted to these seasons.

Although the route of transmission has not yet been systematically investigated, it is widely accepted that the transmission of HBoV most likely occurs by smear, or droplet infections or aerosols, and nasal or oral uptake, as described for the majority of "common cold viruses." The transmission route passes through airway excretions but could also happen via the gastrointestinal route, as HBoV is also shedded by stool (Fig. 5.2).

The HBoV seroprevalence is high and reached 95%, and more in children up to the age of 5 years.[99,100] This seroprevalence remains high in most adults,[62,68] but decreases from 96% to 59% in European adults if antibodies against HBoV strains 2–4 were depleted. Thus, in 41% of patients no long term immunity could be generated, supporting the assumption that the virus is able to persist and could also reinfect elderly patients.[101] Surprisingly, HBoV-1 DNA can also be detected in blood and blood products from healthy Chinese blood donors with a lower seropositivity compared to the afore mentioned cohorts.[102]

4 CLINICAL FEATURES

The HBoV-1 infection is clinically indistinguishable from other respiratory infections and can solely be proven by molecular assays. The spectrum of HBoV infections ranges from asymptomatic[53,103,104] to mild upper respiratory infections[53,105–107] up to serious and life threatening lower respiratory tract infections[56,95,108–118] in all age groups.[56,57,95,104,108–121] The immune response against HBoV starts with an IgM response followed by the formation of IgG,[99,100] but no life long immunity is generated in at least 40% of patients due to the original antigenic sin.[62,68,122]

HBoV-1 is able to infect the central nervous system[82,84] and it has been identified as a putative cause of idiopathic lung fibrosis[44] supported by the fact that a set of profibrotic cytokines were upregulated during HBoV infection in adults and their HBoV dependent upregulation was confirmed in cell culture.[123,124] Whereas, HBoV does not induce a clear Th1 or Th2 response.[125] Furthermore, the HBoV dependent regulated cytokines include a subset of cytokines which are known to be involved in several cancer-associated pathways, supporting the hypothesis that HBoV may be associated with long term diseases or even cancerogenesis.[126–128] Although this hypothesis requires further prospective studies, HBoV DNA was detected in lung- and colorectal tumors. Detection of HBoV DNA, eventually combined with persistence, was described besides detection in normal lung tissue,[104] and in lung- and colorectal tumors,[128] in other tissues such as, tonsils,[27,129–131] and myocardium, and may affect further tissues that have not yet been tested for HBoV-positivity.

Lung fibrosis, especially the idiopathic lung fibrosis (IPF) is characterized by a Th2-type dominated immune response in the affected tissue (as reviewed by:[132–134]). The Th2 response in the lung is accompanied by increased expression levels of IL-4, IL-5, IL-10, and IL-13 and is followed by increased levels—besides others—of CCL17 (TARC) and CCL5 (RANTES). Moreover, fibrosis is related to expression of TNF and IL-8, and it is worth mentioning that the neutralization of TARC leads to a

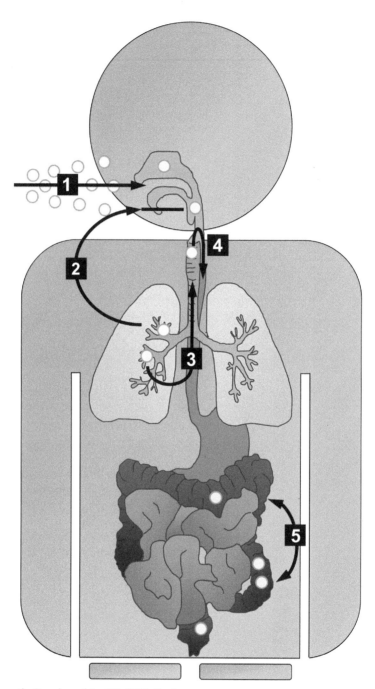

FIGURE 5.2 Schematic Overview of the HBoV Life Cycle

1, entry through the nasopharyngeal space; 2, infection of the lung; 3 and 4, swallowing of the expectorated infectious secretion; 5, infection of the gastrointestinal tract. Additionally, the virus spreads via the bloodstream and causes classical viraemia (not indicated).

reduction of fibrosis in the animal model.[133,135] In addition, an elevation of the TARC/IP-10 ratio is also characteristic for fibrosis and was previously discussed as a marker for IPF.[136]

Moreover, a so far unique follow up case in which the infection/reactivation of HBoV occurred between two episodes of BAL sampling, and in which the fibrosis associated cytokines were expressed in association with the HBoV infection but not before, supports the previously obtained data. This leads to the conclusion, that HBoV colonisation/chronic infection may be at least one trigger that could stimulate airway remodelling. However, it could be argued that in vivo not only the resident airway epithelial cells are involved in the immune response but also additional patient specific factors will contribute to altered profibrotic cytokine profiles. To address this problem, experiments in an air-liquid-interface culture of human airway epithelial cells were performed. These experiments confirmed that the profibrotic cytokines were expressed by the infected cell cultures but were hardly or not at all expressed in mock-infected cells and they reveal that the identified cytokines belong to the initial immune response after HBoV infection (123).

According to the literature the two HBoV proteins VP2 and NP1 seem to influence the regulation of the Interferon beta pathway but the data appear controversially as VP2 upregulates the pathway[137] while NP1 inhibits the IFN-beta production when overexpressed.[138]

5 COINFECTIONS AND PERSISTENCE

Simultaneously with the discovery of HBoV in 2005, multiplexing methods started to become an accepted diagnostic tool and as a consequence detection of multiple infections, especially in respiratory tract diseases, has become a common phenomenon.[53,139–143] Nowadays, multiple infections with up to six pathogens being simultaneously present in a single respiratory sample are frequent[53,139–144] and have misled some researchers to the statement that the human bocavirus, also occurring in asymptomatic patients, is a harmless bystander rather than a pathogen.[145,146] This hypothesis seems to be supported by the fact that for HBoV a formal fulfilment of Koch's modified postulates was not yet possible[147] because no animal model exists so far and also because volunteer transmission trials cannot be recommended based on our current knowledge.[127]

On the contrary, although there is a cohort of asymptomatic carriers,[53,104,139,146,148,149] several studies have shown that HBoV induces clinical symptoms.[50,77,111,112,139,150–156] The asymptomatic viral shedding is meanwhile believed to originate from a long term shedding after an acute infection or from persisting viruses,[26,27,81,122,157–160] most recently confirmed by a long term prospective cohort study.[53,161] Thereby, it was shown that the rate of asymptomatic HBoV infections is similar to the rate of rhinovirus infections, and no one would doubt that rhinoviruses are true pathogens.[53]

Moreover, HBoV induces a serious cytopathic effect in infected cell cultures, which is a typical feature of a pathogen.[14,32,45,157]

6 DIAGNOSTICS

Besides several published homebrew PCRs and real-time PCRs (as reviewed by[3]), numerous commercial assays, such as the Luminex RVP–Assay,[104,162] the Idaho FilmArray,[144,162] or the RespiFinder assay[104] have been developed and released to the market enabling the detection of HBoV from clinical samples. However, multiplexing solely allows us to detect the viral DNA in a respiratory sample without providing the essential information if an active replicative infection underlies the currently clinical episode requiring laboratory testing.[53] As HBoV can be shedded for longer than 3 months after the acute symptomatic phase,[53] a proper diagnostics of human bocavirus requires the proof of active replication, which can be done either by detection of a viremia in the peripheral blood,[77,93,101,122,163–167] or by detection of spliced viral RNA transcripts that were shown to be present exclusively during the active phase of the replication.[168]

7 SUMMARY AND PERSPECTIVE

There is an increasing body of evidence showing that the human bocavirus is a serious pathogen that on the one hand is associated with acute respiratory infections, sometimes with life threatening complications, and on the other hand also could contribute to long term diseases of the airways resulting in lung carcinoma or lung fibrosis. Therefore, it remains crucial to analyze the long-term effects of HBoV infections to identify the mechanisms of HBoV persistence and to determine the host factors for asymptomatic infections, as well as to test the hypothesis that HBoV could trigger the development of lung cancer and fibrosis. In any cases, the proper diagnostics of HBoV require attention and need to be evaluated regarding its interaction with other respiratory viruses that may simultaneously be detected in clinical episodes.

REFERENCES

1. Allander T, Tammi MT, Eriksson M, Bjerkner A, Tiveljung-Lindell A, Andersson B. Cloning of a human parvovirus by molecular screening of respiratory tract samples. *Proc Natl Acad Sci USA* 2005;**102**:12891–6.
2. Cotmore SF, Agbandje-McKenna M, Chiorini JA, et al. Arch Virol. *The family Parvoviridae* 2014; **159**:1239–47.
3. Schildgen O, Muller A, Allander T, et al. Clin Microbiol Rev. *Human bocavirus: passenger or pathogen in acute respiratory tract infections?* 2008;**21**:291–304.
4. Morita E, Sugamura K. Human parvovirus B19-induced cell cycle arrest and apoptosis. *Springer Semin Immunopathol* 2002;**24**:187–99.
5. Morita E, Tada K, Chisaka H, et al. Human parvovirus B19 induces cell cycle arrest at G(2) phase with accumulation of mitotic cyclins. *J Virol* 2001;**75**:7555–63.
6. Bashir T, Rommelaere J, Cziepluch C. In vivo accumulation of cyclin A and cellular replication factors in autonomous parvovirus minute virus of mice-associated replication bodies. *J Virol* 2001;**75**:4394–8.
7. Bashir T, Horlein R, Rommelaere J, Willwand K. Cyclin A activates the DNA polymerase delta -dependent elongation machinery in vitro: a parvovirus DNA replication model. *Proc Natl Acad Sci USA* 2000;**97**:5522–7.
8. Op De Beeck A, Caillet-Fauquet P. The NS1 protein of the autonomous parvovirus minute virus of mice blocks cellular DNA replication: a consequence of lesions to the chromatin? *J Virol* 1997;**71**:5323–9.
9. Oleksiewicz MB, Alexandersen S. S-phase-dependent cell cycle disturbances caused by Aleutian mink disease parvovirus. *J Virol* 1997;**71**:1386–96.
10. Op De Beeck A, Anouja F, Mousset S, Rommelaere J, Caillet-Fauquet P. The nonstructural proteins of the autonomous parvovirus minute virus of mice interfere with the cell cycle, inducing accumulation in G2. *Cell Growth Differ* 1995;**6**:781–7.
11. Bantel-Schaal U, Stohr M. Influence of adeno-associated virus on adherence and growth properties of normal cells. *J Virol* 1992;**66**:773–9.

12. Metcalf JB, Bates RC, Lederman M. Interaction of virally coded protein and a cell cycle-regulated cellular protein with the bovine parvovirus left terminus ori. *J Virol* 1990;**64**:5485–90.

13. Siegl G. Molecular biology and pathogenicity of human and animal parvoviruses. *Behring Inst Mitt* 1990; **85**:6–13.

14. Dijkman R, Koekkoek SM, Molenkamp R, Schildgen O, van der Hoek L. Human bocavirus can be cultured in differentiated human airway epithelial cells. *J Virol* 2009;**83**:7739–48.

15. Chen AY, Cheng F, Lou S, et al. Characterization of the gene expression profile of human bocavirus. *Virology* 2010;**403**:145–54.

16. Tattersall P, Ward DC. Rolling hairpin model for replication of parvovirus and linear chromosomal DNA. *Nature* 1976;**263**:106–9.

17. Cotmore SF, Gottlieb RL, Tattersall P. Replication initiator protein NS1 of the parvovirus minute virus of mice binds to modular divergent sites distributed throughout duplex viral DNA. *J Virol* 2007; **81**:13015–27.

18. Cotmore SF, Tattersall P. Encapsidation of minute virus of mice DNA: aspects of the translocation mechanism revealed by the structure of partially packaged genomes. *Virology* 2005;**336**:100–12.

19. Cotmore SF, Tattersall P. Genome packaging sense is controlled by the efficiency of the nick site in the right-end replication origin of parvoviruses minute virus of mice and LuIII. *J Virol* 2005;**79**:2287–300.

20. Corsini J, Cotmore SF, Tattersall P, Winocour E. The left-end and right-end origins of minute virus of mice DNA differ in their capacity to direct episomal amplification and integration in vivo. *Virology* 2001; **288**:154–63.

21. Cotmore SF, Tattersall P. An asymmetric nucleotide in the parvoviral 3' hairpin directs segregation of a single active origin of DNA replication. *EMBO J* 1994;**13**:4145–52.

22. Cotmore SF, Nuesch JP, Tattersall P. Asymmetric resolution of a parvovirus palindrome in vitro. *J Virol* 1993;**67**:1579–89.

23. Cotmore SF, Nuesch JP, Tattersall P. In vitro excision and replication of 5' telomeres of minute virus of mice DNA from cloned palindromic concatemer junctions. *Virology* 1992;**190**:365–77.

24. Bohmer A, Schildgen V, Lusebrink J, et al. Novel application for isothermal nucleic acid sequence-based amplification (NASBA). *J Virol Methods* 2009;**158**:199–201.

25. Lusebrink J, Schildgen V, Tillmann RL, et al. Detection of head-to-tail DNA sequences of human bocavirus in clinical samples. *PLoS One* 2011;**6**:e19457.

26. Schildgen O, Qiu J, Soderlund-Venermo M. Genomic features of the human bocaviruses. *Future Virol* 2012;**7**:31–9.

27. Kapoor A, Hornig M, Asokan A, Williams B, Henriquez JA, Lipkin WI. Bocavirus episome in infected human tissue contains non-identical termini. *PLoS One* 2011;**6**:e21362.

28. Li L, Pesavento PA, Leutenegger CM, et al. A novel bocavirus in canine liver. *Virol J* 2013;**10**:54.

29. Zhao H, Zhao L, Sun Y, et al. Detection of a bocavirus circular genome in fecal specimens from children with acute diarrhea in beijing, china. *PLoS One* 2012;**7**:e48980.

30. Yang WZ, Huang CP, Duan ZJ. Identification and characterization of porcine bocavirus episomes. *Bing Du Xue Bao* 2012;**28**:418–23.

31. Schnepp BC, Clark KR, Klemanski DL, Pacak CA, Johnson PR. Genetic fate of recombinant adeno-associated virus vector genomes in muscle. *J Virol* 2003;**77**:3495–504.

32. Huang Q, Deng X, Yan Z, et al. Establishment of a reverse genetics system for studying human bocavirus in human airway epithelia. *PLoS Pathog* 2012;**8**:e1002899.

33. Gerspach R, Matz B. Herpes simplex virus-induced "rolling circle" amplification of SV40 DNA sequences in a transformed hamster cell line correlates with tandem integration of the SV40 genome. *Virology* 1989; **173**:723–7.

34. Gerspach R, Matz B. Herpes simplex virus-directed overreplication of chromosomal DNA physically linked to the simian virus 40 integration site of a transformed hamster cell line. *Virology* 1988;**165**:282–5.

35. Schildgen O, Graper S, Blumel J, Matz B. Genome replication and progeny virion production of herpes simplex virus type 1 mutants with temperature-sensitive lesions in the origin-binding protein. *J Virol* 2005;**79**:7273–8.

36. Alazard-Dany N, Nicolas A, Ploquin A, et al. Definition of herpes simplex virus type 1 helper activities for adeno-associated virus early replication events. *PLoS Pathog* 2009;**5**:e1000340.

37. Johansson S, Buchmayer S, Harlid S, et al. Infection with Parvovirus B19 and Herpes viruses in early pregnancy and risk of second trimester miscarriage or very preterm birth. *Reprod Toxicol* 2008; **26**:298–302.

38. Rollin R, Alvarez-Lafuente R, Marco F, et al. Human parvovirus B19, varicella zoster virus, and human herpesvirus-6 in mesenchymal stem cells of patients with osteoarthritis: analysis with quantitative real-time polymerase chain reaction. *Osteoarthritis Cartilage* 2007;**15**:475–8.

39. Rohayem J, Dinger J, Fischer R, Klingel K, Kandolf R, Rethwilm A. Fatal myocarditis associated with acute parvovirus B19 and human herpesvirus 6 coinfection. *J Clin Microbiol* 2001;**39**:4585–7.

40. Thomson BJ, Weindler FW, Gray D, Schwaab V, Heilbronn R. Human herpesvirus 6 (HHV-6) is a helper virus for adeno-associated virus type 2 (AAV-2) and the AAV-2 rep gene homologue in HHV-6 can mediate AAV-2 DNA replication and regulate gene expression. *Virology* 1994;**204**:304–11.

41. Streiter M, Malecki M, Prokop A, et al. Does human bocavirus infection depend on helper viruses? A challenging case report. *Virol J* 2011;**8**:417.

42. Asano Y, Yoshikawa T. Human herpesvirus-6 and parvovirus B19 infections in children. *Curr Opin Pediatr* 1993;**5**:14–20.

43. Bauer HJ, Monreal G. Herpesviruses provide helper functions for avian adeno-associated parvovirus. *J Gen Virol* 1986;**67**(Pt 1):181–5.

44. Windisch W, Schildgen V, Malecki M, et al. Detection of HBoV DNA in idiopathic lung fibrosis, Cologne, Germany. *J Clin Virol* 2013;**58**:325–7.

45. Shen W, Deng X, Zou W, et al. Identification and functional analysis of novel non-structural proteins of human bocavirus 1. *J Virol* 2015;**89**(19).

46. Akinloye OM, Ronkko E, Savolainen-Kopra C, et al. Specific viruses detected in nigerian children in association with acute respiratory disease. *J Trop Med* 2011;**2011**:690286.

47. Albuquerque MC, Pena GP, Varella RB, Gallucci G, Erdman D, Santos N. Novel respiratory virus infections in children, Brazil. *Emerg Infect Dis* 2009;**15**:806–8.

48. Al-Rousan HO, Meqdam MM, Alkhateeb A, Al-Shorman A, Qaisy LM, Al-Moqbel MS. Human bocavirus in Jordan: prevalence and clinical symptoms in hospitalised paediatric patients and molecular virus characterisation. *Singapore Med J* 2011;**52**:365–9.

49. Bastien N, Brandt K, Dust K, Ward D, Li Y. Human Bocavirus infection, Canada. *Emerg Infect Dis* 2006;**12**:848–50.

50. Bharaj P, Sullender WM, Kabra SK, Broor S. Human bocavirus infection in children with acute respiratory tract infection in India. *J Med Virol* 2010;**82**:812–6.

51. Binks MJ, Cheng AC, Smith-Vaughan H, et al. Viral-bacterial co-infection in Australian Indigenous children with acute otitis media. *BMC Infect Dis* 2011;**11**:161.

52. Bubshait DK, Albuali WH, Yousef AA, et al. Clinical description of human bocavirus viremia in children with LRTI, Eastern Province, Saudi Arabia. *Ann Thorac Med* 2015;**10**:146–9.

53. Byington CL, Ampofo K, Stockmann C, et al. Community surveillance of respiratory viruses among families in the Utah Better identification of germs-longitudinal viral epidemiology (BIG-LoVE) study. *Clin Infect Dis* 2015;**61**(8):1217–24.

54. Carrol ED, Mankhambo LA, Guiver M, et al. PCR improves diagnostic yield from lung aspiration in Malawian children with radiologically confirmed pneumonia. *PLoS One* 2011;**6**:e21042.

55. Chieochansin T, Samransamruajkit R, Chutinimitkul S, et al. Human bocavirus (HBoV) in Thailand: clinical manifestations in a hospitalized pediatric patient and molecular virus characterization. *J Infect* 2008;**56**: 137–42.

56. Choi EH, Lee HJ, Kim SJ, et al. The association of newly identified respiratory viruses with lower respiratory tract infections in Korean children, 2000-2005. *Clin Infect Dis* 2006;**43**:585–92.

57. Chow BD, Huang YT, Esper FP. Evidence of human bocavirus circulating in children and adults, Cleveland, Ohio. *J Clin Virol* 2008;**43**:302–6.

58. Chung JY, Han TH, Kim CK, Kim SW. Bocavirus infection in hospitalized children, South Korea. *Emerg Infect Dis* 2006;**12**:1254–6.

59. De Vos N, Vankeerberghen A, Vaeyens F, Van Vaerenbergh K, Boel A, De Beenhouwer H. Simultaneous detection of human bocavirus and adenovirus by multiplex real-time PCR in a Belgian paediatric population. *Eur J Clin Microbiol Infect Dis* 2009;**28**:1305–10.

60. Dina J, Vabret A, Gouarin S, et al. Detection of human bocavirus in hospitalised children. *J Paediatr Child Health* 2009;**45**:149–53.

61. Do AH, van Doorn HR, Nghiem MN, et al. Viral etiologies of acute respiratory infections among hospitalized Vietnamese children in Ho Chi Minh City, 2004-2008. *PLoS One* 2011;**6**:e18176.

62. Endo R, Ishiguro N, Kikuta H, et al. Seroepidemiology of human bocavirus in Hokkaido prefecture, Japan. *J Clin Microbiol* 2007;**45**:3218–23.

63. Esposito S, Bosis S, Niesters HG, et al. Impact of human bocavirus on children and their families. *J Clin Microbiol* 2008;**46**:1337–42.

64. Essa S, Owayed A, Altawalah H, Khadadah M, Behbehani N, Al-Nakib W. The prevalence of human bocavirus, human coronavirus-NL63, human metapneumovirus, human polyomavirus KI and WU in respiratory tract infections in Kuwait. *Med Princ Pract* 2015;**24**:382–7.

65. Fabbiani M, Terrosi C, Martorelli B, et al. Epidemiological and clinical study of viral respiratory tract infections in children from Italy. *J Med Virol* 2009;**81**:750–6.

66. Foulongne V, Olejnik Y, Perez V, Elaerts S, Rodiere M, Segondy M. Human bocavirus in French children. *Emerg Infect Dis* 2006;**12**:1251–3.

67. Furuse Y, Suzuki A, Kishi M, et al. Detection of novel respiratory viruses from influenza-like illness in the Philippines. *J Med Virol* 2010;**82**:1071–4.

68. Guido M, Zizza A, Bredl S, et al. Seroepidemiology of human bocavirus in Apulia, Italy. *Clin Microbiol Infect* 2012;**18**:E74–6.

69. Heydari H, Mamishi S, Khotaei GT, Moradi S. Fatal type 7 adenovirus associated with human bocavirus infection in a healthy child. *J Med Virol* 2011;**83**:1762–3.

70. Hindiyeh MY, Keller N, Mandelboim M, et al. High rate of human bocavirus and adenovirus coinfection in hospitalized Israeli children. *J Clin Microbiol* 2008;**46**:334–7.

71. Hustedt JW, Christie C, Hustedt MM, Esposito D, Vazquez M. Seroepidemiology of human bocavirus infection in Jamaica. *PLoS One* 2012;**7**:e38206.

72. I PM, Nelson EA, Cheuk ES, Leung E, Sung R, Chan PK. Pediatric hospitalization of acute respiratory tract infections with Human Bocavirus in Hong Kong. *J Clin Virol* 2008;**42**:72–4.

73. Jacques J, Moret H, Renois F, Leveque N, Motte J, Andreoletti L. Human Bocavirus quantitative DNA detection in French children hospitalized for acute bronchiolitis. *J Clin Virol* 2008;**43**:142–7.

74. Kaplan NM, Dove W, Abu-Zeid AF, Shamoon HE, Abd-Eldayem SA, Hart CA. Human bocavirus infection among children, Jordan. *Emerg Infect Dis* 2006;**12**:1418–20.

75. Kesebir D, Vazquez M, Weibel C, et al. Human bocavirus infection in young children in the United States: molecular epidemiological profile and clinical characteristics of a newly emerging respiratory virus. *J Infect Dis* 2006;**194**:1276–82.

76. Klinkenberg D, Blohm M, Hoehne M, et al. Risk of rotavirus vaccination for children with SCID. *Pediatr Infect Dis J* 2015;**34**:114–5.

77. Korner RW, Soderlund-Venermo M, van Koningsbruggen-Rietschel S, Kaiser R, Malecki M, Schildgen O. Severe human bocavirus infection, Germany. *Emerg Infect Dis* 2011;**17**:2303–5.

78. Krakau M, Brockmann M, Titius B, et al. Acute human bocavirus infection in MDS patient, Cologne, Germany. *J Clin Virol* 2015;**69**:44–77.

79. Lin F, Zeng A, Yang N, et al. Quantification of human bocavirus in lower respiratory tract infections in China. *Infect Agent Cancer* 2007;**2**:3.

80. Medici MC, Tummolo F, Albonetti V, Abelli LA, Chezzi C, Calderaro A. Molecular detection and epidemiology of astrovirus, bocavirus, and sapovirus in Italian children admitted to hospital with acute gastroenteritis, 2008–2009. *J Med Virol* 2012;**84**:643–50.

81. Meriluoto M, Hedman L, Tanner L, et al. Association of human bocavirus 1 infection with respiratory disease in childhood follow-up study, Finland. *Emerg Infect Dis* 2012;**18**:264–71.

82. Mitui MT, Tabib SM, Matsumoto T, et al. Detection of human bocavirus in the cerebrospinal fluid of children with encephalitis. *Clin Infect Dis* 2012;**54**:964–7.

83. Monteny M, Niesters HG, Moll HA, Berger MY. Human bocavirus in febrile children, The Netherlands. *Emerg Infect Dis* 2007;**13**:180–2.

84. Mori D, Ranawaka U, Yamada K, et al. Human bocavirus in patients with encephalitis, Sri Lanka, 2009-2010. *Emerg Infect Dis* 2013;**19**:1859–62.

85. Niang MN, Diop OM, Sarr FD, et al. Viral etiology of respiratory infections in children under 5 years old living in tropical rural areas of Senegal: The EVIRA project. *J Med Virol* 2010;**82**:866–72.

86. Obuchi M, Yagi S, Oguri A, Takizawa T, Kimura H, Sata T. Outbreak of human bocavirus 1 infection in young children in Toyama, Japan. *Jpn J Infect Dis* 2015;**68**:259–61.

87. Pierangeli A, Scagnolari C, Trombetti S, et al. Human bocavirus infection in hospitalized children in Italy. *Influenza Other Respi Viruses* 2008;**2**:175–9.

88. Pozo F, Garcia-Garcia ML, Calvo C, Cuesta I, Perez-Brena P, Casas I. High incidence of human bocavirus infection in children in Spain. *J Clin Virol* 2007;**40**:224–8.

89. Qu XW, Duan ZJ, Qi ZY, et al. Human bocavirus infection, People's Republic of China. *Emerg Infect Dis* 2007;**13**:165–8.

90. Redshaw N, Wood C, Rich F, Grimwood K, Kirman JR. Human bocavirus in infants, New Zealand. *Emerg Infect Dis* 2007;**13**:1797–9.

91. Santos N, Peret TC, Humphrey CD, et al. Human bocavirus species 2 and 3 in Brazil. *J Clin Virol* 2010;**48**:127–30.

92. Smuts H, Hardie D. Human bocavirus in hospitalized children, South Africa. *Emerg Infect Dis* 2006;**12**:1457–8.

93. Soderlund-Venermo M, Lahtinen A, Jartti T, et al. Clinical assessment and improved diagnosis of bocavirus-induced wheezing in children, Finland. *Emerg Infect Dis* 2009;**15**:1423–30.

94. Souza EL, Ramos JG, Proenca-Modena JL, et al. Human bocavirus in very young infants hospitalized with acute respiratory infection in northeast Brazil. *J Trop Pediatr* 2010;**56**:125–7.

95. Sung CC, Chi H, Chiu NC, et al. Viral etiology of acute lower respiratory tract infections in hospitalized young children in Northern Taiwan. *J Microbiol Immunol Infect* 2011;**44**:184–90.

96. Tan BH, Lim EA, Seah SG, et al. The incidence of human bocavirus infection among children admitted to hospital in Singapore. *J Med Virol* 2009;**81**:82–9.

97. Volz S, Schildgen O, Klinkenberg D, et al. Prospective study of Human Bocavirus (HBoV) infection in a pediatric university hospital in Germany 2005/2006. *J Clin Virol* 2007;**40**:229–35.

98. Weissbrich B, Neske F, Schubert J, et al. Frequent detection of bocavirus DNA in German children with respiratory tract infections. *BMC Infect Dis* 2006;**6**:109.

99. Zhao LQ, Qian Y, Zhu RN, et al. Seroprevalence of antibody against human bocavirus in Beijing, China. *Zhonghua Er Ke Za Zhi* 2008;**46**:111–4.

100. Karalar L, Lindner J, Schimanski S, Kertai M, Segerer H, Modrow S. Prevalence and clinical aspects of human bocavirus infection in children. *Clin Microbiol Infect* 2010;**16**:633–9.

101. Kantola K, Hedman L, Arthur J, et al. Seroepidemiology of human bocaviruses 1-4. *J Infect Dis* 2011;**204**:1403–12.

102. Li H, He M, Zeng P, et al. The genomic and seroprevalence of human bocavirus in healthy Chinese plasma donors and plasma derivatives. *Transfusion* 2015;**55**:154–63.

103. Chonmaitree T, Alvarez-Fernandez P, Jennings K, et al. Symptomatic and asymptomatic respiratory viral infections in the first year of life: association with acute otitis media development. *Clin Infect Dis* 2015;**60**:1–9.

104. Kaur J, Schildgen V, Tillmann R, et al. Low copy number detection of HBoV DNA in BAL of asymptomatic adult patients. *Future Virol* 2014;**9**:715–20.

105. Costa E, Rodriguez-Dominguez M, Clari MA, Gimenez E, Galan JC, Navarro D. Comparison of the performance of 2 commercial multiplex PCR platforms for detection of respiratory viruses in upper and lower tract respiratory specimens. *Diagn Microbiol Infect Dis* 2015;**82**:40–3.

106. Chen KF, Blyn L, Rothman RE, et al. Reverse transcription polymerase chain reaction and electrospray ionization mass spectrometry for identifying acute viral upper respiratory tract infections. *Diagn Microbiol Infect Dis* 2011;**69**:179–86.

107. Cotton M, Innes S, Jaspan H, Madide A, Rabie H. Management of upper respiratory tract infections in children. *S Afr Fam Pract* 2008;**50**:6–12.

108. Zeng SZ, Xiao NG, Zhong LL, Yu T, Zhang B, Duan ZJ. Clinical features of human metapneumovirus genotypes in children with acute lower respiratory tract infection in Changsha, China. *J Med Virol* 2015;**87**(11):1839–45.

109. Li L, Chen ZR, Yan YD, et al. Detection of human bocavirus in nasopharyngeal aspirates versus in broncho-alveolar lavage fluids in children with lower respiratory tract infections. *J Med Virol* 2015;**88**(2):211–5.

110. Ghietto LM, Camara A, Zhou Y, et al. High prevalence of human bocavirus 1 in infants with lower acute respiratory tract disease in Argentina, 2007-2009. *Braz J Infect Dis* 2012;**16**:38–44.

111. Deng Y, Gu X, Zhao X, et al. High viral load of human bocavirus correlates with duration of wheezing in children with severe lower respiratory tract infection. *PLoS One* 2012;**7**:e34353.

112. Arnott A, Vong S, Rith S, et al. Human bocavirus amongst an all-ages population hospitalised with acute lower respiratory infections in Cambodia. *Influenza Other Respi Viruses* 2012;**7**(2):201–10.

113. Pavia AT. Viral infections of the lower respiratory tract: old viruses, new viruses, and the role of diagnosis. *Clin Infect Dis* 2011;**52**(Suppl 4):S284–9.

114. Zhang LL, Tang LY, Xie ZD, et al. Human bocavirus in children suffering from acute lower respiratory tract infection in Beijing Children's Hospital. *Chin Med J* 2008;**121**:1607–10.

115. Schenk T, Huck B, Forster J, Berner R, Neumann-Haefelin D, Falcone V. Human bocavirus DNA detected by quantitative real-time PCR in two children hospitalized for lower respiratory tract infection. *Eur J Clin Microbiol Infect Dis* 2007;**26**:147–9.

116. Kleines M, Scheithauer S, Rackowitz A, Ritter K, Hausler M. High prevalence of human bocavirus detected in young children with severe acute lower respiratory tract disease by use of a standard PCR protocol and a novel real-time PCR protocol. *J Clin Microbiol* 2007;**45**:1032–4.

117. Freymuth F, Vabret A, Dina J, Petitjean J, Gouarin S. Techniques used for the diagnostic of upper and lower respiratory tract viral infections. *Rev Prat* 2007;**57**:1876–82.

118. Ma X, Endo R, Ishiguro N, et al. Detection of human bocavirus in Japanese children with lower respiratory tract infections. *J Clin Microbiol* 2006;**44**:1132–4.

119. Liu WK, Chen DH, Liu Q, et al. Detection of human bocavirus from children and adults with acute respiratory tract illness in Guangzhou, southern China. *BMC Infect Dis* 2011;**11**:345.

120. Midilli K, Yilmaz G, Turkoglu S, et al. Detection of human bocavirus DNA by polymerase chain reaction in children and adults with acute respiratory tract infections. *Mikrobiyol Bul* 2010;**44**:405–13.

121. Kupfer B, Vehreschild J, Cornely O, et al. Severe pneumonia and human bocavirus in adult. *Emerg Infect Dis* 2006;**12**:1614–6.

122. Li X, Kantola K, Hedman L, Arku B, Hedman K, Soderlund-Venermo M. Original antigenic sin with human bocaviruses 1-4. *J Gen Virol* 2015;**96**(10):3099–108.

123. Eichhorn V, Khalfaoui S, Pieper M, et al. Human bocavirus infection induces cytokine expression associated with lung fibrosis. In: Hansmann M-L, ed. 99 Jahrestagung der Deutschen gesellschaft für Pathologie eV. Frankfurt am Main, 28.-31. Mai 2015 Der Pathologe; 2015:SO-040 p. 82.

124. Khalfaoui S, Eichhorn V, Karagiannidis C, et al. Lung infection by human bocavirus induces the release of profibrotic mediator cytokines in vivo and in vitro. *PLoS ONE* 2015;**11**(1):e0147010. doi: 10.1371/journal.pone.0147010.

125. Kumar A, Filippone C, Lahtinen A, et al. Comparison of Th-cell immunity against human bocavirus and parvovirus B19: proliferation and cytokine responses are similar in magnitude but more closely interrelated with human bocavirus. *Scand J Immunol* 2011;**73**:135–40.

126. Chen H, Chen XZ, Waterboer T, Castro FA, Brenner H. Viral infections and colorectal cancer: a systematic review of epidemiological studies. *Int J Cancer* 2015;**137**:12–24.

127. Schildgen V, Khalfaoui S, Schildgen O. Human Bocavirus: from common cold to cancer? Speculations on the importance of an episomal genomic form of human bocavirus. *Rev Med Microbiol* 2014;**25**:113–8.

128. Schildgen V, Malecki M, Tillmann RL, Brockmann M, Schildgen O. The human bocavirus is associated with some lung and colorectal cancers and persists in solid tumors. *PLoS One* 2013;**8**:e68020.

129. Gunel C, Kirdar S, Omurlu IK, Agdas F. Detection of the Epstein–Barr virus, human bocavirus and novel KI and KU polyomaviruses in adenotonsillar tissues. *Int J Pediatr Otorhinolaryngol* 2015;**79**:423–7.

130. Clement N, Battaglioli G, Jensen RL, et al. Prevalence of human bocavirus in human tonsils and adenoids. *Emerg Infect Dis* 2009;**15**:1149–50.

131. Lu X, Gooding LR, Erdman DD. Human bocavirus in tonsillar lymphocytes. *Emerg Infect Dis* 2008;**14**:1332–4.

132. Bagnato G, Harari S. Cellular interactions in the pathogenesis of interstitial lung diseases. *Eur Respir Rev* 2015;**24**:102–14.

133. Keane MP. The role of chemokines and cytokines in lung fibrosis. *Eur Respir Rev* 2008;**17**:151–6.

134. Gauldie J, Jordana M, Cox G. Cytokines and pulmonary fibrosis. *Thorax* 1993;**48**:931–5.

135. Belperio JA, Dy M, Murray L, et al. The role of the Th2 CC chemokine ligand CCL17 in pulmonary fibrosis. *J Immunol* 2004;**173**:4692–8.

136. Kishi M, Miyazaki Y, Jinta T, et al. Pathogenesis of cBFL in common with IPF? Correlation of IP-10/TARC ratio with histological patterns. *Thorax* 2008;**63**:810–6.

137. Luo H, Zhang Z, Zheng Z, et al. Human bocavirus VP2 upregulates IFN-beta pathway by inhibiting ring finger protein 125-mediated ubiquitination of retinoic acid-inducible gene-I. *J Immunol* 2013;**191**:660–9.

138. Zhang Z, Zheng Z, Luo H, et al. Human bocavirus NP1 inhibits IFN-beta production by blocking association of IFN regulatory factor 3 with IFNB promoter. *J Immunol* 2012;**189**:1144–453.

139. Martin ET, Kuypers J, McRoberts JP, Englund JA, Zerr DM. Human bocavirus 1 primary infection and shedding in infants. *J Infect Dis* 2015;**212**:516–24.

140. Balada-Llasat JM, LaRue H, Kelly C, Rigali L, Pancholi P. Evaluation of commercial ResPlex II v2.0, MultiCode-PLx, and xTAG respiratory viral panels for the diagnosis of respiratory viral infections in adults. *J Clin Virol* 2011;**50**:42–5.

141. Pham NT, Trinh QD, Chan-It W, et al. A novel RT-multiplex PCR for detection of Aichi virus, human parechovirus, enteroviruses, and human bocavirus among infants and children with acute gastroenteritis. *J Virol Methods* 2010;**169**:193–7.

142. Spyridaki IS, Christodoulou I, de Beer L, et al. Comparison of four nasal sampling methods for the detection of viral pathogens by RT-PCR-A GA(2)LEN project. *J Virol Methods* 2009;**156**:102–6.

143. Arden KE, McErlean P, Nissen MD, Sloots TP, Mackay IM. Frequent detection of human rhinoviruses, paramyxoviruses, coronaviruses, and bocavirus during acute respiratory tract infections. *J Med Virol* 2006;**78**:1232–40.

144. Loeffelholz MJ, Pong DL, Pyles RB, et al. Comparison of the film array respiratory panel and prodesse real-time PCR assays for detection of respiratory pathogens. *J Clin Microbiol* 2011;**49**:4083–8.

145. Nawaz S, Allen DJ, Aladin F, Gallimore C, Iturriza-Gomara M. Human bocaviruses are not significantly associated with gastroenteritis: results of retesting archive DNA from a case control study in the UK. *PLoS One* 2012;**7**:e41346.

146. Martin ET, Taylor J, Kuypers J, et al. Detection of bocavirus in saliva of children with and without respiratory illness. *J Clin Microbiol* 2009;**47**:4131–2.

147. Williams JV. Deja vu all over again: Koch's postulates and virology in the 21st century. *J Infect Dis* 2010;**201**:1611–4.

148. von Linstow ML, Hogh M, Hogh B. Clinical and epidemiologic characteristics of human bocavirus in Danish infants: results from a prospective birth cohort study. *Pediatr Infect Dis J* 2008;**27**:897–902.

149. Garcia-Garcia ML, Calvo C, Pozo F, et al. Human bocavirus detection in nasopharyngeal aspirates of children without clinical symptoms of respiratory infection. *Pediatr Infect Dis J* 2008;**27**:358–60.

150. Ursic T, Krivec U, Kalan G, Petrovec M. Fatal human bocavirus infection in an 18-month-old child with chronic lung disease of prematurity. *Pediatr Infect Dis J* 2015;**34**:111–2.

151. Ghietto LM, Majul D, Ferreyra Soaje P, et al. Comorbidity and high viral load linked to clinical presentation of respiratory human bocavirus infection. *Arch Virol* 2015;**160**:117–27.

152. Campbell AP, Guthrie KA, Englund JA, et al. Clinical outcomes associated with respiratory virus detection before allogeneic hematopoietic stem cell transplant. *Clin Infect Dis* 2015;**61**:192–202.

153. Akturk H, Sik G, Salman N, et al. Atypical presentation of human bocavirus: severe respiratory tract infection complicated with encephalopathy. *J Med Virol* 2015;**87**:1831–8.

154. Kim JS, Lim CS, Kim YK, Lee KN, Lee CK. Human bocavirus in patients with respiratory tract infection. *Korean J Lab Med* 2011;**31**:179–84.

155. Haidopoulou K, Goutaki M, Damianidou L, Eboriadou M, Antoniadis A, Papa A. Human bocavirus infections in hospitalized Greek children. *Arch Med Sci* 2010;**6**:100–3.

156. Beder LB, Hotomi M, Ogami M, et al. Clinical and microbiological impact of human bocavirus on children with acute otitis media. *Eur J Pediatr* 2009;**168**:1365–72.

157. Deng X, Li Y, Qiu J. Human bocavirus 1 infects commercially available primary human airway epithelium cultures productively. *J Virol Methods* 2014;**195**:112–9.

158. Lehtoranta L, Soderlund-Venermo M, Nokso-Koivisto J, et al. Human bocavirus in the nasopharynx of otitis-prone children. *Int J Pediatr Otorhinolaryngol* 2012;**76**:206–11.

159. Schenk T, Maier B, Hufnagel M, et al. Persistence of human bocavirus DNA in immunocompromised children. *Pediatr Infect Dis J* 2011;**30**:82–4.

160. Martin ET, Fairchok MP, Kuypers J, et al. Frequent and prolonged shedding of bocavirus in young children attending daycare. *J Infect Dis* 2010;**201**:1625–32.

161. Storch GA. Plethora of respiratory viruses and respiratory virus data. *Clin Infect Dis* 2015;**61**(8):1225–7.

162. Babady NE, Mead P, Stiles J, et al. Comparison of the Luminex xTAG RVP fast assay and the Idaho Technology FilmArray RP assay for detection of respiratory viruses in pediatric patients at a cancer hospital. *J Clin Microbiol* 2012;**50**:2282–8.

163. Nascimento-Carvalho CM, Cardoso MR, Meriluoto M, et al. Human bocavirus infection diagnosed serologically among children admitted to hospital with community-acquired pneumonia in a tropical region. *J Med Virol* 2012;**84**:253–8.

164. Don M, Soderlund-Venermo M, Hedman K, Ruuskanen O, Allander T, Korppi M. Don't forget serum in the diagnosis of human bocavirus infection. *J Infect Dis* 2011;**203**:1031–2 author reply 2-3.

165. Hedman L, Soderlund-Venermo M, Jartti T, Ruuskanen O, Hedman K. Dating of human bocavirus infection with protein-denaturing IgG-avidity assays-Secondary immune activations are ubiquitous in immunocompetent adults. *J Clin Virol* 2010;**48**:44–8.

166. Don M, Soderlund-Venermo M, Valent F, et al. Serologically verified human bocavirus pneumonia in children. *Pediatr Pulmonol* 2010;**45**:120–6.

167. Kantola K, Hedman L, Allander T, et al. Serodiagnosis of human bocavirus infection. *Clin Infect Dis* 2008;**46**:540–6.

168. Christensen A, Dollner H, Skanke LH, Krokstad S, Moe N, Nordbo SA. Detection of spliced mRNA from human bocavirus 1 in clinical samples from children with respiratory tract infections. *Emerg Infect Dis* 2013;**19**:574–80.

CIRCULATION OF RESPIRATORY PATHOGENS AT MASS GATHERINGS, WITH SPECIAL FOCUS ON THE HAJJ PILGRIMAGE

P. Gautret*,**, S. Benkouiten*,**

*Aix Marseille University, Research Unit on Emerging Infectious and Tropical Diseases (URMITE), Marseille, France
**University Hospital Institute for Infectious Diseases (Méditerranée Infection), Marseille, France*

1 INTRODUCTION

Mass gathering (MG) medicine is a new field that includes risk assessment, surveillance, and early response to health threats associated with MGs.[1] A MG is usually defined as more than a specified number of persons at a specific location, for a specific purpose, for a defined period of time, and which may greatly vary between different MGs.[2] The number of participants that classifies an event as a MG is arbitrary. Some guidelines specify any gathering to be a MG when more than 1000 attendees are present. Others require the attendance of as many as 25,000 attendees to qualify. However, outbreaks occur irrespective of the size of a gathering. Outbreaks have been reported in settings such as weddings, private parties, and other events involving fewer than 1000 people, but also in large religious MGs attended by millions of pilgrims or sport, sociocultural, political, or commercial large events.[3] According to the World Health Organization, a MG should be defined as *"any occasion that attracts sufficient numbers of people to strain the planning and response resources of the community, city or nation hosting the event."*[2] These MGs can be planned or not, and they may be sporadic or recurrent. A major public health concern in relation to MGs is the international circulation of infectious diseases, and the spread between participants and to the population of the nation hosting the event.[2,4] A number of infectious diseases have been caused or have the potential to cause illnesses and deaths at MGs. The risk and pattern of diseases at MGs are influenced by the features of the event, notably its duration and location, particular activities, and also the participants' characteristics, including their immunity to infectious agents. The first recognition of the consequences of the unique nature of infectious disease for MGs was for food-borne illness arising from person-to-person transmission, with historical reports of cholera outbreaks in the context of the Hajj Muslim pilgrimage and in the Kumb Mela Hindu pilgrimage.[4] By contrast, the majority of outbreaks occurring at MG events reported over the last few decades resulted from respiratory transmission. Respiratory transmission generally requires close contact and is thus more common in MGs with overcrowded living conditions. Intercontinental outbreaks of *Neisseria meningitidis* have been described in the context of various MGs, notably at the Hajj between 1987 and 2001.[4] Over the last 15 years, the rapid international spread of influenza and other respiratory pathogens in association with MGs have been reported on many occasions. It is now clear that one of the major health risks at modern international MGs is the acquisition of respiratory pathogens, with further spread to origin countries through returning attendees.

In this chapter, we review the available data about the circulation of respiratory pathogens at MGs, with a special focus on the Hajj Muslim pilgrimage which has been better studied.

2 RESPIRATORY TRACT INFECTIONS AT THE HAJJ

The annual Hajj pilgrimage to Mecca in the Kingdom of Saudi Arabia (KSA) is the fifth pillar of Islam. The pilgrimage is mandatory for all adult Muslims with physical and financial capacity, at least once in their lifetime. Therefore, 2–3 million Muslims from over 180 countries across the globe gather each year in Mecca for the Hajj. The Hajj and its rituals are physically demanding. Although the Hajj rituals only take 1 week, many pilgrims gather in the KSA for the month-long Hajj season. Upon arrival in Mecca, the holiest city in Islam, most Hajj pilgrims begin their visit by performing the *"Umrah"* (also known as the minor pilgrimage). The *Umrah* is not compulsory but is highly recommended. It can be undertaken at any time of the year and include the *"Tawaf"* (circumambulation seven times in an anticlockwise direction) of the *"Kaaba,"* known as *"Tawaf al-Umrah,"* and *"Sa'I"* (seven trips between two small mountains, *"Al-Safa"* and *"Al-Marwah"*) inside *"Al-Masjid al Haram"* (ie, "The Sacred Mosque"). The Hajj, which retraces the footsteps of the Prophet Mohammed over approximately 1 week, is performed continuously from the 8th to the 13th of *"Dhul Hijja"* (ie, "the month of Hajj"), the last month of the Islamic lunar calendar. As this Islamic lunar calendar is 11 days shorter than the Gregorian calendar, the exact dates of Hajj vary from year to year. As part of the Hajj rituals, pilgrims have to visit different sacred places located outside the city of Mecca, including Mina, where they spend the night from 8th to 9th *Dhul Hijja*; the plain of Arafat for the *"standing ceremony"*, the culminating experience of the Hajj, which lasts on 9th *Dhul Hijja* from after dawn to slightly after sunset; and Muzdalifah where they stay from after sunset on 9th *Dhul Hijja* to after dawn on 10th *Dhul Hijja*. Back at Mina, on 10th *Dhul Hijja*, the pilgrims perform the ritual of *"Jamarat"* (stoning the columns symbolizing the Devil) by throwing seven stones at only the largest of the three pillars in the four-level *Jamarat* Bridge. After the ritual of *Jamarat*, animals are slaughtered in slaughterhouses at Mina, marking the first day of *"Eid al-Adha"* (ie, "Festival of the Sacrifice"). On the same or the following day, the pilgrims return to Mecca for a second circumambulation of the *Kaaba*, known as *"Tawaf al-Hajj"* or *"Tawaf al-Ifadah"*, and the *"Sa'I,"* and then hasten back to Mina to again perform the stoning of the Devil ritual (but this time by throwing seven stones at each of the three pillars) on the 11th and 12th *Dhul Hijja*. The pilgrims are then allowed to return to Mecca after this stoning, but they can also prolong their stay in Mina for another day (the 13th *Dhul Hijja*) to perform the same process of stoning of the pillars as of 11th and 12th *Dhul Hijja*. Finally, pilgrims can leave Mecca after a final circumambulation of the *Kaaba*, known as *"Tawaf al-Wada"* (ie, "the farewell circumambulation"). However, while not required as part of the Hajj, most pilgrims extend their trips to the city of Medina to visit Islam's second holiest site, *"Al-Masjid Al-Nabawi"* (ie, "the Mosque of the Prophet"), which contains Muhammad's tomb.

The Hajj presents a major public health and infection control challenges, both for the Saudi authorities, as well as for the national authorities of the countries of origin of the Hajj pilgrims.[5] In addition to physical exhaustion, sleep deprivation, and extreme heat (in Mecca during October, the average temperature is greater than 38°C during the day and greater than 25°C at night), which increase the susceptibility of pilgrims to airborne infections, inevitable overcrowding for a short period in housing and ritual sites, especially when performing the circumambulation of the *Kaaba* inside the Sacred Mosque in Mecca, with up to eight pilgrims per square meter near the *Kaaba*,[6] when using the pedestrian tunnels leading to the *Jamarat* Bridge in Mina, and in the Mina camp, this is approximately a 3-km^2 area where pilgrims are accommodated in tents, some with up to 50–100 people,[7] greatly increases the risk of acquiring and spreading infectious diseases during the pilgrims' stay, especially airborne diseases. Moreover, returning Hajj pilgrims may contribute to the international spreading of these diseases.

2.1 SYNDROMIC SURVEILLANCE DATA

The "Hajj cough" is considered by pilgrims almost de rigueur.[8] Early reports from the 1978 Hajj season indicated that upper tract respiratory infections formed the bulk of the workload of medical teams attending pilgrims.[9] Recent data indicated that 61% of 4136 ill pilgrims consulting at Mina primary health structures suffered respiratory tract infections.[10] Respiratory tract infection was the leading cause of medical admissions in Saudi hospitals during the Hajj, accounting for 57% in a study, among 160 patients in various hospitals in Mina and Arafat and for 74% in another study, among 16,232 outpatients at an hospital in Medina.[11,12] During the Hajj, pneumonia accounted for 22–39% of 1268 medical admissions in tertiary care structures in Mecca, Mina and Arafat and for 46% of 16,232 medical admissions in a tertiary care hospital in Medina.[11–14] Pneumonia accounted for 22–27% of 505 admissions in intensive care units where they were responsible for 55% of sepsis.[15–17] The incidence of pneumonia evaluated in cohorts of Iranian pilgrims was 3.4 per 100,000 pilgrims in 2005.[18]

Results of cohort surveys evidenced attack rates of respiratory symptoms of 53% among domestic Saudi pilgrims and of, respectively, 71% and 92% among foreign pilgrims from Iran and Indonesia.[19–21] In a survey conducted among French pilgrims participating to the Hajj in 2012–2014, the prevalence of cough was 81% and a high proportion presented with associated sore throat (91%), rhinitis (79%), and hoarseness (63%). Myalgia was reported in 48% of cases and subjective fever in 47%. The incubation time of respiratory symptoms was about 8 days and 52% of pilgrims presenting with a cough during their stay were still symptomatic on return. Among pilgrims with a cough, 69% took antibiotics.[22] Consumption of antibiotics by Hajj pilgrims suffering mild respiratory symptoms is frequent including 54% of patients consulting at various primary health care centers in Mina,[10] 95–99% patients consulting at the Ear, Nose, and Throat clinic of a Hospital in Mecca,[23,24] 72% of a cohort of Iranian pilgrims and 94% of in cohort of Indonesian pilgrims.[19–25]

2.2 ISOLATION OF RESPIRATORY PATHOGENS IN ILL HAJJ PILGRIMS

2.2.1 Viruses

Recent studies using PCR tools were conducted from 2005 through 2014 among a total of 1784 pilgrims who were suffering from upper respiratory tract infections, influenza like illness, lower tract respiratory infection, or pneumonia. Definition of syndromes differed according to authors. Sample size varied from 7 to 555 individuals. These surveys were conducted in ill pilgrims from various nationalities recruited at tertiary care hospitals and primary health care centers in Saudi Arabia or at Mina encampment or when returning in home country. The types of samples included were throat swabs, nasal swabs, nasopharyngeal swabs, sputum, bronchoalveolar, and nasopharyngeal aspirates, depending on the studies.[7,26–33] The viruses most commonly isolated from symptomatic patients during the Hajj are rhinovirus (3 out of 10 ill pilgrims), followed by influenza virus (1 out of 10) and coronaviruses (1 out of 10); other viruses (1 out of 10, grouped) being less frequently isolated, including adenovirus, parainfluenza virus, respiratory syncitial virus, and enterovirus by decreasing frequency. The ongoing Middle East Syndrome Coronavirus (MERS-CoV) epidemic in Saudi Arabia has prompted several countries to establish an enhanced surveillance system to rapidly detect and investigate possible cases of MERS-CoV infection among travelers to from the Middle East, including Hajj and Umrah pilgrims. Results from England, France, Canada, and US revealed that influenza A and B were the viruses most frequently isolated from hospitalized patients followed by rhinovirus; non-MERS coronaviruses, para-influenzae virus, adenovirus, respiratory syncytial virus, enterovirus, and metapneumovirus were less frequently isolated.[34–37] From 2012 through 2014, the surveillance data failed to evidence any case of MERS among Hajj pilgrims and only eight Umrah-associated MERS cases where identified over an estimated 20 million pilgrims who visited Mecca during this period.[38]

2.2.2 Bacteria

There are a few available recent studies addressing the role of bacteria in respiratory tract infections at the Hajj. Data gathered from 1983 pilgrims suffering pneumonia or lower tract respiratory infections between 1991 through 2013. Definition of syndromes differed according to authors. Sample size varied from 38 to 713 individuals. These surveys were conducted in ill pilgrims from various nationalities recruited at hospitals in Saudi Arabia; one was conducted among ill pilgrims arriving at an airport in Saudi Arabia. Types of samples included were serum, sputum and bronchoalveolar and nasopharyngeal aspirates, depending on the studies. Pathogens were identified by conventional culture methods, serology or PCR in one study. The bacteria most commonly isolated from symptomatic patients during the Hajj are *Haemophilus influenzae* (2 out of 10 ill pilgrims), followed by *Streptococcus pneumoniae* (1–2 out of 10), *Mycobacterium tuberculosis* (1 out of 10), *Pseudomonas aeruginosa* (1 out of 10), *Klebsiella pneumoniae* (1 out of 10), and *Chlamydophila pneumoniae* (1 out of 10). Other bacteria were less frequently isolated or investigated including by decreasing frequency *Moraxella catarrhalis*, *Enterococcus sp.*, *Legionella pneumophila*, *Acinetobacter baumannii*, *Staphylococcus aureus*, *Mycoplasma pneumoniae*, *Bordettela pertussis*, *Escherichia coli*, and *Stenotrophomonas maltophilia*.[17,21,32,39–42]

2.3 SYSTEMATIC SCREENING OF RESPIRATORY PATHOGENS IN HAJJ PILGRIMS BEFORE AND AFTER THE PILGRIMAGE

2.3.1 Viruses

From 2003 to 2013, the prevalence of respiratory viruses in cohorts of Hajj pilgrims was systematically investigated independently on the presence of respiratory symptoms in 11 epidemiological studies.[25,35,43–53] However, the design of these studies was also diverse, and included cohorts and longitudinal studies of national pilgrims recruited before they departed from or after they return to their home country, or international pilgrims arriving to and departing from the King Abdulaziz International Airport in Jeddah or at Mina encampment. In these studies, pilgrims were tested using different samples, including throat swabs, nasal swabs, nasopharyngeal swabs, or serum.

A range of respiratory viruses have been detected among pilgrims attending the Hajj pilgrimage. Influenza was the most frequently investigated. The mean prevalence of influenza (detected by PCR methods) was 2.1% among arriving pilgrims and 3.6% among departing pilgrims, with most strains identified as influenza A. In one UK study, a 38.3% seroconversion rate (mostly due to influenza H3N2) was observed in pilgrims returning from Hajj. However, in an Iranian study, the seroconversion rate was 3.6%. A difference in vaccination coverage may explain in part the difference in the reported seroconversion rates between the two studies. Respiratory syncytial virus prevalence was investigated in seven studies, six of which were PCR-based. In PCR-based studies, the mean prevalence of respiratory syncytial virus among pilgrims was 0.5% before and 1.2% after the Hajj. The prevalence of rhinovirus and adenovirus were investigated in six studies. For rhinovirus, the mean prevalence increased from 6.9% before to 19.7% after the Hajj. For adenovirus, the mean prevalence was 0.5% before and 0.6% after the Hajj. Coronaviruses prevalence were also investigated in seven studies, with the mean prevalence rising markedly from 1.2% before to 13.8% after the Hajj, with most infections due to coronavirus 229-E. MERS coronavirus was never isolated. Metapneumovirus and parainfluenza virus carriage were investigated by PCR methods in five studies. The mean prevalence of metapneumovirus was 0.4% before and 0.3% after the Hajj, and that of parainfluenza virus was 1% before and 1.6% after the Hajj. Bocavirus was investigated in four PCR-based studies and detected in only a few arriving pilgrims. Enterovirus was investigated in three PCR-based studies, and the mean prevalence was 0.7% among arriving and 1.1% among departing pilgrims. As far as is known, cytomegalovirus and parechovirus were investigated in only two studies, and were never isolated.

2.3.2 Bacteria

Unfortunately, while numerous studies investigating the carriage of numerous viral pathogens in healthy pilgrims and in pilgrims suffering from respiratory symptoms have been conducted in the past, most of the studies addressing bacterial carriage in Hajj pilgrims have been limited to *N. meningitidis* because of the international *N. meningitidis* serogroup W135 outbreak that occurred in 2001 and 2002.[54]

During the 2013 Hajj season, nasal specimens were prospectively collected from a large multinational cohort of pilgrims from 13 countries and tested for *S. pneumoniae, H. influenzae, K. pneumoniae, S. aureus, Coxiella burnetii, B. pertussis, M. pneumoniae, L. pneumophila, S. pyogenes, Salmonella* spp., *Pneumocystis jirovecii,* and *C. pneumoniae*.[53] In this study, the overall prevalence of bacteria increased from 15.4% before the Hajj to 31% after the Hajj, due to the significant acquisition of *S. pneumoniae, H. influenzae,* and *S. aureus*. The overall acquisition rate of bacteria was 28.3% (12.0%, 11.4% and 7.5%, respectively, for *S. pneumoniae, H. influenzae,* and *S. aureus*). *K. pneumoniae* and *C. burnetii* were also acquired during the stay in KSA by a low proportion of the pilgrims (0.1–3.9%). None of the pilgrims tested positive for *B. pertussis, M. pneumoniae, S. pyogenes, L. pneumophila, Salmonella* spp., *C. pneumoniae,* or *P. jirovecii* at any point during the study period.

Nasopharyngeal pneumococcal carriage prevalence was investigated during the 2011 and 2012 Hajj seasons with four large cross-sectional studies conducted among a cohort of pilgrims sampled using nasopharyngeal swabs at the beginning of Hajj in Mecca (within 2 days of arrival in Mecca) and a cohort of pilgrims sampled at the end of the Hajj in Mina (at any time after arrival in Mina).[55] This study showed an overall pneumococcal carriage prevalence of 6.0% among pilgrims and an increase in pneumococcal carriage from the beginning of Hajj to the end of the Hajj (4.4% versus 7.5%, respectively), particularly of conjugate vaccine serotypes and antibiotic nonsusceptible strains. During the 2012 Hajj season, nasal specimens were collected from a French cohort of pilgrims before departing from France and before leaving the KSA after the 2012 Hajj season. The results from this study demonstrated the acquisition of *S. pneumoniae* nasal carriage in returning Hajj pilgrims, with a prevalence of 7.3% before departing for the KSA and 19.5% before leaving the KSA. None of the participants tested positive for *B. pertussis* or *M. pneumoniae* at any point during the study period.[56] These results were confirmed the following year in another study conducted among a second French cohort of pilgrims departing from France for the 2013 Hajj season, which have reported a *S. pneumoniae* pharyngeal carriage of 50% before departing for the KSA and 62% before leaving the KSA.[50] As part of this second study, nasal samples were also tested for *S. aureus*, thus demonstrating a high rate (22.8%) of nasal carriage of *S. aureus* among pilgrims on return from the Hajj, with a significant increase of the emerging clonal complex 398.[57] In an earlier study, a low rate of methicillin resistant *S. aureus* (MRSA) carriage was noticed among pilgrims attending the 2004 Hajj pilgrimage. In fact, of pilgrims screened, 20.6% were positive for *S. aureus* of which only 1.5% were MRSA.[58]

In a 2005 study, the increased risk of acquiring *M. tuberculosis* infection during the 2002 Hajj season was shown by measuring the immune response to TB antigens, prior to departure and 3 months after return from the Hajj pilgrimage, using a whole-blood assay (Quanti-FERON TB assay) among a cohort of Singaporean pilgrims, who attended a pre-Hajj meningococcal vaccination campaign.[59] Thus, among those pilgrims who were negative prior to the Hajj, 10% had a significant rise in immune response to TB antigens. In a previous prospective seroepidemiological study, 1.4% of adult Singaporean pilgrims were found to have acquired pertussis (defined as prolonged cough and a 14-fold increase in the level of immunoglobulin G to whole-cell pertussis antigen) during their pilgrimage.[60]

3 RESPIRATORY TRACT INFECTIONS AT OTHER MASS GATHERINGS

A number of outbreaks of respiratory tract infections have been reported in various other MG settings.

3.1 RELIGIOUS MASS GATHERINGS

In 2010, in Pakpattan, Pakistan, during the annual celebration of the Urs of Baba Farid, attended by an estimated 500,000 people, the prevalence of acute respiratory illness showed a 25-fold increase while the population increased by a factor of 3.[61] Respiratory tract infections due to influenza (100 cases) were recorded during World Youth Day 2008 in Australia (500,000 participants).[62] Another influenza outbreak (38 cases) was recorded during the Itzapalapa Passion Play celebration in Mexico, 2009 (>2 million participants).[63,64] A measles outbreak (34 cases) occurred at a church gathering in the US, 2005 (500 participants) among unvaccinated participants and another was recorded during the 2010 Taizé festival in France (3500 participants), affecting 27 individuals mainly unvaccinated subjects including secondary and tertiary cases mainly in Germany.[65,66] In 2006, an outbreak of mumps with 214 identified patients occurred after an Easter festival in Austria.[67] A syndromic survey conducted during the 2013 Kumb Mela pilgrimage in India (estimated 100 million participants), evidenced a peak in upper respiratory tract infection symptoms just after the bathing day.[68] Of 412,703 patients who attended to outpatient departments of the hospitals, respiratory infections accounted for 70% of illnesses.[69]

3.2 SPORT MASS GATHERINGS

We found some examples of outbreak at sport events. One involved 36 individuals participating in the Winter Olympiad in Salt Lake City (2002) (1.5 million participants) who tested positive for influenza among 188 individuals presenting with influenza like illness.[70] Special Olympic World Games are reserved for athletes with disabilities; after the Minneapolis gathering in 1991 (20000 participants) a small outbreak of measles (25 cases) was recorded involving not only athletes but also volunteers and a few spectators, most of whom had not been previously vaccinated.[71] Another small measles outbreak (7 cases) occurred subsequent to an international youth sport event in Pennsylvania in 2007 (265,000 participants), and again the majority had previously received no or one dose of vaccine.[72]

3.3 FESTIVALS AND PRIVATE MASS GATHERINGS

Person-to-person transmission of influenza, measles, and mumps viruses has been recorded in the context of large scale open air festivals. Outbreaks of respiratory tract infections due to influenza A(H1N1) pdm09 (totalizing 71 cases) were recorded during the year 2009 at various music festivals in Belgium, Hungary, and Serbia (100,000–4,000,000 participants each).[73–75] Outbreaks of measles occurred at various MGs (totalizing 281 cases), including a large wedding in Spain (2010),[76] a music festival in Germany in 2011 (50,000 participants),[77] an international dog show in Slovenia in 2014 (11,000 participants) resulting in a chain of transmission in Italia,[78,79] and at the Disney theme parks in California in 2014–2015 (24 million annual attendance) resulting in a multistate outbreak.[80,81] Finally, an outbreak of mumps (77 cases) was recorded at an annual village festival in Spain in 2006 (>4500 participants).[82]

4 CONCLUSIONS

The Hajj cough is very common, probably resulting of crowded conditions during ritual performance. It affects all individuals whatever their comorbidities, age, and adherence to preventive measures against respiratory infections including use of face mask, hand hygiene, social distancing, and vaccination.[22] Vaccination against influenza may be considered, although contradictive results have been obtained.[83] A metaanalysis of observational studies demonstrated a significant effect of vaccination on laboratory-proven influenza although the overall effect of the vaccine on disease is yet to be seen and controlled studies exploring the efficacy of influenza vaccine among attendees of MGs are not available.[84] Studies are necessary to evaluate knowledge, attitudes, and practices of Hajj pilgrims regarding vaccination. Randomized controlled trials are needed to assess the efficacy of vaccines and improve the vaccination coverage in this vulnerable population.[83]

Nonpharmaceutical methods, such as face mask use, hand washing or use of hand gel, cough etiquette, social distancing, can be theoretically effective to mitigate the interhuman transmission of respiratory viruses [85] and are consequently recommended to Hajj pilgrims by national public health agencies [86] However, given the limited and inconclusive evidence of their effectiveness at the Hajj, prospective cohort studies are required to confirm whether or not such interventions are effective in interrupting or reducing the spread of respiratory pathogens at the Hajj.[87] A trial is being conducted to provide evidence on the efficacy of face masks in preventing viral respiratory tract infections among pilgrims. However, the adherence to face mask use by Hajj pilgrims remains a challenge.[87] By contrast, and despite their poor knowledge of the usefulness of hand hygiene in preventing respiratory tract infections, hand washing compliance of pilgrims is quite good.[87] Pending further rigorous studies on their effectiveness, we recommend effective hand hygiene practices, notably the use of alcohol-based hand gel, since it requires less time than classical hand washing, and act more rapidly. Social distancing measures are not realistic given the high density to which pilgrims are exposed at holy sites.[87]

Transmission of respiratory pathogens at the Hajj Muslim pilgrimage in the Kingdom of Saudi Arabia is highly frequent, resulting in high acquisition rates of virus and bacteria carriage. The Hajj may, therefore, contribute to the globalization of common respiratory pathogens resulting from the cross-contamination of participants harboring pathogens that easily spread among pilgrims. Numerous viral and bacterial clinical infections of the upper and lower respiratory tract do occur during the pilgrimage leading to a high number of in- and outpatients in Saudi medical structures. Over the past 10 years, the emergence of several viruses including influenza AH1N1, severe acute respiratory syndrome (SARS)-CoV, and MERS-CoV have been a great concern for the international medical community.[5] However, no major outbreaks have occurred at the Hajj, up to now. We must remember that aside from the highly publicized exotic and emerging pathogens, we must be alert to the circulation of common pathogens, which silently cause much more casualties than the exotic newcomers which occupy the forefront of the stage and get all the headlines.[88]

Although outbreaks are less frequently reported in or after MGs outside the Hajj and Umrah pilgrimages, they do sometimes occur. Most common were vaccine preventable diseases, mainly measles and influenza, while outbreaks of mumps have rarely been recorded. Almost all measles outbreaks at MGs have occurred among unvaccinated or incompletely vaccinated individuals.[84] However, the effectiveness of vaccination in this context remains to be evaluated. It is also noticeable that many outbreaks occurring at MGs result in international spread of communicable diseases.

Despite extensive surveillance, reports of outbreaks of respiratory tract infections at large sport events including the Olympics, the FIFA, and EURO football cups are scant. This is likely because the crowd density is much lower and because collective housing is not common at these events. Large music festivals, by contrast, share some characteristics with the Hajj regarding crowd density and housing conditions in tent camps. The younger age of the participants at music festivals likely account for the distinct pattern of pathogens responsible for outbreaks in this context, compared to that of the Hajj.[89] Other large religious MGs, like the Kumb Mela, have been poorly studied and the data concerning the health status of the pilgrims and potential health risks are scarce.[90]

Hopefully, public health research projects, developed in the context of international scientific collaboration, will elucidate the dynamics of communicable diseases transmission during the MGs and the consequence for their international spread. Large multinational cohort surveys are needed to better assess the risk of respiratory pathogens acquisition at MGs. The role of host factor including vaccination status, underlying chronic diseases age, and so on, and the role of environmental factors on the transmission of viruses and bacteria also warrant investigation. Interventional studies addressing the effectiveness of preventive measures are needed. The results of these studies will allow the implementation of evidence-based recommendations on prevention strategy including notably vaccination, hand hygiene, and face mask use. Investigations are also needed to evaluate the relationship link between communication and behavior in order to better adapt preventive message according to needs of specific communities.

REFERENCES

1. Memish ZA, Stephens GM, Steffen R, Ahmed QA. Emergence of medicine for mass gatherings: lessons from the Hajj. *Lancet Infect Dis.* 2012;**12**:56–65.
2. World Health Organization. Communicable disease alert and response for mass gatherings. Technical workshop. Geneva, Switzerland, April 29–30, 2008. www.who.int/csr/resources/publications/WHO_HSE_EPR_2008_8c.pdf.

3. Kok J, Blyth CC, Dwyer DE. Mass gatherings and the implications for the spread of infectious diseases. *Future Microbiol.* 2012;**7**:551–3.
4. Abubakar I, Gautret P, Brunette GW, Blumberg L, Johnson D, Poumerol G, et al. Global perspectives for prevention of infectious diseases associated with mass gatherings. *Lancet Infect Dis.* 2012;**12**:66–74.
5. Memish ZA, Zumla A, Alhakeem RF, Assiri A, Turkestani A, Al Harby KD, et al. Hajj: infectious disease surveillance and control. *Lancet* 2014;**383**:2073–82.
6. Alnabulsi H, Drury J. Social identification moderates the effect of crowd density on safety at the Hajj. *Proc Natl Acad Sci USA* 2014;**111**:9091–6.
7. Barasheed O, Almasri N, Badahdah AM, Heron L, Taylor J, McPhee K, et al. Pilot randomised controlled trial to test effectiveness of facemasks in preventing influenza-like illness transmission among Australian Hajj Pilgrims in 2011. *Infect Disord Drug Targets* 2014;**14**:110–6.
8. Shafi S, Booy R, Haworth E, Rashid H, Memish ZA. Hajj: health lessons for mass gatherings. *J Infect Public Health* 2008;**1**:27–32.
9. Paterson FW. Letter from abu dhabi: welfare of the Hajj. *Br Med J* 1980;**280**:1261–2.
10. Al-Zahrani AG, Choudhry AJ, Al-Mazroa MA, Turkistani AH, Nouman GS, Memish ZA. Pattern of diseases among visitors to Mina health centers during the Hajj season, 1429 H (2008 G). *J Infect Public Health* 2012;**5**:22–34.
11. Al-Ghamdi SM, Akbar HO, Qari YA, Fathaldin OA, Al-Rashed RS. Pattern of admission to hospitals during muslim pilgrimage (Hajj). *Saudi Med J* 2003;**24**:1073–6.
12. Yousuf M, Al-Saudi DA, Sheikh RA, Lone MS. Pattern of medical problems among Haj pilgrims admitted to King Abdul Aziz hospital, Madinah Al-Munawarah. *Ann Saudi Med.* 1995;**15**:619–21.
13. Al-Harbi MA. Pattern of surgical and medical diseases among pilgrims attending Al-Noor Hospital, Makkah. *J Famy Commun Med* 2000;**7**:21–4.
14. Madani TA, Ghabrah TM, Al-Hedaithy MA, Alhazmi MA, Alazraqi TA, Albarrak AM, et al. Causes of hospitalization of pilgrims in the Hajj season of the Islamic year 1423 (2003). *Ann Saudi Med* 2006;**26**:346–51.
15. Madani TA, Ghabrah TM, Albarrak AM, Alhazmi MA, Alazraqi TA, Althaqafi AO, et al. Causes of admission to intensive care units in the Hajj period of the Islamic year 1424 (2004). *Ann Saudi Med* 2007;**27**:101–5.
16. Baharoon S, Al-Jahdali H, Al Hashmi J, Memish ZA, Ahmed QA. Severe sepsis and septic shock at the Hajj: etiologies and outcomes. *Travel Med Infect Dis* 2009;**7**:247–52.
17. Mandourah Y, Al-Radi A, Ocheltree AH, Ocheltree SR, Fowler RA. Clinical and temporal patterns of severe pneumonia causing critical illness during Hajj. *BMC Infect Dis* 2012;**12**:117.
18. Meysamie A1, Ardakani HZ, Razavi SM, Doroodi T. Comparison of mortality and morbidity rates among Iranian pilgrims in Hajj 2004 and 2005. *Saudi Med J* 2006;**27**:1049–53.
19. Deris ZZ, Hasan H, Sulaiman SA, Wahab MS, Naing NN, Othman NH. The prevalence of acute respiratory symptoms and role of protective measures among Malaysian Hajj pilgrims. *J Travel Med* 2010;**17**:82–8.
20. Al-Jasser FS, Kabbash IA, Al-Mazroa MA, Memish ZA. Patterns of diseases and preventive measures among domestic hajjis from Central, Saudi Arabia. *Saudi Med J* 2012;**33**:879–86.
21. Razavi SM, Sabouri-Kashani A, Ziaee-Ardakani H, Tabatabaei A, Karbakhsh M, Sadeghipour H, et al. Trend of diseases among Iranian pilgrims during five consecutive years based on a Syndromic Surveillance System in Hajj. *Med J Islam Repub Iran* 2014;**27**:179–85.
22. Gautret P, Benkouiten S, Griffiths K, Sridhar S. The inevitable Hajj cough: surveillance data in French pilgrims, 2012–2014. *Travel Med Infect Dis* 2015;**13**(6):485–9.
23. Al-Herabi AZ. Road map of an ear, nose, and throat clinic during the 2008 Hajj in Makkah, Saudi Arabia. *Saudi Med J* 2009;**30**:1584–9.
24. Al-Herabi AZ. Impact of pH1N1 influenza A infections on the otolaryngology, head and neck Clinic during Hajj, 2009. *Saudi Med J* 2011;**32**:933–8.
25. Imani R, Karimi A, Habibian R. Acute respiratory viral infections among Tamattu' Hajj pilgrims in Iran. *Life Sci J* 2013;**10**:449–53.

26. Rashid H, Shafi S, Haworth E, El Bashir H, Ali KA, Memish ZA, et al. Value of rapid testing for influenza among Hajj pilgrims. *Travel Med Infect Dis* 2007;**5**:310–313.

27. Rashid H, Shafi S, Booy R, El Bashir H, Ali K, Zambon M, et al. Influenza and respiratory syncytial virus infections in British Hajj pilgrims. *Emerg Health Threats J* 2008;**1**:e2.

28. Rashid H, Shafi S, Haworth E, El Bashir H, Memish ZA, Sudhanva M, et al. Viral respiratory infections at the Hajj: comparison between UK and Saudi pilgrims. *Clin Microbiol Infect* 2008;**14**:569–74.

29. Alborzi A, Aelami MH, Ziyaeyan M, Jamalidoust M, Moeini M, Pourabbas B, et al. Viral etiology of acute respiratory infections among Iranian Hajj pilgrims, 2006. *J Travel Med* 2009;**16**:239–42.

30. Moattari A1, Emami A, Moghadami M, Honarvar B. Influenza viral infections among the Iranian Hajj pilgrims returning to Shiraz, Fars province. *Iran. Influenza Other Respir Viruses* 2012;**6**:e77–9.

31. Barasheed O, Rashid H, Alfelali M, Tashani M, Azeem M, Bokhary H, et al. Viral respiratory infections among Hajj pilgrims in 2013. *Virol Sin* 2014;**29**:364–71.

32. Memish ZA, Almasri M, Turkestani A, Al-Shangiti AM, Yezli S. Etiology of severe community-acquired pneumonia during the 2013 Hajj-part of the MERS-CoV surveillance program. *Int J Infect Dis* 2014;**25**:186–90.

33. Aberle JH, Popow-Kraupp T, Kreidl P, Laferl H, Heinz FX, Aberle SW. Influenza A and B Viruses but Not MERS-CoV in Hajj Pilgrims, Austria, 2014. *Emerg Infect Dis* 2015;**21**:726–7.

34. German M, Olsha R, Kristjanson E, Marchand-Austin A, Peci A, Winter AL, Gubbay JB. Acute respiratory infections in travelers returning from MERS-CoV-affected areas. *Emerg Infect Dis* 2015;**21**(9):1654–6.

35. Gautret P, Charrel R, Benkouiten S, Belhouchat K, Nougairede A, Drali T, et al. Lack of MERS coronavirus but prevalence of influenza virus in French pilgrims after 2013 Hajj. *Emerg Infect Dis* 2014;**20**:728–30.

36. Thomas HL, Zhao H, Green HK, Boddington NL, Carvalho CF, Osman HK, et al. Enhanced MERS coronavirus surveillance of travelers from the Middle East to England. *Emerg Infect Dis* 2014;**20**:1562–4.

37. Shahkarami M, Yen C, Glaser C, Xia D, Watt J, Wadford DA. Laboratory testing for Middle East respiratory syndrome coronavirus, California, USA, 2013–2014. *Emerg Infect Dis* 2015;**21**:1664–6.

38. Sridhar S, Brouqui P, Parola P, Gautret P. Imported cases of Middle East respiratory syndrome: an update. *Travel Med Infect Dis* 2015;**13**:106–9.

39. El-Sheikh SM1, El-Assouli SM, Mohammed KA, Albar M. Bacteria and viruses that cause respiratory tract infections during the pilgrimage (Haj) season in Makkah, Saudi Arabia. *Trop Med Int Health* 1998;**3**:205–9.

40. Alzeer A, Mashlah A, Fakim N, Al-Sugair N, Al-Hedaithy M, Al-Majed S, et al. Tuberculosis is the commonest cause of pneumonia requiring hospitalization during Hajj (pilgrimage to Makkah). *J Infect* 1998;**36**:303–6.

41. Asghar AH, Ashshi AM, Azhar EI, Bukhari SZ, Zafar TA, Momenah AM. Profile of bacterial pneumonia during Hajj. *Indian J Med Res* 2011;**133**:510–3.

42. Abdulrahman NK, Chaudhry AJ, Al Mazroa M. Etiology of upper respiratory tract infection among international pilgrims arriving for Hajj 2010 G. *Saudi Epidemiol Bull* 2011;**19**:14–5.

43. El Bashir H, Haworth E, Zambon M, Shafi S, Zuckerman J, Booy R. Influenza among UK pilgrims to Hajj, 2003. *Emerg Infect Dis* 2004;**10**:1882–3.

44. Kandeel A1, Deming M, Elkreem EA, El-Refay S, Afifi S, Abukela M, et al. Pandemic (H1N1) 2009 and Hajj Pilgrims who received Predeparture Vaccination, Egypt. *Emerg Infect Dis* 2011;**17**:1266–8.

45. Memish ZA, Assiri AM, Hussain R, Alomar I, Stephens G. Detection of respiratory viruses among pilgrims in Saudi Arabia during the time of a declared influenza A(H1N1) pandemic. *J Travel Med* 2012;**19**:15–21.

46. Gautret P, Charrel R, Belhouchat K, Drali T, Benkouiten S, Nougairede A, et al. Lack of nasal carriage of novel corona virus (HCoV-EMC) in French Hajj pilgrims returning from the Hajj 2012, despite a high rate of respiratory symptoms. *Clin Microbiol Infect* 2013;**19**:e315–7.

47. Ziyaeyan M, Alborzi A, Jamalidoust M, Moeini M, Pouladfar GR, Pourabbas B, et al. Pandemic 2009 influenza A (H1N1) infection among 2009 Hajj Pilgrims from Southern Iran: a real-time RT-PCR-based study. *Influenza Other Respir Viruses* 2012;**6**:e80–4.

48. Benkouiten S, Charrel R, Belhouchat K, Drali T, Salez N, Nougairede A, et al. Circulation of respiratory viruses among pilgrims during the 2012 Hajj pilgrimage. *Clin Infect Dis* 2013;**57**:992–1000.
49. Ashshi A, Azhar E, Ayman Johargy A, Asghar A, Momenah A, Turkestani A, et al. Demographic distribution and transmission potential of influenza A and 2009 pandemic influenza A H1N1 in pilgrims. *J Infect Dev Ctries* 2014;**8**:1169–75.
50. Benkouiten S, Charrel R, Belhouchat K, Drali T, Nougairede A, Salez N, et al. Respiratory viruses and bacteria among pilgrims during the 2013 Hajj. *Emerg Infect Dis* 2014;**20**:1821–7.
51. Memish ZA, Assiri A, Almasri M, Alhakeem RF, Turkestani A, Al Rabeeah AA, et al. Prevalence of MERS-CoV nasal carriage and compliance with the Saudi health recommendations among pilgrims attending the 2013 Hajj. *J Infect Dis* 2014;**210**:1067–72.
52. Annan A, Owusu M, Marfo KS, Larbi R, Sarpong FN, Adu-Sarkodie Y, et al. High prevalence of common respiratory viruses and no evidence of Middle East Respiratory Syndrome Coronavirus in Hajj pilgrims returning to Ghana, 2013. *Trop Med Int Health* 2015;**20**:807–12.
53. Memish ZA, Assiri A, Turkestani A, Yezli S, Al Masri M, Charrel R, et al. Mass gathering and globalization of respiratory pathogens during the 2013 Hajj. *Clin Microbiol Infect* 2015;**21**:e571–9.
54. Al-Tawfiq JA, Zumla A, Memish ZA. Respiratory tract infections during the annual Hajj: potential risks and mitigation strategies. *Curr Opin Pulm Med* 2013;**19**:192–7.
55. Memish ZA, Assiri A, Almasri M, Alhakeem RF, Turkestani A, Al Rabeeah AA, Akkad N, Yezli S, Klugman KP, O'Brien KL, van der Linden M, Gessner BD. Impact of the Hajj on pneumococcal transmission. *Clin Microbiol Infect* 2015;**21**:77e8–77e11.
56. Benkouiten S, Gautret P, Belhouchat K, Drali T, Salez N, Memish ZA, et al. Acquisition of *Streptococcus pneumoniae* carriage in pilgrims during the 2012 Hajj. *Clin Infect Dis* 2014;**58**:e106–9.
57. Verhoeven PO, Gautret P, Haddar CH, Benkouiten S, Gagnaire J, Belhouchat K, et al. Molecular dynamics of *Staphylococcus aureus* nasal carriage in Hajj pilgrims. *Clin Microbiol Infect* 2015;**21**:650e5–8.
58. Memish ZA, Balkhy HH, Almuneef MA, Al-Haj-Hussein BT, Bukhari AI, Osoba AO. Carriage of *Staphylococcus aureus* among Hajj pilgrims. *Saudi Med J* 2006;**27**:1367–72.
59. Wilder-Smith A, Foo W, Earnest A, Paton NI. High risk of *Mycobacterium tuberculosis* infection during the Hajj pilgrimage. *Trop Med Int Health* 2005;**10**:336–9.
60. Wilder-Smith A, Earnest A, Ravindran S, Paton NI. High incidence of pertussis among Hajj pilgrims. *Clin Infect Dis* 2003;**37**:1270–2.
61. Hassan S, Imtiaz R, Ikram N, Baig MA, Safdar R, Salman M, et al. Public health surveillance at a mass gathering: urs of Baba Farid, Pakpattan district, Punjab, Pakistan, December 2010. *East Mediterr Health J* 2013;**19**(Suppl. 2):S24–8.
62. Blyth CC, Foo H, van Hal SJ, Hurt AC, Barr IG, McPhie K, et al. Influenza outbreaks during World Youth Day 2008 mass gathering. *Emerg Infect Dis* 2010;**16**:809–15.
63. Zepeda HM, Perea-Araujo L, Zarate-Segura PB, Vázquez-Pérez JA, Miliar-García A, Garibay-Orijel C, et al. Identification of influenza A pandemic (H1N1) 2009 variants during the first 2009 influenza outbreak in Mexico City. *J Clin Virol* 2010;**48**:36–9.
64. Zepeda-Lopez HM, Perea-Araujo L, Miliar-García A, Dominguez-López A, Xoconostle-Cázarez B, Lara-Padilla E, et al. Inside the outbreak of the 2009 influenza A (H1N1)v virus in Mexico. *PLoS One* 2010;**5**:e13256.
65. Parker AA, Staggs W, Dayan GH, Ortega-Sánchez IR, Rota PA, Lowe L, et al. Implications of a 2005 measles outbreak in Indiana for sustained elimination of measles in the United States. *N Engl J Med* 2006;**355**:447–55.
66. Pfaff G, Lohr D, Santibanez S, Mankertz A, van Treeck U, Schonberger K, Hautmann W. Spotlight on measles 2010: measles outbreak among travellers returning from a mass gatering, Germany, September to October 2010. *Euro Surveill* 2010;**15**(50.) pii: 19750.

67. Schmid D, Holzmann H, Alfery C, Wallenko H, Popow-Kraupp TH, Allerberger F. Mumps outbreak in young adults following a festival in Austria, 2006. *Euro Surveill* 2008;13 pii: 8042.
68. Singh BP, Mudera CP. Mass gathering health care. *Popul Health Manag* 2014;**17**:316–7.
69. Cariappa MP, Singh BP, Mahen A, Bansal AS. Kumbh Mela 2013: healthcare for the millions. *Med J Armed Forces India* 2015;**71**:278–81.
70. Gundlapalli AV, Rubin MA, Samore MH, Lopansri B, Lahey T, McGuire HL, et al. Influenza, Winter Olympiad, 2002. *Emerg Infect Dis* 2006;**12**:144–6.
71. Ehresmann KR, Hedberg CW, Grimm MB, Norton CA, MacDonald KL, Osterholm MT. An outbreak of measles at an international sporting event with airborne transmission in a domed stadium. *J Infect Dis* 1995;**171**:679–83.
72. Centers for Disease Control, Prevention (CDC). Multistate measles outbreak associated with an international youth sporting event—Pennsylvania, Michigan, and Texas, August–September 2007. *MMWR Morb Mortal Wkly Rep* 2008;**57**:169–73.
73. Gutiérrez I, Litzroth A, Hammadi S, Van Oyen H, Gérard C, Robesyn E, et al. Community transmission of influenza A (H1N1)v virus at a rock festival in Belgium, 2–5 July 2009. *Euro Surveill* 2009;**14** pii = 19294.
74. Botelho-Nevers E, Gautret P, Benarous L, Charrel R, Felkai P, Parola P. Travel-related influenza A/H1N1 infection at a rock festival in Hungary: one virus may hide another one. *J Travel Med* 2010;**17**:197–8.
75. Ristic M, Seguljev Z, Nedeljkovic J, Ilic S, Injac D, Dekic J. Importation and spread of pandemic influenza virus A(H1N1) in autonomous province of Vojvodina in preepidemic period. *Med Pregl* 2010;**63**:502–5.
76. Lopez HB, Laguna SJ, Marin R, I, Gallardo G, V, Perez ME, Mayoral Cortes JM. Spotlight on measles 2010 An ongoing outbreak of measles in an unvaccinated population in Granada,Spain, October to November 2010 an ongoing outbreak of measles in an unvaccinated population in Granada, Spain, October to November 2010. *Euro Surveill* 2010; 15, pii: 19746.
77. Santibanez S, Prosenc K, Lohr D, Pfaff G, Jordan Markocic O, Mankertz A. Measles virus spread initiated at international mass gatherings in Europe, 2011. *Euro Surveill* 2014;**19** pii: 20891.
78. Grgič-Vitek M, Frelih T, Ucakar V, Fafangel M, Jordan Markocic O, Prosenc K, et al. An outbreak of measles associated with an international dog show in Slovenia, November 2014. *Euro Surveill* 2015;**20**:pii:21012.
79. Filia A, Riccardo F, Del Manso M, D'Agaro P, Magurano F, Bella A. Regional contact points for measles surveillance. Measles outbreak linked to an international dog show in Slovenia - primary cases and chains of transmission identified in Italy, November to December 2014. *Euro Surveill* 2015;**20**:pii: 21050.
80. McCarthy M. Measles outbreak linked to Disney theme parks reaches five states and Mexico. *BMJ* 2015;**350**:h436.
81. Zipprich J, Winter K, Hacker J, Xia D, Watt J, Harriman K. Centers for Disease Control and Prevention (CDC). Measles outbreak—California, December 2014–February 2015. *MMWR Morb Mortal Wkly Rep* 2015;**64**:153–4.
82. Gerstel L, Lenglet A, Garcia CM. Mumps outbreak in young adults following a village festival in the Navarra region, Spain, August 2006. *Euro Surveill* 2006;**11** pii 3078.
83. Alqahtani AS, Rashid H, Heywood AE. Vaccinations against respiratory tract infections at Hajj. *Clin Microbiol Infect* 2015;**21**:115–27.
84. Alqahtani AS, Alfelali M, Arbon P, Booy R, Rashid H. Burden of vaccine preventable diseases at large events. *Vaccine* 2015;**33**(48):6552–63.
85. Jefferson T, Del Mar CB, Dooley L, Ferroni E, Al-Ansary LA, Bawazeer GA, et al. Physical interventions to interrupt or reduce the spread of respiratory viruses. *Cochrane Database Syst Rev* 2011;**7** CD006207.
86. Alqarni H, Memish ZA, Assiri AM. Health conditions for travellers to Saudi Arabia for the pilgrimage to Mecca (Hajj)—2015. *J Epidemiol Glob Health* 2015;**6**(1):7–9.
87. Benkouiten S, Brouqui P, Gautret P. Non-pharmaceutical interventions for the prevention of respiratory tract infections during Hajj pilgrimage. *Travel Med Infect Dis* 2014;**12**:429–42.

88. Charrel RN, Hajj. Umrah, and other mass gatherings: which pathogens do you expect? Beware of the tree that hides the forest! *Travel Med Infect Dis* 2014;**12**:418–9.

89. Botelho-Nevers E, Gautret P. Outbreaks associated to large open air festivals, including music festivals, 1980 to 2012. *Euro Surveill* 2013;**18**:20426.

90. Sridhar S, Gautret P, Brouqui P. A comprehensive review of the Kumbh Mela: identifying risks for spread of infectious diseases. *Clin Microbiol Infect* 2015;**21**:128–33.

INDOOR AIR POLLUTION DUE TO MYCOFLORA CAUSING ACUTE LOWER RESPIRATORY INFECTIONS

T. Dubey

TBON-LAB, Investment Blvd. Hayward, CA, United States

1 INTRODUCTION

Allergy and asthma are common health issues among humans and they may not get attention until they turn into acute respiratory infections. Mostly respiratory diseases are developed due to continuous exposure to an unhealthy or polluted environment, either indoor or outdoor locations.

Prolonged breathing time in polluted air is mostly possible inside a building, which can be a home or work place (with few exceptions such as agriculture fields and construction zones where field workers are exposed for more than 4 h duration). Normally the preliminary symptoms of allergy are ignored unless they develop into chronic respiratory infections.

Asthma is a common disorder that afflicted the health of 24.6 million persons, including children, in the United States during 2009.[1] According to physicians the disease becomes severe when sensitivity toward mold becomes very high.

1.1 DIRECT ASSOCIATION OF FUNGI WITH DEVELOPMENT OF ASTHMA

Primary reasons suspected for such infections are bacteria and virus, while mold as pioneer invader is still a questionable factor directly associated with such infections. There is a big discussion going on to prove that fungal spores accumulating in breathing air can lead to severe respiratory diseases.

Research published by Knutsen et al.[2] has explained the details of fungal species responsible for asthma. They found direct association between increased fungal exposure and loss of asthma control. Jaakkolla et al.[3] found that increased sensitivity for *Aspergillus* and *Cladosporium* species increased the risk for adult-onset asthma. Harley et al.[4] found that children exposed to ascospores and basidiospores during first 3 years of childhood are at risk of asthma. The reason for increased sensitivity for fungi can be a combination of genetic and environmental factors.

Respiratory diseases can be due to outdoor mold spore exposure or due to indoor mold spores accumulated due to a damp environment. Recent reviews from the United States,[5] Europe,[6] and World Health Organization[7] have also confirmed a damp indoor environment as a major factor in the development of respiratory diseases.

Fungal distribution inside buildings may vary depending on the presence of favorable temperature, moisture, and nutritional sources. The presence of mold spores are best represented in house dust collected by HEPA filter. Dust formation takes place due to accumulation of airborne organic and inorganic particulate matter originating from multiple indoor and outdoor sources. House dust consists of a fibrous mixture of hairs, skin cells, dust mites, textile fibers with a particle size range of 10^{-3} to 1 mm (ibid.).

Besides many other components, the fungi are known to be major microbiota of house dusts. Scotts[8] reported from his research on fungi from house dust that outdoor mycoflora act as major supplier of indoor phylloplane spores when wind speed and precipitation favor the release of spores. A large magnitude of fungal spores in house dust can be either amplified within the dust itself or imported from surrounding indoor amplification site or source. The indoor sources include household items varying from small pieces of rotten food from refrigerator, indoor house plant, soil from shoe closets, decaying wood surface, wooden picture frames, or a house pet.[9-11] Table 7.1 shows different environment conditions from residential and industrial/occupational areas responsible for promoting the fungal growth to cause HP and related allergies to exposed persons.

1.2 CHARACTERIZATION OF FUNGI

Fungi are defined as eukaryotic, filamentous, and mostly spore bearing organisms. They grow on plants and animals as saprophytic or parasitic micro- or macroorganisms. Fungal spores vary in morphology and are ubiquitous. Approximately one million fungal species are present in the environment.

Many fungal species are associated with human diseases such as asthma, allergies, respiratory diseases, and skin infections. Respiratory allergy can vary from 6% among general population to 30% among atopic individuals. Toxic metabolites produced by fungal cells can cause allergic manifestations such as asthma, rhinitis, allergic bronchopulmonary mycoses, hypersensitive pneumonitis (HP), and allergic bronchopulmonary aspergilosis (ABPA).

1.3 HISTORICAL BACKGROUND

Airborne fungal spores were first associated with allergy attacks among patients by Storm Van Leeuwen.[12] He suspected that fungi in the home and mattresses of a patient in Holland who suffered from asthma could be causative agents. Similar studies in Germany and the United States have indicated that airborne fungi could cause asthma and hay fever in many patients. An incident of asthma in a patient who was exposed to fungal spores from the bark of maple logs was reported by Towey et al.[13]

Another case of asthma attack was noted by Cobe[14] in a greenhouse worker who inhaled the spores of *Cladosporium* from a tomato plant.

Gambale et al.[15] studied the airborne fungi of Sao Paulo state of Brazil and reported *Cladosporium, Dreschlera, Fusarium, Alternaria, Penicillium, Aspergillus,* and hyphal fragments as most common fungal genera.

1.4 DISEASES CAUSED BY MOLDS IN HUMANS

More information is required to track the pathway of individual mold spore type from various stages:

1. Inhaled by a person;
2. Production of specific toxin or allergen in specific area of respiratory tract;
3. Major symptoms generated due to specific allergen;
4. Cause the asthma; and finally,
5. Its proper treatment by using appropriate antigen.

Table 7.1 Various Sources of Allergens Produced by Fungal Species to Cause Hypersensitive Pneumonitis

Source Domestic/Residential Exposure	Disease	Allergen
Excessive moisture: Leakage area, kitchen sink, ceiling, heating ventilation air conditioning filters (HVAC), wall board, shower, basement sewage conditions, Sauna	Humidifier air conditioner (from contaminated water), *Cephalosporium*-hypersensitive pneumonitis (HP), Sauna taker's lung, hot tub lung	*Penicillium expansum, Aureobasidium pullulans, Thermophilic actinomycetes, Acremonium (=Cephalosporium), Cladosporium* sp.
Spoiled food: Refrigerated food-moldy cheese, rotting fruits, and vegetables	Cheese washer's lung	*Penicillium casei*
Decoration: House plants, live Christmas trees, wooden frames, fish aquariums	Asthma, allergies, Housewife's lung	*Alternaria, Aspergillus, Cladosporium, Aurebasidium pullulans*
Pets: Dogs, cats, and fishes	General asthma, Fisherman's lung	*Paecilomyces* sp., *Saprolegnia* sp.
Clothes and shoes in closet, sports items/balls	Aspergillosis, Sportsman's allergies	*Aspergillus* sp. Smut/Myxomycetes
Outside vegetation: trees, mushrooms, fruits and vegetables, soil	Lycoperdonosis	*Lycoperdon* sp./Puffballs, *Aspergillus* sp., *Penicillium* sp.
Travel: Items inside car, parks/garden soil: bird droppings, heavy dust	Japanese Summer house-HP, Valley fever	*Trichosporon* sp., *Coccidioides imitis*
Occupational/Industrial Exposure	**Disease**	**Allergen**
Agricultural: Compost, field equipments, vehicles, storage of grain, hay and silage, and sugarcane workers	Farmers lung, compost lung, Begassosis	*Aspergillus fumigatus*, thermophilic fungi, actinomycetes, and other soil microorganisms
Mushroom workers	Mushroom worker's lung	Micropolyspora faeni is a bacteria associated with mushrooms
Construction workers lung-dust-burrowing animals	Valley fever disease	Valley fever fungus and other soil fungi
Moldy barley and moldy grapes	Malt and wine worker's lung	*Aspergillus fumigatus, Aspergillus clavatus, Botrytis cinera*
Tobacco workers:moldy tobacco	Tobacco worker's disease	*Aspergillus* sp.
Wood workers: Cedar, mahogany and oak dust and pine and spruce pulp	Wood worker's lung	*Alternaria* species and wood dust fungi

This can be difficult task because more than one fungal species are inhaled by the human body, which may be producing similar toxins but may not have similar antigen for treatment.

Due to this state of uncertainty, we need to look for other options such as prevention of exposure of sensitive persons to toxic mold. This can be possible if we know more about the etiology of predominating mold genera inside and outside a building. Air monitoring of quality and quantity of specific mold genera in surrounding atmosphere is a major tool to watch and prevent the accumulation of spores inside a building.

2 ASSOCIATION OF MOLD WITH ASTHMA AND ALLERGIES-MAJOR ASPECTS

This chapter will discuss various aspects of predominating mold species associated with indoor air quality and their role in causing any respiratory problems among humans. Major aspects to explore the association of mold species with asthma and allergies and the mechanism of their development inside human body include the following:

1. *Pathophysiology*: mechanism of development of disease by producing fungal metabolites;
2. *Ecological studies*: various factors causing mold growth inside the building;
3. Methods to study the mold and related problems in indoor and outdoor environments; and,
4. Induced outbreak of mold allergies due to human activity (eg, bringing a live Christmas tree from the outside of a nursery to the inside of a residential or commercial building).

2.1 PATHOPHYSIOLOGY

2.1.1 Toxic metabolites

The major toxic metabolites produced by different groups of fungi responsible for these allergies include:

1. *B-Glucan and Dectin receptor*: glucans are major components of fungi and may form 60% of dry weight of fungi;
2. *Fungal protease*: can damage epithelial tissues to enter mucosal layer;
3. *Chitinase*: can cause increased IgE levels and increase in asthma;
4. *HLA-class II antigen*: which is found in patients with ABPA; and,
5. *IL4RA and IL 13 polymorphisms*: genes containing single nucleotide polymorphism (SNPs), which are associated with severe asthma.

Mycotoxins and volatile organic compounds are produced as a major toxic chemical by fungal cells to cause mycotoxicoses but less is known about their exposure limits to cause the disease. Information available on mycotoxins is insufficient from medical textbooks and can be unrecognized by medical professionals, unless they affect large numbers of people.

In 1999, the World Health Organization[16] published a bulletin to explain the chemical nature of various mycotoxins and the etiology of disease due to production of specific mycotoxin by a mold species. This bulletin also explains the identification of mycotoxin producing fungi, their epidemiological, clinical, and histological nature and their role (when available) in outbreaks of mycotoxicoses resulting from exposure to different fungi.

The various groups of mycotoxins as explained by WHO bulletin are given as follows:

1. *Aflatoxins*: These are produced by different *Aspergillus* species growing on nuts, cereal, and rice if exposed to water or high humidity. These toxins can be harmful to human health. The two major *Aspergillus species* that produce aflatoxins are:
   a. *Aspergillus flavus*: which produces only B aflatoxins; and,

b. *A. parasiticus*: which produces both B and G aflatoxins. Aflatoxins [M.sub.1] and [M.sub.2] are oxidative metabolic products of aflatoxins [B.sub.1] and [B.sub.2] produced by animals following ingestion, and so appear in milk (both animal and human), urine, and feces. Aflatoxicol is a reductive metabolite of aflatoxin [B.sub.1].

Aflatoxins are acutely toxic, immunosuppressive, mutagenic, teratogenic, and carcinogenic compounds. The main target organ for toxicity and carcinogenicity is the liver. The evaluation of epidemiological and laboratory results carried out in 1987 by the International Agency for Research on Cancer (IARC) provided evidence that naturally occurring mixtures of aflatoxins can be carcinogenic in humans. Therefore, they are classified as Group 1 carcinogens, except for aflatoxin [M.sub.1], which is possibly carcinogenic to humans (Group 2B).

2. *Ochratoxins*: These are present as secondary metabolites of *Aspergillus and Penicillium* strains, found on cereals, coffee, bread, and meat and food from animal origin. The most frequent is ochratoxin A, which is also the most toxic. It has been shown to be nephrotoxic, immunosuppressive, carcinogenic, and teratogenic in all experimental animals tested so far.

3. *Trichothecenes*: Currently, 148 Trichothecenes have been isolated but only a few have been found to contaminate food and feed. They are produced mostly by members of the *Fusarium* genus, although other genera [eg, *Trichoderma, Trichothecium, Myrothecium*, and *Stachybotrys* (black mold)] are also known to produce these compounds. Deoxynivalenol (DON), also known as vomitoxin, nivalenol (NIV), diacetoxyscirpenol (DAS) are most frequently observed contaminants. On the other hand T-2 toxin is rare.

In experimental animals, trichothecenes are 40 times more toxic when inhaled than when given orally. In several cases, trichothecene mycotoxicosis was caused by a single ingestion of bread containing toxic flour or rice.

Trichothecenes are also very common in air samples collected during the drying and milling process on farms, in the ventilation systems of private houses and office buildings. Trichothecens may accumulate in residential buildings if the walls have high humidity due to seepage. A higher concentration of mold spores can increase the concentration of Trichothecens and can change a healthy green building into a sick moldy building. This can be proven when the buildings and ventilation systems are thoroughly cleaned and symptoms of airborne toxicosis disappear.

4. *Zearalenone*: Earlier known as F-2, this is produced mainly by *Fusarium graminearum* and related species, principally in wheat and maize. Zearalenone and its derivatives produce estrogenic effects in various animal species and cause infertility, vulval oedema, vaginal prolapse, mammary hypertrophy in females, and atrophy of testes and enlargement of mammary glands in males.

5. *Fumonisins*: These are mycotoxins produced by *Fusarium moniliforme* and related species when they grow in maize. Only Fumonisins B.sub.1and B.sub.2 are of toxicological significance. Disaster due to this mycotoxin Fumonisin [B.sub.1] was best observed during the outbreak of acute food-borne disease in 27 villages of India. The individuals affected were from the poorest social strata, who had consumed maize and sorghum harvested and left in the fields during unseasonable rains. The main features of the disease were transient abdominal pain, borborygmus, and diarrhea, which began half an hour to 1 h following consumption of unleavened bread prepared from moldy sorghum or moldy maize.

2.1.2 TTC concept

The threshold of toxicologic concern (TTC) is based on a known structure-activity relationship. Frawley[17] and Munro[18] emphasized methods of evaluation of safety and characterization of risk to determine the daily chemical intake level that is acceptable for the human body without causing adverse health effects. This was assumed for a 70 year life span (Hardin et al.[19]). The TTC concept was further extended to other inhaled substances by converting a TTC expressed in *microgram/person/day* to an airborne concentration (ng/M^3). The resulting concentration of no toxicologic concern (CONTC) (30 ng/M^3) is considered as a generic airborne concentration that is harmless to human. It was also suggested that mycotoxin levels of dust and fungal spores from agricultural fields have potential to produce mycotoxins greater than CONTC levels. On the other hand, common exposure to mycotoxins inside buildings are below CONTC and harmless.

Unacceptable relationship of airborne mold exposure and clinical manifestations of allergic rhinitis. Bush et al.[20] evaluated various data to establish a clear relationship between mold and respiratory diseases and gave following reasons for the unclear relationship between the mold as causal organism and human respiratory diseases.

1. Presence of IgE antibodies to molds as part of polysensitization among atopic patients (those with allergic asthma, allergic rhinitis, and atopic dermatitis).
2. Lower airway diseases (asthma) have well known allergic responses to inhaled mold antigens.
3. There is no evidence to prove that exposure to outdoor airborne molds can cause allergic rhinitis. In addition, the studies on the contribution of indoor molds to upper airway allergy are even less compelling.
4. Exposure to airborne molds is not sufficient to prove them to be a factor in atopic dermatitis, urticaria, angioedema, or anaphylaxis.

It was also suggested that patients with suspected mold allergy should be evaluated by means of a standard clinical method for evaluation of potential allergies (skin or blood testing) for IgE antibodies in response to appropriate mold antigens.

2.1.3 Role of prick test

According to Menezes et al.,[21] the skin test is a deliberate and controlled exposure to a suspected allergen conducted mainly to confirm clinical atopic sensitivity. The skin test has recently become a widespread diagnostic aid. A basic premise of skin test is that reaginic antibody will fix to the skin mast cell similar to that occurring in other target organs. It involves an interaction between the allergen and IgE fixed to mast cells in the skin with liberation of chemical mediators, which results in local erythema and wheal formation.

In a clinical review by the European Respiratory Society, Kousha et al.[22] emphasized the fact that air born fungi are spread through atmospheric air and if we know the ecological distribution of anemophilous fungi in a city then specific treatment of allergic manifestations induced by inhaled allergens can be easier. Their use in an individual's allergy is widespread. The authors used the prick test in 50 patients suffering from asthma and rhinitis and 10 healthy persons with no resp. allergy as control. Fungal allergens extracts were prepared from 10 of the most predominant mold species found in air by using sodium bicarbonate. All ten of the predominant fungi could provoke skin test reactivity in individuals with resp. allergy in a specific geographic location. *Aspergillus, Alternaria, and Dreschlera* were positive in all 50 patients. Against this, all of the healthy people used as control tests were negative.

2.1.4 Allergic bronchopulmonary aspergillosis

Allergic bronchopulmonary aspergillosis (ABPA) is a hypersensitivity reaction to *Aspergillus* fumigatus antigens. This is generally seen in patients with atopy, asthma, or cystic fibrosis. Kousha et al. have also reported that invasive pulmonary aspergillosis occurs primarily in severely immunodeficient patients and that the number of such patients have increased.

The effect of inhalation of *A. fumigatus* spores to different types of sensitive patients include:

1. *Normal host-no sequel;*
2. *Patients with cavity lung disease*: aspergilloma;
3. *Chronic lung disease or mild immune-compromised host*: chronic necrotizing aspergillosis;
4. *Immunocompromised host*: invasive pulmonary aspergillosis; and,
5. Patients with asthma, cystic fibrosis, and atopy can lead to allergic reaction.

2.1.5 Sensitization to ABPA

Sensitization to *Aspergillus* antigens is an important phenomenon in asthmatics, especially those with atopy but this also increases the risk of more severe airflow obstruction and more prescriptions for oral corticosteroids. These observations also suggest that it is crucial to screen asthmatic patients for sensitization to *Aspergillus* antigens and to monitor these patients more closely and exclude the presence of ABPA.

ABPA is usually suspected on clinical grounds. The diagnosis is confirmed by radiological and serological testing. Other symptoms include elevated IgE total serum and sputum cultures reveal *Aspergillus* spp. Elevated Serum IgE could be used as a marker for flare-ups and response to therapy.[23]

The pathogenesis of ABPA is not completely understood. There is no clear correlation between *Aspergillus* load in the environment and the development of ABPA.[24] More than one immune responses may be involved,[25] such as *Aspergillus*-specific IgE-mediated type I hypersensitivity reactions[5], specific IgG-mediated type III hypersensitivity reactions,[6] and abnormal T-lymphocyte responses.[7]

2.2 ECOLOGICAL STUDIES

Direct association between mold species and human disease caused due to inhalation of mold spores is a big challenge among medical mycologists. One of the major reasons behind this fact is lack of sufficient data to create a standard for mold concentration inside the buildings.

Various efforts are being made to furnish more reliable data from different habitats to prove that fungal species can be one of the strongest initiators in causing the upper and lower respiratory diseases of the human body.

Current research data suggest the following:

- Various mold spores present in outdoor air can act as source of indoor mold exposure;
- The occurrence of various mold species in different geographical regions play a major role in outbreak of allergy season of the area;
- Different mold genera are associated with specific substrates inside a building and serve as source of allergies to sensitive persons;
- High moisture level supports mold growth and plays a significant role in converting a healthy green building into a moldy and sick building; and,
- Mold spore concentration collected from indoor environment must be compared with same from outdoor environment and remediation should be planned accordingly.

In the following section we describe the major ecological studies on mold spores associated with indoor air quality.

2.2.1 Studies by James Scott

James Scott in his thesis[8] studied the mycoflora from indoor household dust. According to his observations, the development of allergies to fungi follow the biological process similar to other allergens such as pollen, dust, dandruff and so on. The airborne spores of fungal species such as *Alternaria, Aspergillus,* and *Cladosporium* are found throughout the world and can cause allergic rhinitis and allergic asthma. Fungal spores are especially important in the tropics where climatic conditions are favorable for growth and sporulation. Higher concentrations of spores can cause an increased incidence of allergic diseases. He isolated *Aspergillus* and *Penicillium* as predominant fungal genera collected as vacuum dust. These species were reported as major sources of asthma and allergies to the exposed persons. He also used molecular PCR techniques to identify different species of *Penicillium* collected from house dusts. These fungal species can be present on different surfaces, including carpet, and were collected as house dust where only small-sized spores with longer viability can survive to accumulate and cause respiratory diseases.

Scott also studied the characterization of fungal diversity of house dust and found major predominating fungal genera as *Aspergillus, Alternaria, Chaetomium, Fusarium, Myrothesium, Paecilomyces, Penicillium, Stachybotrys,* and *Tolypocladium.* He also emphasized analysis of household dust as a major storehouse for dominant indoor fungal taxa such as *Penicillium* and its role in causing the human respiratory diseases.

Aspergillus and *Penicillium* types are accumulated in house dust because they are small in size (2–5 um) and can easily penetrate the dust bags of vacuum cleaner to regenerate and accumulate in carpet dust.

There were noted correlations between positively associated dust borne fungal species and their ecological similarities, as described by Pope and coworkers.[26] They reviewed asthma trends and observed that magnitude of allergen exposure increased the potential for allergic sensitization, and suggested an increase in asthma morbidity and mortality as reflected by hospital admission and statistics.

Many of these molds are passive and allochthonous arising from different outdoor habitats. These mold genera also form specific assemblage of species from dust microform of many ecologically homogenous groups,[8] as shown:

1. *Phylloplane (leaf surface) molds*: *Allernaria alternata, Cladosporium cladosporioides, Epicoccum, Aspergillus versicolor*;
2. *Xerophilic molds*: *Eurotium herbarum, Wallemia saba.* They are present due to activities within indoor environment, such as food and clothes;[27,28]
3. Soil borne fungi, present due to indoor movement through footwear, transferred by pets or by potted ornamental plants[29]: *Trichoderma viride, Penicillium citrienigrun*[30,31];
4. Assemblage of fungal species as contaminants of water damaged building material: *A. versicolor, Aspergillus ustus, Chaetomium globosum, Penicillium aurentiogriseum, Penicillium brevicompactum, Penicillium chrysogenum,* and *Stachybotrys chartarum.*

Accumulation of small-spored fungi also found a disproportionate high levels of *Aspergillus* and *Penicillium* in household dust which could be due to morphology, long viability, thin wall, and globose shape.[32,33]

2.2.2 Studies by Dubey

Dubey and Amal[34] noted that many household items from daily use can provide the substratum for mold growth and accumulation of spores. These materials are used in residential and occupational buildings and can increase the spore concentrations and cause allergies to sensitive persons. They focused on the colonization of indoor household items by *Aspergillus* species and used viable bulk culture method. An affinity of *Aspergillus* species with 14 different household substrates was noted, which included fiber, wall material, wall paper, wooden material, leather, textiles, and so on. CFU counts of different species of *Aspergillus* were recorded from 374 samples collected from different buildings of Northern California during 2003. *A. versicolor* was most predominant species followed by *Aspergillus niger, A. fumigatus,* and *A. flavus.* Wall material, wall paper, and fiber/textile were preferred by 80% of species. More allergic mold from wall material also suggested a need for thorough inspection of wall materials while looking for the source of mold growth.

A summary of various items serving as possible mold spore source from residential and occupational environments is presented in Table 7.1.[8,9,11,34-38]

Dubey[39] collected data on ecology and distribution of mold spores from 18 different locations of Northern California. This was a preliminary survey on distribution of mold genera from outdoor environments of 18 different cities of Northern California (Table 7.2).

The basic idea behind this survey was the assumption that mapping of outdoor mold concentrations can help to find the source of indoor mold accumulation in specific season of a location.

Collection of data from different locations can be used to create a mold forecast system to warn the high alert for allergy season.

Their survey also indicated that local environmental factors such as vegetation, rainfall, and wind–velocity support the predominance of specific mold species in the area. High spore counts of predominating fungal species from certain regions could be due to the favorable temperature, moisture, shape and size of spores, and rate of multiplication. Majority of locations showed basidiospores as most dominating spore type followed by *Cladosporium, Aspergillus/Penicillium* type, and unidentified ascospores.

2.2.3 Studies by Baxter

Baxter's team of investigators had evaluated several thousand buildings from Southern California during 1994–2001, and classified as clean and sick buildings.[40] He also gave an illustrated account of standard procedures to study the air-mycoflora of San Diego in Southern California. His studies included the surveys of several residential and commercial buildings for moisture levels, water intrusion, and mold concentrations. These data were used to classify the buildings as healthy and sick buildings (Table 7.3).

Currently the American Industrial Hygiene Association and the American Conference of Governmental Industrial Hygienists[41] (ACGIH) recommend that indoor sampling data should be compared with the local outdoor environment and/or a control environment. Both of these organizations also acknowledge the absence of standards or acceptable limits for the concentration of fungi due to the absence of data sufficient to establish them. Baxter compared short-term mold spore data collected from well-characterized residential and commercial buildings that have been classified based on predetermined visual inspection criteria. These results were compared with outdoor sampling results that have also been collected on a short-term basis at entry and exit points to buildings, along with ground level outdoor data collected during routine indoor air quality investigations.

Table 7.2 Distribution of Predominating Mold Species From 18 Different Cities of Northern California

Date	City	Ascospores Spores/M³	Aspergillus/ Penicillium Type Spores/M³	Basidiospores Spores/M³	Cladosporium Spores/M³	Total Counts Spores/M³
May 14	Alameda		800	1,750	1,000	5,150
Aug. 14	Brisbane	1,650		1,750	900	5,800
Jun. 14	Concord		1,250	700	1,300	4,050
Aug. 15	Concord	500		1,000	1,650	3,750
May 15	Crescent City	4,300	2,000	8,750		16,550
Apr. 15	Danville			2,100	2,250	5,700
Apr. 14	Dublin			1,250	1,000	3,400
Jul. 15	Dublin			1,200	1,300	3,250
Aug. 15	Eureka		2,500		1,250	5,500
Jul. 15	Eureka		1,500		1,750	3,700
May 15	Eureka		4,750		500	6,200
Feb. 14	Fremont	2,150		2,200	1,200	6,850
Mar. 15	Fremont	6,000		10,000	1,250	21,000
Jul. 15	Hayward		2,000	750	1,750	5,350
May 15	Hayward	1,100		1,200	1,700	4,700
Aug. 14	Hayward	7,640		4,770	2,640	20,500
Mar. 15	Las Gatos		1,000		3,200	7,400
Apr. 15	Livermore	1,000		2,200	1,500	4,400
Aug. 10	Livermore	1,000	1,150		1,250	4,500
May 15	Napa	1,200			6,400	9,350
Aug. 14	Oakland	750		1,250	1,100	4,150
Jun. 15	Oakland		2,250	2,000	1,700	7,500
Apr. 15	Oakland			3,000	1,500	6,300
May 14	Petaluma		1,250	3,000	1,750	7,700
Jun. 14	Pleasanton		1,350	3,350	850	6,600
Jul. 15	Redwood City		2,000		1,500	6,400
Aug. 15	San Francisco	300		1,650	1,300	3,650
Jun. 15	San Francisco		3,250	680	550	4,650
Jun. 15	San Francisco		700	1,250	1,750	4,850
Feb. 14	Sunnyvale		500	1,000	1,350	3,500
Total Observations	30	12	16	23	29	
	% Occurrence	40	53.3	76.6	96.6	
	% Dominance	0.08	16.6	52.1	37.9	

Table 7.3 Baxter's Classification of Buildings Based on Moisture Levels			
Classification	**No. of Buildings**	**No. of Samples**	**Total Number of Mold Species**
Residential buildings (clean)	19	55	16
Residential buildings (water stained)	30	108	17
Residential buildings (mold growth)	77	230	18
Commercial buildings (clean)	37	107	18
Commercial buildings (mold growth)	27	76	16

Baxter et al. [39].

Based on complete visual inspection and comprehensive building history, different buildings were classified under the following groups:

1. "Clean" commercial office and nonindustrial workplace environments;
2. "Moldy" commercial office and nonindustrial workplace environments showing mold growth;
3. "Clean" residential single-family dwellings and apartments;
4. Residential single-family dwellings and apartments showing water staining only;
5. "Moldy" residential single-family dwellings and apartments, showing mold growth.

2.2.3.1 Criteria for building classification
To classify a building as "clean," it has to satisfy all of the following conditions:

1. No history of flooding was observed;
2. No evidence of moisture intrusion;
3. No history of sewer backups;
4. No visible mold growth;
5. No interior surface moisture measurements exceeding 15% (equivalent wood scales); and,
6. Causes for air quality complaint were other than water intrusion or fungal growth were suspected.

The criteria for "Water stained" buildings were classified as if they met the following interior conditions (Table 7.3)

a. Water staining was observed in single to multiple locations beyond the allowances for the clean building category; and,
b. No visual evidence of mold growth was observed.

Criteria for "Mold Growth" classification: interior visible mold growth on areas totaling greater than one square foot (0.09 m^2) was required.

2.3 METHODS TO STUDY
The study of mold concentration from air requires the following steps:

1. Collection of air samples;
2. Preparation and microscopic analysis of spore trap samples;

3. Identification and counting of spores;
4. Calculation of LOD (limit of detection); and,
5. Interpretation of spore count data.

2.3.1 Collection of sample

Spore trap as slit impaction sampling is currently the most commonly used, and fastest method for the collection of both nonviable and culturable airborne mold spores. Because of the size selection criteria associated with this type of collection device, particles larger than 2.7 μm in aerodynamic diameter are collected with at least 50% collection efficiency. This spore trap method is capable of identification and counting of fungal spores to classify them as genus or morphologically similar groups present in the air regardless of its viability. The method is helpful to decide whether fungal concentration inside a building are "normal" or "atypical" if remediation procedure for mold contaminated environment was satisfactory.

Equipment: The most commonly used devices are the Zefon Air-O-Cell (Zefon International, Ocala, Fla.), Allergenco sampler (San Antonio, Texas), and Burkard personal sampler (Burkard Manufacturing Co., Rickmansworth, Hertfordshire, UK).[42]

Regional outdoor data has been reported through the National Allergy Bureau Aeroallergen Network of the American Academy of Allergy, Asthma, and Immunology.[43] Baxter's data were collected from the rooftops of multistory buildings and over long time intervals (ie, 5–7 days) by using the Burkard 7-day sampler.

Indoor Data: In contrast, the majority of the data collected for the evaluation of buildings (indoors and outdoors) were based on short-term impaction samplers These devices are frequently used to collect 5–10-min samples at entry and exit points or fresh air intakes and supply locations of heating and air-conditioning systems.

2.3.2 Preparation and microscopic analysis of spore trap samples

This requires aseptical techniques to remove the impaction slide and prepare for microscopic identification and counting through an efficient microscope with known field diameter.

2.3.3 Identification and counting of spores

The area of deposition of an Air-O-Cell sample measures approximately 14.5 by 1.1 mm and visually resembles penicillin. This device is an impaction sampler and is not similar to filter sampling devices, the particle deposition density varies significantly along the width of the area from the center to each edge. Accurate counting requires mold spores to be counted in a series of slices, or traverses, perpendicular to the long axis of the rectangular particle deposition area, the diameter of the microscope field of view is measured and used to establish the width of the pathway covered by performing each traverse. The percentage of the sample analyzed is then calculated by multiplying the width of each traverse times the number of traverses, divided by the actual length of the deposition trace.

When using the Nikon Labophot-2 microscope at a magnification of 400×, and counting 8–10 traverses, this corresponds to analyzing 25% of the entire trace.

- A minimum of four traverse widths separated each analyzed traverse. Analysis stopping rules similar to NIOSH Method 7400 can be employed,[44] if spore concentrations are relatively high.

- Only spores found within the microscopic field of view or crossing the border of the microscope field of view by more than 50% of the spore area can be counted.
- A 100% analysis of a trace is highly recommended by researchers to evaluate the mold concentrations in hospitals, children's daycare centers, or in senior livings areas where risk of mold exposure is very high.[45]

Spore Identification Procedures: Mold spore identification is based on comparison with known slide reference standards and reference atlases.[46–48] Percentage of unknown or unclassified spores can vary from 5–15% of the total counts of a trace.

2.3.4 Calculation of the limit of detection

Spore concentrations are calculated according to the following equation:

$$Cp = L \times P \ (1) \ DN \ QT/1000$$

where Cp = concentration of particles per cubic meter of air (cts/m^3), P = number of particles counted L = length of entire deposition trace (mm), D = microscope field of view and traverse width (mm), N = number of sample traverses counted, Q = sample flow rate in liters per minute (L/min), T = time (min)

Measurements for the limit of detection (LOD) can vary depending on the sample air volume and the density of particles on the slide. The typical mold spore detection limit ranged from 40 spores/m^3 to over 3000 spores/m^3 depending on air sample volume and actual total spore density. In Baxter's experiment, high detection limits occurred when the 100-spore count stopping point was reached in only a few traverses. Thus, LOD values for a single fungal group and building can vary by fivefold or more. The best approach was to use the constant value of zero

Measurements of values below <LOD and concentration mean values within buildings.

Baxter's team made one to ten individual measurements per building. In almost all cases, sample sizes were less than 3, and in the majority of buildings only one measurement was made. In cases where a single measurement was made for the entire building, it was always collected from the room or office closest to the suspected origin of the air quality complaint or observed damage condition. When more than one measurements were made in a building, the arithmetic average for each spore category was used to represent that building. Values less than LOD were assigned a zero and included in the average.

Ideally, a more accurate method such as maximum likelihood estimation would be used to estimate these censored or non detect values; however, the small sample sizes (usually <3) per building limit this approach. Alternatively, a value such as LOD/$(2)^{0.5}$ would be used as being more representative of the values that exist in the distribution below the LOD.

2.3.5 Interpretation of spore count data

This is the most challenging part of the whole procedure, mainly due to the lack of a standard acceptable limit of spore concentration. There is also a great degree of variation in ecology and etiology of individual fungal species due to daily or seasonal variations in surrounding atmosphere viz. moisture level, temperature, nutrition, and wind velocity. Any of these factors can become a controlling factor in sudden increase or decrease of specific spore concentrations in target building. Finally, this may lead us to determine if a building is clean or sick and to plan for remediation accordingly.

2.4 INDUCED OUTBREAK OF MOLD ALLERGIES DUE TO HUMAN ACTIVITY—CASE STUDY

Dubey[35] reported a comparison of indoor and outdoor mold genera from residential and commercial buildings during Christmas season when natural Christmas trees were placed inside the buildings and various activities were established around the tree. This was a special survey of mold concentration during the peak allergy season and Christmas holidays when natural Christmas trees are brought inside the residential and commercial buildings for decorations and more indoor activities are involved. The main purpose was to see if increased rate of asthma and allergy in holiday season was due to the high concentration of mold spores.

Air samples were collected from outdoor nurseries as a source of origin of the Christmas trees. These data were compared with same, collected from inside the buildings at three different intervals as: a week before setting the tree, during the presence of tree, and week after removal of the tree.

Major mold species included *Aspergillus, Aureobasidium pullulance, Cladosporium, Fusarium* and *Penicillium*. A higher concentration of colony forming units was noted during the presence of Christmas tree than before and after placing the tree. Similar genera were recorded from outdoor air of nursery area as the source of Christmas tree. A possible relationship was proposed between increased number of asthma and allergy cases during holiday season and high spore concentration of these allergic molds during the same period (Tables 7.4 and 7.5).[35]

Table 7.4 Fungal Occurrence During the Presence and Absence of Christmas Tree From a Residential Building

Sampling Time	Before Setting of Tree	Before Setting of Tree	During Presence of Tree	During Presence of Tree	After Removal of Tree	After Removal of Tree
Location	Indoor	Outdoor	Indoor	Outdoor	Indoor	Outdoor
Spore/CM^3	690	4200	9700	4300	1200	4500
CFU/CM^3	20	150	423	640	60	350
Total fungal genera	4	7	10	8	5	6
Predominant fungal genera	*Basidiospores, Cladosporium, Penicillium* sp.	*A. versicolor, Ascospores, Basidiospores, Cladosporium, Chaetomium, Chrysosporium, Pythium* sp.	*Aspergillus fumigatus, Aspergillus nigrum, Aureobasidium pullulans, Cladosporium, Chaetomium, Chrysosporium*	*Aspergillus nigrum, Basidiospores, Cladosporium, Epicoccum, Fusarium, Pythium, Penicillium*	*Aspergillus nigrum, Cladosporium, Chaetomium, Penicillium*	*A. versicolor, Basidiospores, Cladosporium, Fusarium*

Table 7.5 Fungal Occurrence During the Presence and Absence of Christmas Tree From a Commercial Building

Sampling Time	Before Setting of Tree	Before Setting of Tree	During Presence of Tree	During Presence of Tree	After Removal of Tree	After Removal of Tree
Location	Indoor	Outdoor	Indoor	Outdoor	Indoor	Outdoor
Spore/CM^3	740	6800	4000	9700	2200	6300
CFU/CM^3	60	680	150	320	110	450
Total fungal genera	6	7	12	9	5	6
Predominant fungal genera	*Aspergillus, Ascospore, Basidiospore, Cladosporium, Penicillium, smut, rust*	*Alternaria, A. versicolor, Ascospore, Basidiospore, Cladosporium, Penicillium, smut, rust*	*Aspergillus nigrum, Aspergillus fumigatus, Aureobasidium pullulans, Chaetomium, Cladosporim, Mucor, Penicillium, T. viride*	*Alternaria, Aspergillus versicolor, Cladosporium, Mucor, Penicillium, smut, rust*	*Alternaria, Aspergillus fumigatus, A. versicolor, Aureobasidium pullulans, Cladosporium, Mucor, Penicillium*	*Alternaria, Aspergillus versicolor, Cladosporium, Mucor, Penicillium, smut, rust*

Dubey [35].

3 CONCLUSIONS

There is strong evidence of increased lower respiratory diseases among humans caused by different fungal species. These fungi produce toxic metabolites which can be allergic to human and cause respiratory diseases. The pathophysiological mechanism of action of such allergens have been reported for these allergens but there are insufficient data to explain a clear relationship between many common indoor fungi and their allergic reactions produced in the human body. Researchers indicate that more spores will produce more toxins but there is no standard number of different spores that will produce a toxin level harmful to the human body. In this uncertainty of standards, the comparative studies between spore counts of outdoor and indoor spores have added the data to support the evidence that fungal species from various environmental conditions can invade inside, accumulate and cause asthma and allergies to sensitive persons if exposed for a longer duration.

Mold is becoming more important with global warming and increase in CO_2 levels. Chances are that there will be a further increase in mold spore concentrations due to high CO_2 levels, which can lead to higher number of sensitive persons for asthma.[46] However, the treatments for these diseases are not completely under control due to a lack of research data.

The WHO guidelines also suggest in their concluding remarks that the relationships between dampness, microbial exposure, and health effects cannot be estimated.[7] It is a difficult task to give any medical advice or recommendation for any specific threshold numbers or acceptable levels of contamination by microorganisms. However, it is recommended that dampness and mold-related problems must be remedied because they increase the risk of hazardous exposure to microbes and chemicals.

REFERENCES

1. AAAAI. Centers for Disease Control and Prevention: vital signs. In Asthma Statistics; 2011-May. Available from: http://www.aaaai.org/about-the-aaaai/newsroom/asthma-statistics.aspx.
2. Knutsen AP, Bush RK, Demain JG, Denning DW, Dixit A, Fairs A, et al. Fungi and allergic lower respiratory tract diseases. *J Allergy Clin Immunol* 2012;**129**(2):280–91 quiz 292–293.
3. Jaakkola MS, Ieromnimon A, Jaakkola JJ. Are atopy and specific IgE to mites and molds important for adult asthma? *J Allergy Clin Immunol* 2006;**117**:642–8.
4. Harley KG, Macher JM, Lipsett M, Duramad P, Holland NT, Prager SS, et al. Fungi and pollen exposure in the first months of life and risk of early childhood wheezing. *Thorax* 2009;**64**:353–8.
5. Cockrill BA, Hales CA. Allergic bronchopulmonary aspergillosis. *Ann Rev Med* 1999;**50**:303–16.
6. Knutsen AP, Slavin RG. *In vitro* T cell responses in patients with cystic fibrosis and allergic bronchopulmonary aspergillosis. *J Lab Clin Med* 1989;**113**:428–35 Medline.
7. Chauhan B, Santiago L, Kirschmann DA, et al. The association of HLA-DR alleles and T cell activation with allergic bronchopulmonary aspergillosis. *J Immunol* 1997;**159**:4072–6.
8. Scott JA. Studies on indoor fungi. A thesis submitted in conformity with the requirements for the degree of Doctor of Philosophy in Mycology, Graduate Department of Botany in the University of Toronto; 2001. Available from: http://WWW.sporometrics.com/Thesis.htm
9. Burge HA, Solomon WR, Muilenberg ML. Indoor plantings as allergen exposure sources. *J Allergy Clin Immunol* 1982;**70**:101–8.
10. Summerbell RC, Staib F, Dales R, Nolard N, Kane J, Zwanenburg H, et al. Ecology of fungi in human dwellings. *J Med Veterin Mycol* 1992;**30**(Suppl.1):279–85.
11. Dubey T. Colonization of indoor household items by *Aspergillus* species. Poster presented in AIHA annual meeting; 2003.
12. Storm Van Leeuwen W. Weitere untersuchunger uber asthma und Klima Klin. *Wochenschr* 1925;**4**:1294.
13. Towey JW, Sweeney HC, Huron WH. Severe bronchial asthma apparently due to fungus spores found in maple bark. *JAMA* 1932;**99**:453.
14. Cobe HM. Asthma due to a mold hypersensitivity due to *Cladosporium*. *J Allergy* 1932;**3**:389–91.
15. Gambale W, Purchio A, Paula CR. Influência de fatores abióticos na dispersão aérea de fungos na cidade de São Paulo. *Brasil Rev Microbiol* 1983;**14**:204–14.
16. Peraica M, Radic B, Lucic A, Pavlovic M. Diseases caused by molds in humans. Bulletin of the World Health Organization 1999. http://www.mold-survivor.com/diseases_caused_by_molds_in_huma.html.
17. Frawley JP. Scientific evidence and common sense as a basis for food-packaging regulations. *Food Cosmet Toxicol* 1967;**5**:293–308.
18. Munro IC. Safety assessment procedures for indirect food additives an overview. Report of a workshop. *Regul Toxicol Pharmacol* 1990;**12**:2–12.
19. Hardin BD, Robins CR, Fallah P, Kelman BJ. The Concentration of no toxicologic concern (CONTC) and airborne mycotoxins. *J Toxicol Env Health* 2009;**72**:585–98 Part A.
20. Bush RK, Portnoy JM, Saxon A, Terr AI, Wood RA. The medical effects of mold exposure. *J Allergy Clin Immunol* 2006;**117**(2):326–33.
21. Menezes EA, Carvalho PG, Trindade ECPM, Sobrinho GM, Cunha FA, Castro FFM. Airborne fungi causing respiratory allergy in patients from Fortaleza, Ceara, Brazil. *Jornal Brasileiro de Patologia e Medicina*

Laboratorial 2004;**40**(2.) http://www.scielo.br/scielo.php?pid=s1676-24442004000200006&script=sci_arttext

22. Kousha M, Tadi R, Soubani AO. Pulmonary aspergillosis. a clinical review. *Eur Respir Rev* 2011;**20**(121): 156–74.

23. Greenberger PA, Patterson R. Diagnosis and management of allergic bronchopulmonary aspergillosis. *Ann Allergy* 1986;**56**(6):444–8.

24. Tillie-Leblond I, Tonnel AB. Allergic bronchopulmonary aspergillosis. *Allergy* 2005;**60**(8):1004–13.

25. Wang JL, Patterson R, Rosenberg M, Roberts M, Cooper BJ. Serum IgE and IgG antibody activity against *Aspergillus fumigatus* as a diagnostic aid in allergic bronchopulmonary aspergillosis. *Am Rev Respir Dis* 1978;**117**:917–27.

26. Pope AM, Patterson R, Burge H. *Indoor allergens: assessing and controlling adverse health effects.* Washington, DC: National Academy Press; 1993.

27. Pitt JI, Hocking AJ. *Fungi and food spoilage.* 2nd ed. Gaithersburg, Maryland: Chapman and Hall; 1999.

28. Samson RA, van Reenen-Hoekstra ES. *Introduction to food-borne fungi.* 3rd ed. Baarn, The Netherlands: Centraal bureau voor Schimmelcultures; 1988.

29. Summerbell RC, Krajden S, Kane J. Potted plants in hospitals as reservoirs of pathogenic fungi. *Mycopathologia* 1989;**106**:13–22.

30. Barron GL. *The genera of hyphomycetes from soil.* Baltimore: Williams and Wilkins; 1968.

31. Domsch KH, Gams W, Anderson TH. *Compendium of Soil Fungi,* vol. 1. London: Academic Press; 1980.

32. Flannigan B, Miller JD. Health implications of fungi in indoor environments-an overview. In: Samson RA, Flannigan B, Flannigan ME, Verhoeff AP, Adan OCG, Hoekstra ES, editors. *Health implications of fungi in indoor environments.* Amsterdam: Elsevier; 1994 Air Quality Monographs 2.

33. Scott JA, Straus NA, Wong B. Heteroduplex fingerprinting of *Penicillium brevicompactum* from house dust. In: Bioaerosols, Fungi, Mycotoxins, E., Johanning, editors. Proceedings of the Third International Conference on Fungi, Mycotoxins and Bioaerosols, September 23–25, 1998. Saratoga Springs, New York. Albany, New York: Eastern New York Occupational and Environmental Health Center. 1999, p. 335–342.

34. Dubey T, Amal K. Distribution mapping of indoor fungi from various locations in Northern California. Poster Presentation Annual Conference, AIHA; 2003.

35. Dubey T. Distribution of outdoor fungi in different cities of Northern California. Part of presentation for UC Berkeley Extension; 2015.

36. Baxter DM, Perkins JL, McGhee CR, Seltzer JM. A regional comparison of mold spore concentrations outdoors and inside "clean" and "mold contaminated" Southern California buildings. *J Occup Environ Hyg* 2005;**2**:8–18.

37. American Conference of Governmental Industrial Hygienists (ACGIH). *Bioaerosols: assessment and control.* Cincinnati, Ohio: ACGIH; 1999.

38. Allergenco, Inc. *Owner's manual and user's cuide,* Air Sampler (MK-2). San Antonio: Allergenco, Inc; 1992.

39. American Academy of Allergy, Asthma, and Immunology: National Allergy Bureau Aeroallergen Network. Available from: http://www.aaaai.org/NAB/index.cfm?p=default

40. National Institute of Occupational Safety and Health (NIOSH): Method 7400. Asbestos and other fibers by PCM. *Manual of analytical methods,* 4th ed. Cincinnati, Ohio: NIOSH, Department of Health and Human Services; 1994.

41. Dubey T, Hobeck H. Distribution of mold spores on a trace inside a spore trap. Presentation in IAQA annual meeting; 2006.

42. Smith EG. *Sampling and identifying allergenic pollens and molds.* San Antonio: Blewstone Press; 1990.

43. St-Germain G. *Identifying filamentous fungi.* Belmont, California: Star Publishing Company; 1996.

44. Wang CK, Zabel RA. *Identification manual for fungi from utility poles in the Eastern United States.* Lawrence, Kansas: Allen Press, Inc; 1990.

45. Dubey T. Mold spores from Christmas tree. Poster presentation in Mycological Society of America (MSA) annual meeting on July 28th, 2009. Number: P2EP038.

46. WHO Guidelines for Indoor Air Quality. Dampness and mould. Heseltine E, Rosen J, editors. www.euro.who.int/-data/assets/pdf file/0017/4332645.pdf; 2009.

47. Hunninghake GW, Richardson HB. Hypersensitivity, pneumonitis and eosinophilic pneumonias. *Harrison's principles of internal medicine.* 14th ed. New York: McGraw- Hill; 1998. p. 1426–9 Chapter 253.

48. Park HS, Jung KS, Kim SO, Kim SJ. Hypersensitivity pneumonitis induced by *Penicillium expansum* in a home environment. *Clin Exp Allergy* 1994;**24**:383–5.

IS THERE A LINK BETWEEN ENVIRONMENTAL ALLERGENS AND PARASITISM?

8

I. Postigo, J.A. Guisantes, J. Martínez

University of The Basque Country, Department of Immunology, Microbiology and Parasitology, Faculty of Pharmacy and Laboratory of Parasitology and Allergy, Research Center Lascaray, Paseo University, Vitoria, Spain

1 INTRODUCTION

Over recent decades, several epidemiologic studies performed in different parts of the world have demonstrated a significant increase in the prevalence of atopy and have consequently demonstrated an increase in the probability of type I allergic diseases involving asthma, hay fever, atopic dermatitis, or allergic gastroenteritis.[1]

The reason and mechanism by which allergy and parasitism have consistent overlaps that allow them to be studied under a similar framework begins with the formulation of the "hygiene hypothesis," as well as the immunological basis supporting this hypothesis.

The "hygiene hypothesis" was originally postulated to explain the outbreak of allergy, which has presently expanded to an epidemic proportion.[2] The hypothesis postulates that a lack of early exposure to different germs or parasites (clean environments) could have a direct relationship with the development of allergic diseases. It seems that an idle immune system is more prone to an inappropriate immune response, leading to immunopathological effects such as allergy or autoimmunity depending on the type and nature of the antigens to which exposure occurs.[3,4]

The "hygiene hypothesis" has been a subject of disagreement ever since the term was first proposed.[5] The text suggests that allergic diseases (hay fever and eczema) are less common in children from larger families because they are exposed to more infectious agents through their siblings than are children from families with only one child, where transmission of infectious agents is more limited.

The "hygiene hypothesis" was interpreted as an explanation for the increased prevalence of allergic diseases in Western countries, with the assumption that opportunities for infection were reduced through the higher standards of hygiene achieved in those countries.

Improved hygiene is believed to mediate its effect through decreased exposure to infectious agents in early life. Recent studies highlight the importance of the gastrointestinal microbial environment in the development of allergic diseases. In particular, infection with hepatitis A, *Helicobacter pylori*, and *Toxoplasma* in individuals living in temperate climates, and geo-helminths in those living in endemic

The Microbiology of Respiratory System Infections. http://dx.doi.org/10.1016/B978-0-12-804543-5.00008-7

areas, have been shown to be associated with a reduced-risk of atopic manifestations.[6] Because the hygiene hypothesis has significant gaps, other concepts have been proposed in an attempt to create new descriptions to reduce these gaps. In particular, the "microbiome hypothesis" and the "biodiversity hypothesis" have emerged.

Current studies suggest that features of the home, medical practices, and cleanliness behaviors are all involved in the hygiene effect in some way. Traditional markers associated with the protective microbial environment have been supplanted by "culture-independent microbiome science," distinguishing the characteristics of potentially protective microbiomes from pathologic features.[3] Rapidly declining biodiversity may contribute to the rapidly increasing prevalence of allergies in urban populations worldwide. According to the "biodiversity hypothesis," reduced contact between people and natural environmental features and biodiversity may adversely affect the human commensal microbiota and its immunomodulatory capacity.[7]

Cohabitation with different allergenic sources in early life, such as pets, several sources of microbial infections, and other domestic animals could have an additive protective effect on the development of allergies and asthma via similar and shared mechanisms. Revisiting the "hygiene hypothesis" using more accurate descriptions based on the preventive microbe intervention platforms, termed the "microbiome theory," it is possible to reinforce and reformulate the "hygiene hypothesis."[3]

There is evidence that exposure to some infections can promote either allergy development or allergy protection, depending on several external and internal factors. The chemical or biological products of the environment, the type of microbiota existing in each place, the timing and bioburden of infectious agents, the characteristics of the intestinal flora, the parasitic infections or the genetic susceptibility of the host can all play an important role in the future development of atopy and allergic diseases.[8]

These relationships are the starting point that allow an introduction to immunology and molecular biology to better understand all of these apparently interrelated phenomena.

Translating the "hygiene hypothesis" into immunological terms, and on the basis of the aforementioned aspects related to this theory, it can be suggested that these processes begin in uterus and during prenatal life, as well as in early postnatal developmental stages, and could represent an opportunity for allergy-preventing environmental factors. The current knowledge of the cellular and molecular mechanisms used to explain the events of the hygiene hypothesis includes changes in the fine balance between the Th1, Th2, and regulatory T cell responses. These responses can become activated or not depending on the way of activation and the activation status of the innate immune cells.[9]

T cell-mediated immunity is an adaptive process of developing antigen-specific T lymphocytes to eliminate viral, bacterial, or parasitic infections, as well as malignant cells. This type of immunity can also be involved in the aberrant recognition of self-antigens leading to autoimmune inflammatory diseases, or some special antigens defined as "allergens" which lead to allergic diseases.

In fact, T cell-mediated immunity is the central element of the adaptive immune system and includes a primary response by naïve T cells, effector functions by activated T cells, and the persistence of antigen-specific memory T cells. Moreover, T cell-mediated immunity is a key part of a complex and coordinated response that includes other effector cells such as macrophages, natural killer cells, mast cells, basophils, eosinophils, and neutrophils.

Initially, two major functional T helper subpopulations were distinguished by their cytokine profiles: Th1 cells enhance proinflammatory cell-mediated immunity and were shown to induce delayed-type hypersensitivity (DTH) and mediate the response to some protozoans such as *Leishmania* and

Trypanosoma. Th2 cells promote noninflammatory immediate immune responses and have been shown to be essential in B cell production of IgG, IgG-4,[10] IgA, and IgE.

Undoubtedly, the introduction of the Th1/Th2 paradigm to the natural and adaptive immune response was a tipping point to better understand the evolution of the immune response against different pathogens and their possible immunopathological effects.[11]

However, additional cellular T cell subsets have been defined in recent years, contributing to a better understanding of different immunopathological facts not explainable by Th1/Th2 cell interactions.

Naïve T cells are the most homogenous representatives of unactivated CD4+ (T helper) and CD8+ (T cytotoxic) subsets. Once their activation occurs, they can be distinguished by their cytokine profiles. Thus, activated T helper cells can be subdivided into Th1, Th2, Th17, and Treg subsets based on the profile of cytokines they produce. In addition, there are subsets of regulatory T (Treg) cells that add complexity to T cell heterogeneity.

2 ALLERGENS

2.1 ALLERGENS AS CAUSE OF SENSITIZATION

According to the World Health Organization/European Academy of Allergy and Clinical Immunology (WHO/EAACI) allergy definitions, allergens are defined as "antigens that cause allergy." Most allergens reacting with IgE and IgG antibodies are proteins, often with carbohydrate side chains, but in certain circumstances, pure carbohydrates have been postulated to be allergens. In rare instances, low molecular weight chemicals, such as, isocyanates and anhydrides acting as haptens are still referred to as allergens for their capacity to induce IgE antibodies. In the case of allergic contact dermatitis, the classical allergens are low molecular weight chemicals, such as, chromium, nickel, and formaldehyde, which react with T cells.[12]

Other definitions have also used that incorporate new parameters into the mentioned definition. Chapman et al.[13] defined allergens as "environmental agents that induce IgE-mediated immediate hypersensitivity," which includes the term "environment."

However, the potential for environmental and food allergens to become allergens would be restricted to those proteins that are structurally related to a limited range of metazoan parasite proteins and should be sufficiently different from the host proteins as to display an IgE response and its dependent effector mechanisms.[14] Fitzsimmons and Dunne[14] used another definition that includes the term "nonparasitic": "Allergens are nonparasitic antigens capable of stimulating a type I hypersensitivity reactions in atopic individuals"; the reference to parasitism is included in this definition.[14]

2.2 ALLERGENS AS TYPE I ALLERGIES' DIAGNOSES AND TREATMENT TOOLS

The worldwide prevalence of allergic diseases is rising dramatically in both developed and developing countries. The WHO estimates that 10–40% of the global population, depending on the countries, is atopic and is at risk of suffering some type of allergic disease, such as, asthma, hay fever, allergic conjunctivitis, allergic gastrointestinal disorders, or allergic dermatitis.[15]

To date, the only specific treatment applied in this type of disease has been hyposensitization with allergen extracts, which has been used empirically since the early 20th century.

However, despite the time elapsed after its implantation, it remains unclear which factors are involved in the action of these treatments and how they induce tolerance to allergens leading to type I hypersensitivity. This type of tolerance, which does not always occur consistently, is discussed in the context of different types of medicine.

With the recent application of genomics and proteomics methods, the knowledge about individual, well-characterized allergens has increased the expectations of the diagnosis and the specific treatment of allergic diseases mediated by IgE.[16-24] These advances have also improved the availability of well-characterized tools to study the mechanism of action for how these treatments induce allergen tolerance.[25-28]

First, characterizations of allergenic sources have greatly improved. The development of specific tools to detect and quantify individualized allergens in biological products that are made of complex, antigenic mosaics has allowed for the detailed study of allergen compositions.[29-30] Second, individualized allergens themselves, their hypoallergenic variants, or relevant peptides may constitute the therapeutic products.[18]

Starting in 2001, a consortium (CREATE Project) involving basic and clinical researchers, biotechnological industries that produce allergens, and regulatory agencies was created to standardize the materials and methods used to develop these products. The aim of this work was the production of recombinant allergens to use as universal references, after they were compared with their homologous native forms and to define the most appropriate methods for quantifying individualized allergens.[31] In recent studies on the molecular basis of allergy, contrary to traditional thinking, most allergens have a narrow range of functionality. These allergens are found in a restricted number of protein families.[32] The biochemical classification system for protein families reveals more than 12,000 protein families, and approximately only 200 families contain allergenic proteins.

2.3 THE BIOLOGICAL FUNCTION AND NATURE OF ALLERGENS

According to the data included in the Allergome database,[33] there are now approximately 2500 allergenic sources that are well-defined as allergens, and the number of these well-identified allergens (excluding isoforms and epitopes) is just over 3000. From a molecular taxonomy perspective, the Database of Allergen Families[34] shows that of the 12,273 existing protein families (March, 2011), only 186 families include allergens. There are approximately 1000 allergens that could be assigned to a specific protein family, and only 10% include allergens with unknown or unidentified biological functions. Only a few families such as the prolamins, polcalcines, profilins, tropomyosins, cupins, and PR10-related proteins include 26% of the aforementioned 1100 allergens. The allergens that belong to these few families are a highly homogeneous group of proteins. They are present in many different allergenic sources that form the most important group of allergens to cause the cross-reactivity phenomenon. Conversely, there are species-specific or family-specific allergens in more phylogenetically restrictive protein families. Some of these allergens such as the major allergen from *Alternaria alternata* (Alt a 1), define a new family of proteins.[32,35-37]

Most allergens are proteins. The intrinsic properties of proteins make them undistinguishable from the proteins that stimulate other immunoglobulin isotypes (IgA, IgG, or IgM), which constitute the conventional antigens.

As mentioned previously, within the substantial heterogeneity of the existing protein families, only a few contain allergenic proteins, and, of all the protein domains, only 2.1% include allergenic proteins.[33]

If so, apart from the classic factors that influence antigenicity, such as the molecular size, concentration, solubility, stability, biochemical activity, and phylogenetic proximity, what would cause a protein to become an allergen? Or put in another way, what makes an allergen?

It is clear that the answers to these questions lie in how the immune system responds to different antigens that are capable of stimulating the immune system and in the strategies that are used to respond to interactions with each antigen.

Because allergenicity cannot be defined by the common rules for allergen-IgE binding sites, a complete set of old and new concepts related to answering the question, "What makes an allergen?" should be formulated in the future.

2.4 DEFINING THE CONCEPT OF ALLERGENS

Taking into account that the immune response against helminthic infections are predominantly Th2, which mainly involves cytokines, such as IL-3, Il-4, IL-10 and IL-13, which are implicated in the increased levels of IgE antibodies, eosinophils, basophils, and mast cells, and that the allergic phenotype has a very similar pattern of immune response, several authors have suggested that both immunological processes could have the same or a similar origin.[10,38–42]

On the other hand, Fitzsimmon et al.[43] in their excellent review reported homologous protein families between allergen proteins and some metazoan parasite proteins. Some of the top 10 allergen families (tropomyosins, PR-1, lipocalins, profilins, and serine-proteases) have members in helminth species, and many of them have IgE-binding ability. Other proteins included in this top-ten list, such as prolamins and expansins, have not been identified in helminth species to date.

Currently, all available data about the structure, biochemical, and biological functions of the different allergens allow the clustering of these factors into allergen families, and certainly, these characteristics should be part of the definition of allergens.[44]

The phylogeny of allergen sources and allergen taxonomic markers can also contribute to a better understanding of the allergen concept.[45,46]

Currently, it is very difficult to discriminate between allergenic and non-allergenic proteins, considering the classic concepts of major, minor or non-allergenic proteins. In previous years, increasing knowledge about the innate immune system has modified our understanding about the protein allergenicity concept.

Pathways of innate immune activation appear central to the contribution of allergenicity. The intrinsic properties of allergens seem to activate both the innate and adaptive immune system. Recently, it has been demonstrated that some allergens possess intrinsic adjuvant properties to stimulate innate immunity. The adjuvant properties appear to contribute to the allergic sensitization in atopic or sensitized individuals.[44,46,47] These adjuvant properties are mainly mediated by protease activity, carbohydrate residues, and lipids, which interact with the innate immune system to promote allergic Th2 responses.

The question, "What makes an allergen?" can be answered by not only characteristics defined by the protein itself and other adjuvant components capable of activating Th2 responses, but also the origin and evolution of the immune response to these proteins in each form of parasitism. Allergenicity in susceptible subjects is a complex and multifactorial phenomenon that can be explained only by the combination of all of the aforementioned factors.

3 HELMINTH PARASITES AND ALLERGY

"Helminth" is a nontaxonomic general term meaning "worm." In Helminthology, associated with medical health, two major phyla of parasitic worms are recognized: the Nematoda (round worms), which are subdivided into two main classes, Adenophorea and Secernentea; and the Platyhelminthes (flatworms), which are subdivided into two main classes, Cestoda (tapeworms) and Trematoda (flukes).

Among the helminthiasis, the most common human infections are those caused by geohelminths (intestinal parasites also known as soil-transmitted helminths). Highly prevalent parasites such as *Ascaris lumbricoides*, *Trichuris trichura* and hookworms belong to this family. Recent global estimates indicate that approximately 3.5 billion people are infected with one or more of these common nematode parasites.[48]

Other important helminth parasitosis includes Schistosomiasis, which are transmitted by cercariae swimming in water, and Filariasis, which are transmitted by arthropods, although both of these have a more restricted distribution.

Fluke infections are also important helminthiasis transmitted by metacercariae in water or foods. Their adults are established in the bile ducts, the lungs or the intestine, depending on the species.

Fasciola hepatica is a liver fluke that causes Fascioliasis, a zoonotic infection that affects approximately 50 million people worldwide; over 180 million are at risk of infection in both developed and undeveloped countries.[49]

Despite the fact that "immunity against helminths" and "parasites and allergy" are subjects that have been exhaustively reported in the literature, only a limited number of references about the "parasitic origins of the allergic response" can be found, even in discussions on the evolutionary relationship of allergens.[14]

It is unanimously accepted that the immune response to helminth parasites and the immunological basis of allergy occur under the same rules.

Most allergen or helminth antigen-specific CD4+ human T cell clones have a Th2 phenotype, whereas the majority of T cell clones specific to microbial antigens or antigens responsible for type IV hypersensitivity exhibit a Th1 phenotype.

The selective or preferential activation of CD4+ T cell subsets secreting a defined pattern of cytokines is the key to defining the means by which the immune response shifts towards protection or towards immunopathology.[50,51]

However, the Th1/Th2 paradigm cannot adequately explain the development of certain inflammatory responses. A new subset of T cells called Th17 cells represent another independent group of T cells with specific functions that can eliminate certain extracellular pathogens, which are presumably not adequately handled by Th1 or Th2 cells. The major function of Th17 cells has been described in the induction of autoimmune tissue inflammation.

Recently, Th9 cells were also proposed as a subset that develops under the influence of IL-4 and TGF-β to produce IL-9. There are now several subsets that may have the potential to produce immunological disease.

Th9 cells participate in the lesions of many diseases, such as allergic inflammation, tumors, and parasitic infections.[52] The current understanding of the contribution of Th9 cells to both effective immunity and immunopathological disease could provide important support for the future development of treatments for allergic and autoimmune diseases.[53]

In summary, although the evidence for the polarized cytokine secretion profiles of Th1 and Th2 is indisputable, several recent studies have shown a more complex pattern of cytokine interactions in different models of the immune response, including autoimmune models that are inconsistent with the simple dichotomy paradigm (Table 8.1).[52,54–56]

Additional knowledge of the IgE- mediated response common to allergy and helminths and the immune response and a comparison between immunogens from parasites and allergenic sources targeted by IgE or those involved in the activation of the innate and adaptive immune responses, could provide additional keys to how the host–parasite relationships evolved to induce, suppress, and regulate these interactions. Furthermore, which tools the parasites are using to escape and how the hosts apply immunological resources to achieve a more effective defense against parasite infections remain unanswered questions.

Answers to these questions would undoubtedly be keys for success in the study of immune regulation in response to allergic phenomenon through parasite immune modulators in addition to studying the immunologic origin of allergic diseases. As Fitzsimmons and Dune[14] cite, "parasitology could help to introduce the evolution into the allergy."

Table 8.1 Differentiation of Effector T Cells Subsets

Th Cell Subset	Profile Cytokine Secretion	Effect	Immunopathology
Th1	γ-Interferon IL-2 B-TNF	• Macrophage activation • Cell-mediated immunity • Phagocyte-dependent protective responses	Pathogenesis of organ-specific autoimmune disorders, Crohn's disease, *Helicobacter pylori*-induced peptic ulcer, acute kidney allograft rejection, and unexplained recurrent abortions
Th2	IL-4, IL-5, IL-10, IL-13	• Antibody production • Eosinophil activation • Inhibition of several macrophage functions (providing phagocyte-independent protective responses)	Atopic disorders Helminth infection immune responses
Th 17	IL-17, IL-17F, IL-21, IL-22	• Clearing extracellular pathogens • Tissue inflamation	Autoimmune tissue inflammation
Th 9	IL-9, IL-10, IL-21	• Mast cells activation with effects on the epithelial cells of the lung and gut • Direct effect on regulatory T (Treg) cells, T helper 17 (Th17) cells, and antigen presenting cells (APCs).	Nematode infection immune response Melanoma Allergic inflamation Anti-Tumors immunity Leukimia
Treg[a]	IL-35 IL-10 TGB-β	• Maintenance of peripheral tolerance • Down-modulation of the amplitude of an immune response • Prevention of autoimmune diseases	

Treg participate in all cell-mediated immune responses, directly affecting Th1, Th2, Th17, CTL, and B cell reactions against "self" and "foreign" Ag.
[a]Treg can be divided into different subsets based on the expression of FoxP3 and/or the production of IL-10, TGF-β, and IL-35.

Helminth infections have immunomodulatory effects on antiparasite inflammatory responses in humans, but the results are not conclusive. Helminth infections have been associated with reduced as well as increased prevalence of atopic diseases in different populations, depending on the helminth species,[57] epidemiology and distribution,[58] host genetics,[59,60] and parasite burden. High levels of parasites may induce down-regulation while low burdens may have the opposite effect.[61,62]

There is evidence that some helminths are associated with atopy and/or risk factors for asthma. Thus, *Ascaris, Anisakis,* or *Toxocara,* and asthma have a strong positive association, but others such as *Trichuris* or *Enterobius* do not.[39,63–65] There is also evidence that *Schistosoma mansoni* and hookworm infections induce a protective effect to allergic asthma by decreasing the immune response to allergens and clinical manifestations of asthma.[66,67]

One of the most remarkable characteristics of the many human helminth infections is the preponderance of asymptomatic infections as a positive co-evolutionary result of the host–parasite relationship. This situation is the primary cause of the development of parasite-mediated immune regulation.[38,67–69]

Several studies have shown that helminth infections induce a complex immune regulatory network that involves dendritic cells, Treg cells and systematic elevated levels of IL-10 and possibly TGF-β, B reg cells and activated macrophages[14,38,43,65,67] to achieve chronic asymptomatic conditions.

Several clinical studies have found that certain helminth infections (*Trichuris suis* or *Necator americanus*) protect against the development of aberrant inflammation, but it is clear that the final objective is to identify the parasite-derived immunomodulatory components responsible for protective effects.[65–69]

These components are a complex array of antigens/allergens, adjuvants and different immunomodulators capable of activating the pathway through innate immune cells (DC subsets) by recognition of the various families of receptors (PRR: pattern recognition receptors) such as Toll-like (TL) receptors, Nucleotide-binding Oligomerization Domain (NOD)-like receptors, pattern recognition (RIG)-like receptors, and C-type lectin receptors that allow the recognition of a great variety of pathogens though their characteristic components.[70]

Understanding the mechanisms by which helminths regulate inflammation may potentially lead to the development of strategies focused on the control of undesirable inflammation in allergic and autoimmune diseases.

4 CONCLUDING REMARKS

Both allergic diseases and helminth infections elicit strong Th2 responses, and both allergens and parasite antigens seem to have common molecular and biochemical features that support, at least in part, the similarity of both immune responses. In allergic diseases, the allergen itself triggers the Th2 response but not the regulatory response conducting to the tolerance of the allergenic molecules. In parasitism, however, the immune evasion mechanism of the parasite itself redirects the initial allergic phenomenon to the regulatory responses that allow it to coexist with its host. In this context, it has been understood that in a large number of parasitism models, when the immunological status of the host is normal, the parasitic infections manifest in chronic asymptomatic and even silent forms. However, asthma can have a positive or negative association with helminth infections based on several factors, such as the helminth species, the clinical status of the infection, the host's genetics, or the intensity of the infection.

In order to clarify the evolutionary origin of IgE-mediated allergies and their possible relationships with parasites, it will be necessary to define the common markers that initiate both responses. This knowledge could provide the molecular basis for the possible phylogenetic relationships among such heterogeneous sources as pollen and helminth parasites. The current theory regarding the substances that trigger the activation of immunological reactions in allergy would be reinforced, and it might be determined that immunological responses depend not only on the structure and homology of these substances but also on their biological functions.

Studies of molecular allergy have allowed us to determine the relationship between allergy activation and the development of clinical symptoms. It is currently possible to associate one or a few allergenic molecules with patients' allergenic profiles and to know which molecules are implicated in the risk of developing more acute symptoms. Further studies of individualized allergens in combination with the model of immune regulation observed in helminths would be very useful for developing alternative strategies for the prevention, treatment, and control of inflammatory diseases, including allergic diseases.

REFERENCES

1. Thomsen SF. Epidemiology and natural history of atopic diseases. *Eur Clin Respir J* 2015; **2**: 24642.
2. Pfefferle PI, Teich R, Renz H. The immunological basis of the Hygiene Hypothesis. In: Pawankar, R., Holgate, S.T., Rosenwasser, LJ., editors. *Allergy Frontiers. Epigenetic, Allergens and Risk Factors*. New York; 2009, p: 325-48.
3. Liu AH. Revisiting the hygiene hypothesis for allergy and asthma. *J Allergy Clin Immunol* 2015;**136**:860–5.
4. Rook GAW. Hygiene Hypothesis and autoinmune diseases. *Clinic Rev Allergy Immunol* 2012;**42**:5–15.
5. Strachan DP. Hay fever, hygiene and household size. *BMJ* 1989;**299** 1259-60.
6. Sheikh A, Strachan DP. The hygiene theory: fact or fiction? *Curr Opin Otolaryngol Head Neck Surg* 2004;**12**:232–6.
7. Hanski I, von Hertzen L, Fyhrquist N, Koskinen K, Torppa K, Laatikainen T, et al. Environmental biodiversity, human microbiota, and allergy are interrelated. *Proc Natl Acad Sci* 2012;**109**:8334–9.
8. Fishbein AB, Fuleihan RL. The "hygiene hypothesis" revisited: does exposure to infectious agents protect us from allergy? *Curr Opin Pediatr* 2012;**24**:98–102.
9. Garn H, Renz H. Epidemiological and immunological evidence for the hygiene hypothesis. *Immunobiology* 2007;**212**(6):441–52.
10. Yazdanbakhsh M, Kremser PG, van Ree R. Allergy, parasites and the hygiene hypothesis. *Science* 2002;**296** 490-L 494.
11. Mosmann TR, Coffman RL. TH1 and TH2 cells: different patterns of lymphokine secretion lead to different functional properties. *Annu Rev Immunol* 1989;**7** 145-L 173.
12. Johansson SG, Hourihane JO, Bousquet J, Bruijnzeel-Koomen C, Dreborg S, Haahtela T, et al. EAACI (the European Academy of Allergology and Cinical Immunology) nomenclature task force. A revised nomenclature for allergy. An EAACI position statement from the EAACI nomenclature task force. *Allergy* 2001;**56**:813–24.
13. Chapman MD, Pomés A, Aalbrese RC. Molecular biology of allergens: structure and immune recognition. In: Holgate ST, Rosenwasser LJ, editors. *Allergy Frontiers. Epigenetic, Allergens and Risk Factors*. New York; 2009. p. 265–289.
14. Fitzsimmons CM, Dunne DW. Survival of the fittest: allergology or parasitology? *Trends Parasitol* 2009;**25**:447–51.

15. Pawankar R, Sanchez-Borges M, Bonini S, Kaliner MA. Rhinitis, Conjunctivitis, and Rhinosinusitis. In: Pawankar R, Canonica GW, Holgate ST, Lockey RF, editors. *WAO White Book on Allergy*. Milwaukee; 2011, p. 27–34.

16. Valenta R, Lidholm J, Niederberger V, Hayek B, Kraft D, Grönlund H. The recombinant allergen-based concept of component-resolved diagnostics and immunotherapy (CRD and CRIT). *Clin Exp Allergy* 1999;**29**:896–904.

17. Valenta R, Niespodziana K, Focke-Tejkl M, Marth K, Huber H, Neubauer A, et al. Recombinant allergens: what does the future hold? *J Allergy Clin Immunol* 2011;**127**:860–4.

18. Marth K, Focke-Tejkl M, Lupinek C, Valenta R, Niederberger V. Allergen peptides, recombinant allergens and hypoallergens for allergen-specific immunotherapy. *Curr Treat Options Allergy* 2014;**1**:91–106.

19. Valenta R, Campana R, Marth K, van Hage M. Allergen-specific immunotherapy: from therapeutic vaccines to prophylactic approaches. *J Intern Med* 2012;**272** 144-L 157.

20. Valenta R, Linhart B, Swoboda I, Niederberger V. Recombinant allergens for allergen-specific immunotherapy: 10 years anniversary of immunotherapy with recombinant allergens. *Allergy* 2011;**66**:775–83.

21. Harwanegg C, Laffer S, Hiller R, Mueller MW, Kraft D, Spitzauer S, et al. Microarrayed recombinant allergens for diagnosis of allergy. *Clin Exp Allergy* 2003;**33**:7–13.

22. Valenta R, Kraft D. From allergen structure to new forms of allergen-specific immunotherapy. *Curr Opin Immunol* 2002;**14**:718–27.

23. Mothes N, Valenta R, Spitzauer S. Allergy testing: the role of recombinant allergens. *Clin Chem Lab Med* 2006;**44**:125–32.

24. Valenta R, Kraft D. Recombinant allergens for diagnosis and therapy of allergic diseases. *Curr Opin Immunol* 1995;**7**:751–6.

25. Curin M, Weber M, Thalhamer T, Swoboda I, Focke-Tejkl M, Blatt K, et al. Hypoallergenic derivatives of Fel d 1 obtained by rational reassembly for allergy vaccination and tolerance induction. *Clin Exp Allergy* 2014;**44**:882–94.

26. Baranyi U, Pilat N, Gattringer M, Linhart B, Klaus C, Schwaiger E, et al. Persistent molecular microchimerism induces long-term tolerance towards a clinically relevant respiratory allergen. *Clin Exp Allergy* 2012;**42**:1282–92.

27. Palomares O, Yaman G, Azkur AK, Akkoc T, Akdis M, Akdis CA. Role of Treg in immune regulation of allergic diseases. *Eur J Immunol* 2010;**40**:1232–40.

28. Valenta R, Kraft D. Recombinant allergen molecules: tools to study effector cell activation. *Immunol Rev* 2001;**179**:119–27.

29. Schenk S, Valenta R, Kraft D. Molecular and functional characterization of allergens: basic and practical aspects. *Arb Paul Ehrlich Inst Bundesamt Sera Impfstoffe Frankf AM* 1994;**87**:221–32.

30. Kazemi-Shirazi L, Niederberger V, Linhart B, Lidholm J, Kraft D, Valenta R. Recombinant marker allergens: diagnostic gatekeepers for the treatment of allergy. *Int Arch Allergy Immunol* 2002;**127**:259–68.

31. van Ree R. CREATE partnership. The CREATE Project: development of certified reference material for allergenic products and validation of methods for their quantification. *Allergy* 2008;**63**:310–26.

32. Radauer C, Bublin M, Wagner S, Mari A, Breiteneder H. Allergens are distributed into a few protein families and possess a restricted number of biochemical functions. *J allergy Clin immunol* 2008;**121**:847–52.

33. Allergome. <www.allergome.org>; 2015.

34. Database of Allergen Families. AllFam. <www.meduniwien.ac.at/allergens/allfam>; 2011.

35. Wagner GE, Gutfreund S, Fauland K, Keller W, Valenta R, Zangger K. Backbone resonance assignment of Alt a 1, a unique β-barrel protein and the major allergen of *Alternaria alternata*. *Biomol NMR Assign* 2014;**8**:229–31.

36. Chruszcz M, Chapman MD, Osinski T, Solberg R, Demas M, Porebski PJ, et al. *Alternaria alternata* allergen Alt a 1 a unique β-barrel protein dimer found exclusively in fungi. *J Allergy Clin Immunol* 2012;**130** 241-L 247.

37. Postigo I, Gutiérrez-Rodríguez A, Fernández J, Guisantes JA, Suñén E, Martínez J. Diagnostic value of Alt a 1, fungal enolase and manganese-dependent superoxide dismutase in the component-resolved diagnosis of allergy to pleosporaceae. *Clin Exp Allergy* 2011;**41** 443-L 451.

38. Jackson JA, Friberg IM, Little S, Bradley JE. Review series on heltminths, immune modulation and the Hygiene hypothesis: Immunity agains helminths and immunological phenomena in modern human populations: coevolucionary legacies? *Immunology* 2008;**126**:18–27.

39. Cooper PJ. Interactions between helminth parasites and allergy. *Curr Opin Allergy Clin Immunol* 2009;**9**:29–37.

40. Allen JE, Maizels RM. Diversity and dialogue in immunity to helminths. *Nat Rev Immunol* 2011;**11**:375–88.

41. Gause WC, Wynn TA, Allen JE. Type 2 immunity and wound healing: evolutionary refinement of adaptive immunity by helminthes. *Nat Rev Immunol* 2013;**13**:607–14.

42. Artis D, Maizels RM, Finkelman FD. Forum: immunology: allergy challenged. *Nature* 2012;**484** 459-L 469.

43. Fitzsimmons CM, Falcone FH, Dunne DW. Helminth allergens parasite specific IgE and its protective role in human immunity. *Front Immunol* 2014;**5**:1–12.

44. Scheurer S, Toda M, Vieths S. What makes an allergen? *Clin Exp Allergy* 2015;**45**:1150–61.

45. Gabriel MF, Postigo I, Gutierrez A, Suñen E, Tomaz C, Martinez J. Development of a PCR-based tool for detecting immunologically relevant Alt a 1 and Alt a 1 homologue coding sequences. *Med Mycology* 2015;**53**:636–42.

46. Gabriel MF. Evaluation of Alt a 1 as a specific marker of exposure to fungal allergenic sources and clinical relevance of a manganese-dependent superoxide dismutase and serine protease as new *A. alternata* allergens. PhD Thesis. Universidade da Beira Interior. Portugal; 2015.

47. Deifl S, Bohle B. Factors influencing the allergenicity and adjuvanticity of allergens. *Immunotherapy* 2011;**3**:881–93.

48. Ojha SC, Jaide C, Jinawath N, Rotjanapan P, Baral P. Geohelminths: public health significance. *Infect Dev Ctries* 2014;**8**:5–16.

49. Nyindo M, Lukambagire AH. Fascioliasis: an ongoing zoonotic trematode infection. *Biomed Res Int.* 2015;**2015**, Article ID 786195

50. Romganani S, editor. Th1 and Th2 cells in health and disease. *Chem Immunol*; **63**. Basel Switzerland: Karger AG; 1996.

51. Romagnani S. Type 1 T helper and type 2 T helper cells: functions, regulation and role in protection and disease. *Int J Clin Lab Res* 1991;**21**:152–8.

52. Jia L, Wu C. Differentiation, regulation and function of Th9 cells. In: Sun B, editor. *T helper cell differentiation and their functions. Advances in experimental medicine and biology*. the Netherlands: Springer; 2014. p. 181–207.

53. Kaplan MH, Hufford MM, Olson MR. The development and *in vivo* function of T helper 9 cells. *Nat Rev Immunol* 2015;**15**:295–307.

54. Romagnani S. Th1/Th2 cells. *Inflamm Bowel Dis* 1999;**5**:285–94.

55. Broere F, Apasov SG, Sitkovsky MV, van Eden W. T cell subsets and T cell-mediated immunity. In: Nijkamp FP, Parnham MJ, editors. *Principles of immunopharmacology*. 3rd revised and extended edition. Basel Switzerland: Springer AG; 2011. p. 15–27.

56. Zou W, Regulatory T, cells. Tumor immunity and immunotherapy. *Nature Rev Immunol* 2006;**6**:295–307.

57. Capron M, Trottein F, editor. Parasites and allergy. *Chemical Immunol Allergy*. Basel Switzerland: Karger AG; 2006. p. 90.

58. Heukelbach J, Feldmeier H. Epidemiological and clinical characteristics of hookworm-related cutaneous larva migrans. *Lancet Infect Dis* 2008;**8** 302-L 309.

59. Peisong G, Yamasaki A, Mao XQ, Enomoto T, Feng Z, Gloria-Bottini F, et al. An asthma-associated genetic variant of STAT6 predicts low burden of ascaris worm infestation. *Genes Immun* 2004;**5**:58–62.

60. Moller M, Gravenor MB, Roberts SE, Sun D, Gao P, Hopkin JM. Genetic haplotypes of Th-2 immune signalling link allergy to enhanced protection to parasitic worms. *Hum Mol Genet* 2007;**16**:1828–36.

61. O'Regan NL, Steinfelder S, Venugopal G, Rao GB, Lucius R, Srikantam A, et al. *Brugia malayi* microfilariae induce a regulatory monocyte/macrophage phenotype that suppresses innate and adaptive immune responses. *PLoS Negl Trop Dis* 2014;**8**(10):e3206.

62. Saeed I, Taira K, Kapel CM. *Toxocara canis* in experimentally infected silver and arctic foxes. *Parasitol Res* 2005;**97**:160–6.

63. Al Ghwass MM, El Dash HH, Amin SA, Hussin SS. Intestinal parasitic infections and atopic infections disease in children. A hospital based study. *J Egypt Soc Parasitol* 2015;**45** 413-L 419.

64. Leonardi-Bee J, Pritchard D, Britton J. Asthma and current intestinal parasite infection: systematic review and meta-analysis. *Am J Respir Crit Care Med* 2006;**174**:514–23.

65. Schabussova I, Wiedermann U. Allergy and worms: let's bring back old friends? *Wien Med Wochenschrift* 2014;**164** 382-L 391.

66. Araujo MI, Carvalho EM. Human schistosomiasis decreases immune response to allergen and clinical manifestations of asthma. *Chem Immunol Allergy* 2006;**90**:29–44.

67. Flohr C, Quinnell RJ, Britton J. Do helminth parasites protect against atopy and allergic disease? *Cli Exp Allergy* 2008;**39**:20–32.

68. Wilson MS, Maizels RM, Regulatory T. cells induced by parasites and modulation of allergic responses. *Chem Immunol Allergy* 2006;**90**:176–95.

69. Aranzamendi C, de Bruin A, Kuiper R, Boog CJ, van Eden W, Rutten V, et al. Protection against allergic airway inflammation during the chronic and acute phases of *Trichinella spiralis* infection. *Clin Exp Allergy* 2013;**43**:103–15.

70. Aranzamendi C, Sofronic-Milosavljevic L, Pinelli E. Helminths: immunoregulation and inflammatory diseases—which side are *Trichinella* spp. and *Toxocara* spp. on? *J Parasitol Res* 2013;**2013**:329438.

RESPIRATORY INFECTIONS IN IMMUNOSUPPRESSED PATIENTS

9

S.R. Konduri, A.O. Soubani

*Wayne State University School of Medicine, Division of Pulmonary,
Critical Care and Sleep Medicine, Detroit, Michigan, United States*

1 INTRODUCTION

Alterations of the immune system are common. They could be due to hematological malignances such as acute leukemia, tissue and organ transplantation, primarily hematopoietic stem cell transplantation (HSCT); and chemotherapeutic and immunosuppressive medications used for the treatment of a variety of malignant and nonmalignant conditions. Also, disruption in the anatomical barriers and invasive procedures are other causes of altered immune states.

Pulmonary complications, both infectious and noninfectious, are important cause of morbidity and mortality in immunosuppressed (Fig. 9.1). The incidence of these complications reaches 60% of patients.[1,2] The mortality of pulmonary complications is 40% and approaches 90% in patients who require mechanical ventilation.[3] The etiology of pulmonary disease in immunosuppressed patients may be predicted based on variables such as the type of immunodeficiency, the presence of neutropenia, the prophylactic agents being used, and the temporal relation to the administration of chemotherapy.

2 PATHOPHYSIOLOGY PREDISPOSING PATIENTS TO INFECTIONS

The risk of infection in immunosuppressed patients primarily stems from defects in the number or function of neutrophils, T cells, and B cells. Abnormalities in each of these cell lines increases the risk of specific infections (Fig. 9.2). However the functions of these defense mechanisms are interdependent, so an abnormality in one cell line may compromise the function of other defense mechanisms. Neutropenia is the most frequent immunodeficiency in patients with hematological malignancies. It is seen in patients with acute leukemia and secondary to myelosuppression following chemotherapy. It has been previously shown that the neutrophil function is suppressed in some cancers, which consequently increased risk of infection after chemotherapy.[4] In addition to decreasing the production of neutrophils, de novo cancer,[5–7] chemotherapy and radiation interfere with the chemotactic, phagocytic activity, and the intracellular killing of microorganisms by these cells. Furthermore, corticosteroids, which are part of treatment of many conditions, are well known to cause peripheral neutrophilia by reducing their adherence to endothelial cells and chemotactic properties. Neutrophils are protective

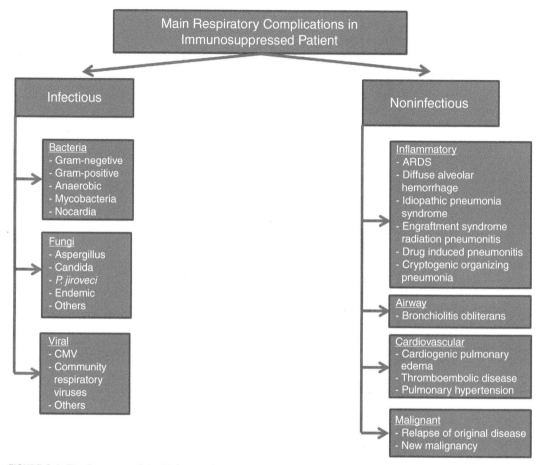

FIGURE 9.1 The Spectrum of the Main Respiratory Complications in Immunosuppressed Patients

against bacterial and fungal infections. The risk of infection in neutropenia increases below an absolute neutrophil count (ANC) of 500 cells/m^3. The risk increases significantly with ANC < 100 cells/m^3. The duration of neutropenia and rate of decline of ANC count also individually pose risk for acquiring infection. In fact, almost all patients with ANC <100/m^3 for more than 3 weeks become febrile. Aspergillosis is one such infection that highlights these risk factors in patients with neutropenia. In a study on patients with *Aspergillus* infections, the authors have noted that the risk of *Aspergillus* infection was increased by 1% with each day of neutropenia for the first 3 weeks, thereafter the risk increases by 4%.[8]

B cells are responsible for the humoral branch of the immune system. They produce immunoglobulins (IgM, IgG, IgE, IgA, and IgD) after appropriate stimulus by the antigen-presenting cells. Abnormalities in their number and function, predispose to infection with encapsulated bacteria such as *Streptococcus pneumoniae*, *Hemophilus influenzae*, and *Staphylococcus aureus*. Lymphoproliferative

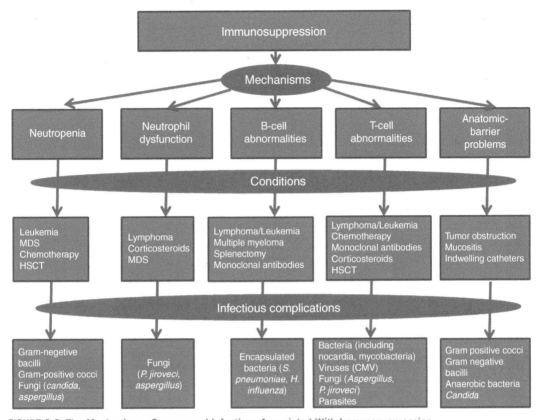

FIGURE 9.2 The Mechanisms, Causes, and Infections Associated With Immunosuppression

malignancies such as chronic lymphocytic leukemia, multiple myeloma, treatment with monoclonal antibodies such as Rituximab, and intensive chemotherapy regimens are some of the other scenarios where B cell abnormalities are seen.

T-cells are responsible for cellular-mediated immunity. They play a vital role by regulating monocyte-macrophage antigen handling, production of cytokines and intracellular pathogen elimination. Abnormalities in T-cell function predispose to infections from organisms such as *Aspergillus* and *Pneumocystis jirovecii*, viruses, *Nocardia* and *Mycobacteria*. Corticosteroids, immunosuppressants such as Tacrolimus, Sirolimus, Azathioprine, chemotherapeutic agents like cyclophosphamide, fludarabine, cladribine, anti-CD52 monoclonal antibodies (Alemtuzumab) impair T-cell function.

Violation of anatomical barriers, resulting from invasion by cancer or by its treatment, also increases risk for infection. For example, mucositis secondary to chemotherapy may predispose to aspiration or translocation of microorganisms from the gastrointestinal tract, vascular catheters inserted for chemotherapy, parenteral nutrition, and transfusion of blood products may be sources of bacterial and fungal infections.

3 BACTERIAL PNEUMONIA

Bacterial pneumonia caused by Gram-positive and Gram-negative organisms is a significant problem in patients with malignancies and is a leading cause of infectious death in these patients. It constitutes about 34% of the infectious episodes in patients with acute leukemia.[9] They account for 15% of all respiratory infections.[10] Up to 60% of patients with neutropenia can develop lung infiltrates during the course of their disease.[11] These bacteria may lead to community and hospital acquired pneumonia.

Factors that generally increase the risk of bacterial pneumonia in immunosuppressed patients include the nature of the immunodeficiency, chemotherapeutic regimen, degree, and duration of neutropenia.[2,12] Other factors that may predispose to bacterial pneumonia include older age, lower performance status,[13] mucositis that increases the risk of aspiration[14] and indwelling catheters that increase the risk of bacteremia and septic emboli to the lungs.[15]

Certain characteristics of immunosuppression may predispose the patient to particular organisms. For example, encapsulated bacteria like *S. pneumoniae, H. influenzae* are seen more frequently in patients with multiple myeloma, chronic lymphocytic leukemia. *Legionella pneumophila* is more frequently seen in patients with lymphoma.

The presentation of bacterial pneumonia in these patients may be subtle, and not infrequently, patients present with neutropenic fever without localizing symptoms and signs due to blunted inflammatory response.[16] Radiological signs may also be subtle and scarce. The chest radiograph usually shows focal consolidation. However, they may be normal. The infiltrates may progress quickly to multifocal or diffuse changes that are compatible with acute respiratory distress syndrome. High resolution computerized tomography of the chest (HRCT) is recommended in febrile neutropenic patients, as it may show pulmonary infiltrates in up to 50% of patients with negative chest radiograph.[17] Many patients need invasive procedures like bronchoscopy for obtaining adequate samples for culture and serological testing. This is mainly due to a lower yield on sputum as a result of oropharyngeal contamination, use of prophylactic antibiotics, and so on.

The organisms frequently isolated are Gram-negative bacteria such as *Pseudomonas aeruginosa, Klebsiella spp, Escherichia coli, Moraxella catarrhalis, H. influenzae, Stenotrophomonas maltophilia*[18] and Gram-positive organisms (patients without a recent antibiotic exposure) like *S. pneumoniae* and *S. aureus.*[19] A 10 year study in a single hospital revealed incidence of Gram-negative pneumonia in 58% of bacteremia cases.[19] The microbiology of pulmonary infections in neutropenic patients is frequently similar to that of hospital acquired pneumonia or ventilator associated pneumonia.

Following HSCT, patients continue to be at risk for bacterial pneumonia post engraftment of the stem cells, albeit less frequently. The risk factors for bacterial pneumonias during this period include the presence of acute or chronic graft versus host disease (GVHD) and the immunosuppressive therapy used to treat these patients. These patients are at risk of infection with encapsulated organisms such as *S. pneumoniae* and *H. influenza.* Gram-negative organism may be the etiology in up to 25% of patients.[12]

Since the introduction of Pneumococcal vaccine, there has been decreased incidence of penicillin resistant and invasive serotypes of *Pneumococci.* However, the incidence of other serotypes is still significant. In a single center study, Pneumococcal pneumonia was seen in 6.5% of patients. Increased incidence was noted in hematological malignancies versus solid tumors. Other risk factors noted were chronic obstructive pulmonary disease and diabetes mellitus. Most of these were healthcare associated infections.[20]

Empiric broad-spectrum antibiotics should be started immediately when bacterial pneumonia is suspected in patients with hematological malignancies. The American Thoracic Society guidelines for the treatment of healthcare associated pneumonia in patients with risk factors such as HSCT or neutropenia, recommend a regimen that includes antipseudomonal cephalosporin or carbepenem, or β-lactam, plus antipseudomonal fluoroquinolone or aminoglycoside, plus linezolid or vancomycin if methicillin resistant *S aureus* is suspected. In patients who develop bacterial pneumonia late following HSCT, coverage of encapsulated organisms with a fluoroquinolone is recommended. The antibiotic regimen should be narrowed if the etiologic agent is identified.[21,22] Institutional resistance pattern needs to be considered when determining the choice of antibiotics.

Legionella pneumophila pneumonia is occasionally reported in patients with hematological malignancies. A study involving 49 patients with this infection reported that lymphopenia, systemic corticosteroids, and chemotherapy as the most common risk factors. It is identified using DFA (direct fluorescent antibody assay). Mortality can be as high as 31%.[23] Macrolide or fluoroquinolone are the primary treatment. A prolonged course of antibiotics may be needed depending on the initial response to treatment.

Stenotrophomonas maltophilia is a growing concern. It is commonly seen in patients with obstructive lung disease, prolonged mechanical ventilation, recent broad-spectrum antibiotic exposure, and neutropenia.[24] Lobar or lobular consolidation with absence of pleural effusion is common. Cavitation is rarely seen. Trimethoprim-sulfamethoxazole (TMP-SMZ) remains the preferred antibiotic in most patients. Ceftazidime, Moxifloxacin, Tigecycline, and Colistin are alternatives.

Nocardia infection has been rarely reported with an incidence ranging between 0.3 and 1.7%.[25,26] The main predisposing factors include neutropenia, acute or chronic GVHD, and lack of prophylaxis with Trimethoprim-sulfamethoxazole. Fifty-six percent of the patients may have pulmonary involvement. The most common radiological findings include nodules with or without infiltrates, cavitation, and empyema. The diagnosis is usually made by Gram staining and modified acid-fast bacillus staining of sputum, bronchoalveolar lavage (BAL), or pleural fluid. Diagnosis can be further confirmed by CT-guided biopsy of pulmonary nodule or surgical lung biopsy. Coinfection with other organisms such as bacteria, *Cytomegalovirus* (CMV) and *Aspergillus* is common and may be seen in up to 36% of the patients.[26]

Treatment of *Nocardia* infection is by prolonged administration of TMP-SMZ. In case of an allergy or side effects associated with sulfa preparations, second line agents that are available include Amikacin, Minocycline, Cephalosporin, or Imipenem. Response to treatment is generally good, and long term survival in patients with nocardiosis is not significantly different from controls.[26] The increased use of TMP-SMZ for prophylaxis against *P. jiroveci* has the added benefit of reducing the incidence of *Nocardia* infections.

Mycobacterium tuberculosis infection is rare in immunosuppressed patients and varies significantly depending on whether the patients lived in an endemic area. In a large cancer center in the United States, the overall rate of active *M. tuberculosis* infection was 0.2 in 1000 new cancer diagnoses. Five out of the 18 patients described had hematological malignancies and 4 were neutropenic.[27] In India, a report of 130 patients with acute leukemia, 9 cases (6.9%) had active tuberculosis[28].

The main risk factors for *M. tuberculosis* in patients with hematological malignancies are allogeneic transplantation, total body irradiation, chronic GVHD, corticosteroid therapy, and residence in endemic areas.[29] The clinical and radiological picture is similar to disease in nonimmunosuppressed patients. However, atypical presentations such as lack of cavitation, rapidly progressive disease, and extra-pulmonary manifestations have been described. The diagnosis of infection is usually made by acid-fast bacillus staining and culture of sputum, BAL, or pleural fluid samples. Rapid diagnosis of

M. tuberculosis may be established by polymerase chain reaction (PCR) test. It is important to check drug susceptibility on positive samples. Treatment is similar to patients without malignancies. Some studies suggest extending treatment to 1 year.[30]

Nontuberculous mycobacterial (NTM) infections are also rare in these patients. NTM infections include *Mycobacterium avium complex, M. kansasii,* and *M. chelonae.* In an epidemiological study of 2856 patients with hematological malignancies at a single center in Taiwan (a tuberculosis endemic area), the prevalence of NTM was about 1.2%.[31] Nontuberculous mycobacteria may colonize the airways of the patients with chronic lung diseases such as bronchiectasis, which may create a challenge in determining the significance of isolating these bacteria from lower respiratory tract samples. The presentation ranges from mild worsening of obstructive lung disease, consolidation, nodules, cavity to disseminated disease.[31] *Mycobacterium avium intracellulare* infection is treated with a regimen that includes Clarithromycin, Azithromycin, Ethambutol, Rifampin, and Rifabutin for 18 months (12 months after negative cultures).[32,33]

4 FUNGAL INFECTIONS

Aspergillus species are ubiquitously found in the environment. They are regularly inhaled into the lower respiratory tract where they transform to short, acutely branching, and septate hyphae. Neutrophils and alveolar macrophages act as the primary defense mechanisms against infection by these organisms. Their activity is further augmented by T-helper cell induced cytokine production (such as TNF-α, interferon-γ, IL-12, and IL-15). As mentioned previously, neutropenia and defects in T-helper cell function also increase vulnerability to invasive disease by *Aspergillus.* Concomitant infection with *Cytomegalovirus* (CMV) in patients who underwent HSCT increases the risk of invasive pulmonary aspergillosis (IPA). The hazards ratio for IPA in this setting increases by 13.3-fold.[34] A retrospective review conducted over a 9-year period, including 385 cases of patients with suspected and documented IPA, revealed that the disease-specific survival rate was 60%. Factors that predicted mortality were allogeneic HSCT, neutropenia, progression of the underlying malignancy, prior respiratory disease, corticosteroids therapy, renal impairment, low monocyte counts, pleural effusion, and disseminated aspergillosis.[35]

IPA usually affects the lower respiratory tract, but occasionally has been reported in extra-pulmonary sites like sinuses, gastrointestinal tract, and skin.[35–39] Patients usually present with symptoms consistent with bronchopneumonia that is unresponsive to antibiotics. In presence of vascular invasion, pleuritic chest pain and hemoptysis have been reported (due to small pulmonary infarcts). IPA is one of the most common causes of hemoptysis in neutropenic patients with cavitation.[40] Patients may also present with seizures, meningitis, epidural abscess, intracranial hemorrhage, ring-enhancing intracranial lesions as result of hematogenous dissemination of the organism to the brain. Skin, kidneys, heart, esophagus, and liver are also occasionally involved.

Diagnosis of IPA in patients begins with identifying at-risk patients based on presence of risk factors. Histopathological diagnosis remains the gold standard for the diagnosis of IPA.[41,42] The presence of septate, acute, branching hyphae, with invasion of native tissues, paired with positive culture from the same site, is diagnostic of IPA. The histopathological findings in neutropenic patients are characterized by scant inflammation, extensive coagulation necrosis associated with hyphal angioinvasion, and high fungal burden. Conversely, in patients with HSCT, there is intense inflammation with neutrophilic infiltration, minimal coagulation necrosis, and low fungal burden.[43]

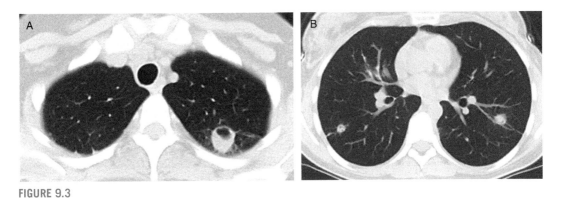

FIGURE 9.3

A and B are Chest CT Scan Views Showing Multiple Pulmonary Nodules, Some are Cavitating in a HSCT Patient With IPA.

The sensitivity and specificity of chest radiograph are low in early stages of the disease. Radiographic abnormalities may include pleural-based infiltrates, rounded opacities, or cavities. Pleural effusions are rare. HRCT has evolved as a useful tool for early diagnosis of IPA. Early implementation of this diagnostic modality has also been shown to favor improved outcomes in these patients.[44,45] It is also a valuable tool in planning of further invasive diagnostic studies like bronchoscopy or surgical lung biopsy. The typical radiological findings include multiple nodules (Fig. 9.3), halo sign, air crescent sign. Halo sign, usually seen in first week, is mainly seen in neutropenic patients. It appears as a zone of low attenuation surrounding a central lesion of higher attenuation owing to hemorrhage around the lesion. Another classic and late radiological sign is the air crescent sign. The crescent-shaped lucency in the region of the original nodule occurs secondary to necrosis. Both the halo and the air crescent signs are neither sensitive nor pathognomic of IPA. In a large study of 236 patients with IPA, 61% of the cases had halo sign. On the other hand, the presence of halo sign portends a good prognosis, especially in patients with hematological malignancies.[46]

The significance of isolating *Aspergillus* species in sputum samples depends on the immune status of the host. Studies have shown a positive predictive value of 80–90% in diagnosing IPA in immunosuppressed patients.[47–49] However, negative sputum studies have been noted in 70% of patients with confirmed IPA.[49,50] Blood cultures are rarely positive in patients with IPA.

Bronchoscopy aids in direct visualization of the bronchial tree and in obtaining BAL for fungal staining, culture, *Aspergillus* antigen assays, and so on. It is specifically useful in patients with diffuse lung involvement. The sensitivity and specificity of BAL in diagnosis of IPA has been reported to be around 50 and 97%, respectively. However, the yield of BAL has also been reported to be lower.[51] Transbronchial biopsies further carry increases the risk of bleeding without adding much to the diagnosis.

Aspergillus fumigatus is the most common cause of IPA. A review of 300 patients with proven IPA, identified *Aspergillus terreus* as the second most common species (23%). Infections by *Aspergillus terreus* were significantly more likely to be nosocomial in origin and resistant to Amphotericin B.[52,53] The triazole antifungal agents have significantly better efficacy against *Aspergillus terreus*. An important part of diagnostic work up for IPA is distinguishing it from infections by other molds like *Scedosporium spp.*, *Pseudallescheria,* and *Fusarium*, which may have similar histological appearance as *Aspergillus*. Hence, it is recommended to include the samples for culture when possible.

Galactomannan (GM) assay is an enzyme-linked immunosorbent (ELISA) based assay that is designed to detect GM, a polysaccharide cell-wall component of the *Aspergillus*. The Food and Drug Administration has approved its application to serum analysis with a threshold value of 0.5 ng/ml. It has been reported that the positive serum GM assay precedes clinical signs and radiographic changes by several days (5–8 days). It may also be useful in assessing the evolution of infection during treatment.[54,55]

A study by Pfeiffer et al., which included 27 studies from 1996–2005 of patients with IPA (defined according to the European Organization for Research on Treatment of Cancer/Mycoses Study Group [EORTC/MSG] criteria), found that serum GM assay to have a sensitivity of 71% and specificity of 89%. The negative predictive value and positive predictive value were 92–98% and 25–62%, respectively.[55]

GM is found in certain food items, β-lactam antibiotics (eg, Piperacillin-Tazobactam), which may contribute to false positive results. The assay itself is species specific, and cannot exclude the concomitant infections by other molds such as *Fusarium*, Zygomycetes, and dematiaceous fungi.

There is accumulating evidence that GM assay may be applicable to body fluids such as BAL, urine, and cerebrospinal fluid. A prospective study of 200 patients with hematological malignancies and profound neutropenia, found that BAL–GM assay had sensitivity of 100% for IPA compared to serum GM assay.[56] Musher et al. showed that incorporating GM assay and quantitative-PCR assay into standard BAL fluid analysis may enhance identification of *Aspergillus* species in patients with hematological malignancies[57]. Factors affecting GM assay in BAL samples was studied by Racil et al. In their study, the sensitivity and specificity of BAL–GM assay was reported to be 78% and 98%, respectively, when combined with serum and bronchial samples. The factors affecting its performance were noted to be neutropenia, BAL standardization and antifungal therapy.[58] Adherence to standard BAL collection technique is paramount for ensuring reliability of the assay findings.

PCR is another tool available for the detection of *Aspergillus* DNA in BAL fluid and serum. The sensitivity has been reported to be between 67% and 100%, and specificity between 55% and 95%.[59] However, the test alone cannot discriminate between colonization and infection. Rienwald et al. conducted the first multicenter, randomized control trial in which they elucidated that combining PCR assay with BAL–GM assay, resulted in sensitivity of 55% and specificity of 100%, respectively. Thus combining the two assays increased the specificity of diagnosing *Aspergillus* without the need for increasing the threshold cut-off value of GM assay, at the cost of sensitivity, to achieve the same specificity.[60] The availability of *Aspergillus* PCR assay is still restricted to highly specialized laboratories.

Detection of serum (1→3)-β-D-glucan, a fungal cell wall constituent has been reported to be a highly sensitive and specific test for invasive deep mycosis, including candidiasis, fusariosis, pneumocystosis and aspergillosis, and could be useful in the immunosuppressed patients. A meta-analysis that included 15 studies, showed the sensitivity and specificity to be 76% and 85%, respectively. Combining with GM assay increased its specificity to 98% for diagnosing IPA. It also showed higher specificity with two consecutive positive results and in patients with hematological malignancies.[61]

The previous assays do not replace the value of clinical signs and radiological data in diagnosing IPA. They solely act as adjunctive data in helping to reach to the diagnosis.

The EORTC/MSG has provided its criteria for the diagnosis of invasive fungal infections (Table 9.1). These criteria are not necessary to consider treatment in clinical practice, rather they are put forward to guide clinical and epidemiological research.

Table 9.1 Fungal Infections Cooperative Group and the National Institute of Allergy and Infectious Diseases Mycoses Study Group (EORTC/MSG) Criteria for the Diagnosis of IPA

Category	Criteria
Proven IPA	1. Microscopic analysis on sterile material: histopathologic, cytopathologic, or direct microscopic examination of a specimen obtained by needle aspiration or sterile biopsy in which hyphae are seen accompanied by evidence of associated tissue damage. 2. Culture on sterile material: recovery of Aspergillus by culture of a specimen obtained by lung biopsy.

Probable IPA
All 3 criteria
	1. Host factors—*1 of the following*
	  a. Recent history of neutropenia (<500 neutrophils/mm^3) for >10 days
	  b. Receipt of an allogeneic stem cell transplant
	  c. Prolonged use of corticosteroids $>/= 0.3$ mg/kg/d of prednisone equivalent for >3 weeks
	  d. Treatment with other recognized T-cell immunosuppressants
	  e. Inherited severe immunodeficiency
	2. Clinical features—*1 of the following 3 signs on CT*
	  a. Dense, well-circumscribed lesion(s) with or without a halo sign
	  b. Air crescent sign
	  c. Cavity
	3. Mycological criteria *1 of the following*
	  a. Direct test (cytology, direct microscopy, or culture) on sputum, BAL fluid, bronchial brush indicating presence of fungal elements or culture recovery *Aspergillus* spp.
	  b. Indirect tests: galactomannan antigen detected in plasma, serum, or BAL fluid.

Possible IPA	Host factors and clinical features. No mycological evidence

The mortality rate of IPA remains high despite initiation of treatment. Treatment should be considered as soon as there is a clinical suspicion for IPA. Prior to introduction of Voriconazole to clinical practice, Amphotericin B was the first line of therapy for IPA. The recommended dose is 1.0–1.5 mg/kg/d. Amphotericin B has serious side-effects including nephrotoxicity, electrolyte disturbances, and hypersensitivity. The lipid-based preparations of Amphotericin B (eg, liposomal Amphotericin B and lipid complex Amphotericin B) have relatively milder side effects. Voriconazole is a broad-spectrum triazole that is currently the treatment of choice for IPA, due to its fungicidal action. In a large prospective, randomized, multicenter trial, comparing Voriconazole to Amphotericin B as the primary therapy for IPA. Voriconazole treatment arm had a higher favorable response rate at week 12 (53% compared with 32% in patients receiving Amphotericin B), and a higher 12-week survival (71% compared with 58%).[62] Voriconazole is available in both intravenous and oral formulations. The recommended starting dose is 6 mg/kg twice daily intravenously, followed by 4 mg/kg/d. Maintenance treatment can be considered after 1 week at 200 mg orally twice daily. The most frequent adverse effects of Voriconazole include visual disturbances described as blurred vision, photophobia, and altered color perception. Liver function test abnormalities and skin reactions are less common side effects.

Posaconazole is another broad-spectrum triazole that is available as an oral and intravenous (recent) formulation. Currently, it is used as salvage therapy in patients with invasive aspergillosis refractory to treatment with Voriconazole or lipid formulation Amphotericin B or in patients who are intolerant to its adverse effects of these medications.[63]

Recently, the FDA has approved the use of Isavuconazonium sulfate, a broad-spectrum triazole for treatment of IPA. A phase III trial involving 516 patients compared Isavuconazole to Voriconazole in treatment of patients with invasive aspergillosis. Isavuconazole was found to be noninferior to Voriconazole for all-cause mortality.[64] It is reported to have slightly lower incidence of adverse effects. Most common adverse effects are gastrointestinal disorders, hypokalemia, and elevated liver enzymes.

Echinocandin derivatives such as Caspofungin, Micafungin, and Anidulafungin are effective agents and can be considered in the treatment of IPA refractory to first line agents or if the patient could not tolerate first line agents.[65] The Infectious Disease Society of America recommends Voriconazole as the primary treatment for IPA, and liposomal Amphotericin B as an alternative regimen. For salvage therapy, agents include lipid-based preparations of Amphotericin B, Posaconazole, Itraconazole, Caspofungin, or Micafungin.[66]

The role of surgical resection although limited, is indicated in cases of massive hemoptysis, or lesions close to the great blood vessels or pericardium, or the resection of residual localized pulmonary lesions in patients with continuing immunosuppression.

Immunomodulatory therapy has been considered to decrease degree of immunosuppression in order to aid in treatment of IPA. Colony stimulating factors such as granulocyte-colony stimulating factors (G-CSF) or interferon-γ have been used. However, current evidence to their use is limited to few randomized studies and case reports. In one such randomized study that included patients receiving chemotherapy for leukemia, prophylactic therapy with GM-CSF led to a lower frequency of fatal fungal infections compared with placebo (1.9% vs 19% respectively) and reduced overall mortality.[67]

Due to high mortality related to invasive aspergillosis, prophylaxis remains an important part of management of patients with risk factors. Patients should be advised to avoid construction areas. Similarly, use of high-efficiency particulate air (HEPA) filtration with or without laminar airflow ventilation is efficient in protecting against IPA. Itraconazole was shown to be effective in preventing fungal infections in neutropenic patients, according to a metaanalysis.[68] Posaconazole has also been shown to be effective prophylactic agent in high risk patients with acute myelogenous leukemia, myelodysplastic syndrome, and GVHD after allogeneic HSCT.

In a Centers for Disease Control sponsored surveillance program for invasive fungal infections that included 886 patients with invasive fungal infection, non-aspergillus mold were responsible for 14% of infections. In addition, *Cryptococcus* infection was found in 4%, endemic fungi like *Coccidiomycoses*, *Blastomyces*, *Histoplasma* in 3%, and *Pneumocystis* in 2% of patients.[69]

Hyaline hyphomycetes such as *Fusarium* and *Scedosporium spp.* are the main non-aspergillus molds that are reported in patients with hematological malignancies. These organisms resemble *Aspergillus* and culture is the only means to distinguish these organisms. Lung is the primary site of infection however they may disseminate to other sites. Corticosteroid therapy and neutropenia are common risk factors. The mortality in patients with pulmonary Fusariosis is as high as 65% within one month of diagnosis.[70] *Scedosporium* has been isolated from the air in hospitals. Patients usually present with thin wall cavities or nodules. Air crescent sign may be present or absent. The patient's history usually denotes that these lesions are refractory to antibiotics and antifungal therapy. The diagnosis is confirmed by isolating the organisms by culture either in blood, BAL or lung tissue. Mortality associated with these fungal infections remains very high, in the range of 70–100%.[71]

Zygomycetes such as *Rhizopus* and *Mucor* are characterized by sparsely septate, broad hyphae with irregular branching. They have a predilection to invade blood vessels with associated tissue destruction due to thrombosis and necrosis. They usually involve the sinopulmonary tree and dissemination to other organs is rare. Although, they are an uncommon cause of invasive fungal infection in immunosuppressed patients (0.5–1.9%), mortality usually reaches 80%.[72]

Cryptococcal infections have been reported to be low due to the widespread use of Fluconazole prophylaxis in patients with hematological malignancies. Patients with profound T-cell depression are at highest risk. The diagnosis of pulmonary cryptococcosis is made by detection of the fungus in BAL or lung tissue samples. The sensitivity and specificity of serum cryptococcal antigen assay is more than 95%. The treatment includes Fluconazole or Itraconazole in milder cases, with use of Amphotericin B in severe cases. Either Fluconazole or Itraconazole is used for maintenance therapy.[73]

Invasive *Candida* infection develops in 11–16% of patients with hematological malignancies. Autopsy studies suggest that 50% of infected patients have pulmonary involvement.[74] Pulmonary infection results from dissemination of invasive candidiasis or candidemia. In a study of 529 patients who died from leukemia or myelodysplastic syndrome and were affected by candidiasis, 45% had lung involvement.[75] Primary *Candida* pneumonia is extremely rare. Histopathological confirmation is necessary in most of the cases. Previous triazole antifungal exposure, renal failure, neutropenia have been associated with higher mortality rates. Lately, there has been a shift in the candida species from *C. albicans* to more resistant species such as *C. glabrata* or *C. krusei*.

Pneumocystis jirovecii pneumonia is caused by inhalation of the aerosolized organisms. Previously, the incidence of *Pneumocystis* pneumonia was up to 20% of patients with hematological malignancies.[76] However the routine prophylaxis with TMX/SMZ has significantly decreased this number. Patients with lymphoproliferative disorders appear to be at highest risk for *P. jiroveci* infection followed by patients treated with the prolonged administration of high dose corticosteroid therapy and purine analogues such as Fludarabine.[77–79]

P. jirovecii pneumonia presents with acute onset, rapidly progressing symptoms of dyspnea, nonproductive cough, and fever. Hypoxemia is usually significant in these patients. Radiologically, there are perihilar interstitial or alveolar infiltrates (Fig. 9.4). Ground glass opacities are seen on HRCT of chest. Lactate dehydrogenase (LDH) levels may be elevated. The diagnosis is usually done on BAL analysis. The sensitivity of this test may be lower than patients with HIV infection due to lower organism burden. Similarly, induced-sputum samples are useful.[80,81]

Mortality has been recently reported to be 20% of patients with malignancy. The need for mechanical ventilation portends poor outcome[82]. Prophylaxis using TMP-SMZ is recommended for patients who are on chronic corticosteroid therapy and following engraftment in allogeneic HSCT recipients for 100 days. However, if a patient has chronic GVHD, then prophylactic treatment is continued as long as the patient is on immunosuppressive therapy.

Endemic mycoses such as *Histoplasmosis, Blastomycosis,* and *Coccidiomycosis* are rare in patients with hematological malignancies.[83] These infections tend to have geographical distribution, and are usually caused by endogenous reactivation of a latent infection. Initial treatment is with high dose Amphotericin B. After controlling the infection, Itraconazole or Fluconazole can be used as a maintenance treatment.

5 CYTOMEGALOVIRUS

CMV infection is an important infection in immunosuppressed patients, especially in posttransplant patients. The predominant site of infection being the gastrointestinal system, however, up to one-third of patients with HSCT may have pulmonary involvement. The incidence of CMV pneumonia has

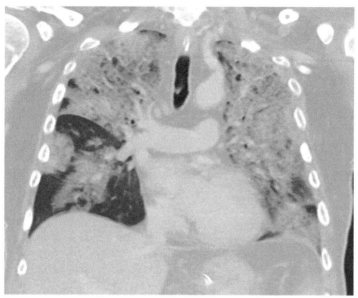

FIGURE 9.4 Reconstructed Chest CT Scan Image Showing Bilateral Pulmonary Infiltrates in a Patient With Acute Myeloid Leukemia and *P. jirovecii* Pneumonia

been noted to be around 3.5% prior to 1997, which has since decreased to 0.8%.[84] The risk of developing CMV pneumonia has been noted to increase with use of T-cell depleting chemotherapy (such as fludarabine, Alemtuzumab, Cytarabine, or high dose cyclophosphamide and corticosteroids), which are used to treat leukemia and lymphoma. It has also been reported in allogeneic HSCT transplant, specifically in seropositive recipients. The risk is least when both donor and recipient are seronegative.[85]

CMV pneumonia presents clinically with acute onset fever, dyspnea, nonproductive cough and pleurisy. The patient may further progress to acute hypoxemic respiratory failure in matter of 2 weeks. Chest imaging may show interstitial pattern of pneumonia or patchy consolidation. HRCT of chest is more sensitive in identifying ground glass opacities, nodular opacities, diffuse or patchy in distribution, and thickening of the interlobular septae. Pleural effusion is also seen in up to 25% of these patients.[86]

The classic histopathological finding of CMV pneumonia is the presence of intracytoplasmic eosinophilic inclusion bodies within the areas of mononuclear, interstitial inflammation on lung biopsy. However, they may be absent early in the disease. These findings may be associated with alveolar epithelial desquamation and hyaline membrane formation. Demonstration of inclusion bodies in the alveolar epithelial cells on cytological examination of the BAL fluid has specificity of 98% for CMV pneumonia, but has low sensitivity, with positive predicted value of 73%.[87]. Rapid culture of the BAL fluid using shell viral technique is highly sensitive (99%) in detecting the virus. On the other hand, it cannot differentiate between viral shedding and invasive disease and, hence, is less specific (67–83%).[87] CMV pp65 antigen assay is based on detection of pp65 antigen in blood neutrophils and as such, its use is limited in neutropenia. However, it is useful in monitoring treatment response. Persistence of antigenemia despite 2 weeks of therapy may suggest antiviral resistance. CMV PCR analysis is a quantitative analysis of viral DNA that is shed in body fluids including blood, urine, cerebrospinal fluid, and alveolar lavage. Detection of viral DNA in alveolar lavage is diagnostic of CMV pneumonia, but is associated with frequent false positivity.[88] Clinicians are often confronted with positive laboratory results, and it is vital to correlate clinical, radiological features with a positive result in order to reliably diagnose CMV infection.

Ganciclovir is currently the standard of treatment for CMV pneumonia. Mortality has been shown to decrease if treatment is initiated prior to onset of respiratory failure. Ganciclovir prophylaxis has also shown reduction in CMV invasive disease and mortality in HSCT patients.[89] It is administered in seropositive patients and in seronegative recipient with a seropositive donor. Prophylaxis is initiated 5 days before engraftment and continued for 100 days after transplantation.

6 COMMUNITY RESPIRATORY VIRUSES

There is limited data on the significance of community respiratory infections in immunosuppressed patients due to seasonal and endemic nature of these infections and possibly due to presence of mild or no symptoms in most of the cases. Commonly identified viruses that may cause serious infections in these patients are *respiratory syncytial virus* (RSV), *parainfluenza, influenza,* and *adenovirus.* In a retrospective analysis of 306 patients with 343 episodes of viral infections, 33% were from *Influenza*, 31% from RSV and 27% from *Parainfluenza* (mainly type 3). About one-third of these patients progressed to having lower respiratory tract infections. High risk patients were identified as age >65 years, severe neutropenia and lymphopenia and allogeneic HSCT. Overall, mortality was 15%.[90]

In lower respiratory tract infections, HRCT of the chest may show patchy ground glass opacities, nodules, bronchial wall thickening or combination of earlier. Pleural effusions are less common. Rapid laboratory diagnosis can be made using direct-(DFA) and indirect-(IFA) fluorescent antibody assays, performed on nasopharyngeal swabs or wash. Sensitivity ranges between 20–52%.[91] Viral cultures usually take 7–14 days to be reported. BAL fluid may also be useful in patients with lower respiratory tract infections. Coinfections with bacteria, *P. jiroveci*, or fungi are commonly seen and these infections should be excluded by the appropriate diagnostic studies.

Infections due to respiratory viruses are usually self-limiting, although lower respiratory infections leading to acute respiratory failure are associated with high mortality. Simple measures such as hand washing, early isolation of patients suspected to have infection, limiting visitors during endemic seasons, and restricting patient contact with symptomatic healthcare personnel or visitors are very effective preventative measures against these infections.

7 CONCLUSIONS

Immunosuppression due to abnormalities in the human's defense mechanisms is common and is associated with a variety of malignant and nonmalignant conditions. Immunosuppressed patients are prone to unique respiratory infections. In recent years there have been important advances in the diagnosis and management of these infections. However, there remains significant challenges and unacceptably high mortality. More research is needed to prevent and minimize the severity and duration of immunosuppression. It is also important to better identify risk factors for these infections. More robust, less invasive diagnostic tests with high sensitivity and specificity are needed. On the other hand, there is still demand for better tolerated and more effective therapeutic agents. The field of respiratory complications in immunosuppressed patients is an exciting one and all indications are that the future will hold more success.

REFERENCES

1. Maschmeyer G, Link H, Hiddemann W, Meyer P, Helmerking M, Eisenmann E, et al. Pulmonary infiltrations in febrile patients with neutropenia. Risk factors and outcome under empirical antimicrobial therapy in a randomized multicenter study. *Cancer* 1994;**73**:2296–304.
2. Soubani AO, Miller KB, Hassoun PM. Pulmonary complications of bone marrow transplantation. *Chest* 1996;**109**:1066–77.
3. Ewig S, Torres A, Riquelme R, El-Ebiary M, Rovira M, Carreras E, et al. Pulmonary complications in patients with haematological malignancies treated at a respiratory ICU. *Eur Respir J* 1998;**12**:116–22.
4. Hubel K, Hegener K, Schnell R, Mansmann G, Oberhäuser F, Staib P, et al. Suppressed neutrophil function as a risk factor for severe infection after cytotoxic chemotherapy in patients with acute nonlymphocytic leukemia. *Ann Hematol* 1999;**78**:73–7.
5. Sone S. Role of alveolar macrophages in pulmonary neoplasias. *Biochim Biophys Acta* 1986;**823**:227–45.
6. Olsen GN, Gangemi JD. Bronchoalveolar lavage and the immunology of primary lung cancer. *Chest* 1985;**87**:677–83.
7. McDonald CF, Atkins RC. Defective cytostatic activity of pulmonary alveolar macrophages in primary lung cancer. *Chest* 1990;**98**:851–85.
8. Gerson SL, Talbot GH, Hurwitz S, Strom BL, Lusk EJ, Cassileth PA. Prolonged granulocytopenia: the major risk factor for invasive pulmonary aspergillosis in patients with acute leukemia. *Ann Intern Med* 1984;**100**:345–51.
9. Bodey GP, Rodriguez V, Chang HY, Narboni G. Fever and infection in leukemic patients—a study of 494 consecutive patients. *Cancer* 1978;**41**:1610–22.
10. Inagaki J, Rodriguez V, Bodey GP. Causes of death in cancer patients. *Cancer* 1974;**33**:568–73.
11. Novakova IR, Donnelly JP, De Pauw B. Potential sites of infection that develop in febrile granulocytopenic patients. *Leuk Lymphoma* 1993;**10**:461–7.
12. Lossos IS, Breuer R, Or R, Strauss N, Elishoov H, Naparstek E, et al. Bacterial pneumonia in recipients of bone marrow transplantation. A five-year prospective study. *Transplantation* 1995;**60**:672–8.
13. Lee JO, Kim DY, Lim JH, Seo MD, Yi HG, Kim YJ, et al. Risk factors for bacterial pneumonia after cytotoxic chemotherapy in advanced lung cancer patients. *Lung Cancer* 2008;**62**:381–4.
14. Labarca JA, Leber AL, Kern VL, Territo MC, Brankovic LE, Bruckner DA, et al. Outbreak of Stenotrophomonas maltophilia bacteremia in allogeneic bone marrow transplant patients: role of severe neutropenia and mucositis. *Clin Infect Dis* 2000;**30**:195–7.
15. Poutsiaka DD, Price LL, Ucuzian A, Chan GW, Miller KB, Snydman DR. Blood stream infection after hematopoietic stem cell transplantation is associated with increased mortality. *Bone Marrow Transplant* 2007;**40**:63–70.
16. Rolston KV. The spectrum of pulmonary infections in cancer patients. *Curr Opin Oncol* 2001;**13**:218–23.
17. Heussel CP, Kauczor HU, Heussel G, Fischer B, Mildenberger P, Thelen M. Early detection of pneumonia in febrile neutropenic patients: use of thin-section CT. *AJR Am J Roentgenol* 1997;**169**:1347–53.
18. Safdar A, Rolston KV. Stenotrophomonas maltophilia: changing spectrum of a serious bacterial pathogen in patients with cancer. *Clin Infect Dis* 2007;**45**:1602–9.
19. Carratalà J, Rosón B, Fernández-Sevilla A, Alcaide F, Gudiol F. Bacteremic pneumonia in neutropenic patients with cancer. *Arch Intern Med* 1998;**158**:868–72.

20. Garcia-Vidal C, Ardanuy C, Gudiol C, Cuervo G, Calatayud L, Bodro M, et al. Clinical and microbiological epidemiology of Streptococcus pneumoniae bacteremia in cancer patients. *J Infect* 2012;**65**:521–7.

21. Niederman MS, Mandell LA, Anzueto A, Bass JB, Broughton WA, Campbell GD, et al. Guidelines for the management of adults with community-acquired pneumonia. Diagnosis, assessment of severity, antimicrobial therapy, and prevention. *Am J Respir Crit Care Med* 2001;**163**:1730–54.

22. Guidelines for the management of adults with hospital-acquired, ventilator-associated, and healthcare-associated pneumonia. *Am J Respir Crit Care Med* 2005;**171**:388–416.

23. Jacobson KL, Miceli MH, Tarrand JJ, Kontoyiannis DP. Legionella pneumonia in cancer patients. *Medicine (Baltimore)* 2008;**87**:152–9.

24. Fujita J, Yamadori I, Xu G, Hojo S, Negayama K, Miyawaki H, et al. Clinical features of Stenotrophomonas maltophilia pneumonia in immunocompromised patients. *Respir Med* 1996;**90**:35–8.

25. Daly AS, McGeer A, Lipton JH. Systemic nocardiosis following allogeneic bone marrow transplantation. *Transpl Infect Dis* 2003;**5**:16–20.

26. van Burik JA, Hackman RC, Nadeem SQ, Hiemenz JW, White MH, Flowers ME, et al. Nocardiosis after bone marrow transplantation: a retrospective study. *Clin Infect Dis* 1997;**24**:1154–60.

27. Aisenberg GM, Jacobson K, Chemaly RF, Rolston KV, Raad II, Safdar A. Extrapulmonary tuberculosis active infection misdiagnosed as cancer: Mycobacterium tuberculosis disease in patients at a Comprehensive Cancer Center (2001–2005). *Cancer* 2005;**104**:2882–7.

28. Mishra P, Kumar R, Mahapatra M, Sharma S, Dixit A, Chaterjee T, et al. Tuberculosis in acute leukemia: a clinico-hematological profile. *Hematology* 2006;**11**:335–40.

29. Ip MS, Yuen KY, Woo PC, Luk WK, Tsang KW, Lam WK, et al. Risk factors for pulmonary tuberculosis in bone marrow transplant recipients. *Am J Respir Crit Care Med* 1998;**158**:1173–7.

30. Yuen KY, Woo PC. Tuberculosis in blood and marrow transplant recipients. *Hematol Oncol* 2002;**20**:51–62.

31. Chen Y, Sheng H, Lai C, Liao H, Huang T, Tsay W, et al. Mycobacterial infections in adult patients with hematological malignancy. *Eur J Clin Microbiol Infect Dis* 2012;**31**:1059–66.

32. Nontuberculous mycobacteria. *Am J Transplant* 2004;**4**(Suppl. 10):42–46.

33. Griffith DE, Aksamit T, Brown-Elliott BA, Catanzaro A, Daley C, Gordin F, et al. An official ATS/IDSA statement: diagnosis, treatment, and prevention of nontuberculous mycobacterial diseases. *Am J Respir Crit Care Med* 2007;**175**:367–416.

34. Fukuda T, Boeckh M, Carter RA, Sandmaier BM, Maris MB, Maloney DG, et al. Risks and outcomes of invasive fungal infections in recipients of allogeneic hematopoietic stem cell transplants after nonmyeloablative conditioning. *Blood* 2003;**102**:827–33.

35. Nivoix Y, Velten M, Letscher-Bru V, Moghaddam A, Natarajan-Amé S, Fohrer C, et al. Factors associated with overall and attributable mortality in invasive aspergillosis. *Clin Infect Dis* 2008;**47**:1176–84.

36. Iwen PC, Rupp ME, Langnas AN, Reed EC, Hinrichs SH. Invasive pulmonary aspergillosis due to Aspergillus terreus: 12-year experience and review of the literature. *Clin Infect Dis* 1998;**26**:1092–7.

37. Young RC, Bennett JE, Vogel CL, Carbone PP, DeVita VT. Aspergillosis. The spectrum of the disease in 98 patients. *Medicine (Baltimore)* 1970;**49**:147–73.

38. Prystowsky SD, Vogelstein B, Ettinger DS, Merz WG, Kaizer H, Sulica VI, et al. Invasive aspergillosis. *N Engl J Med* 1976;**295**:655–8.

39. Allo MD, Miller J, Townsend T, Tan C. Primary cutaneous aspergillosis associated with Hickman intravenous catheters. *N Engl J Med* 1987;**317**:1105–8.

40. Albelda SM, Talbot GH, Gerson SL, Miller WT, Cassileth PA. Pulmonary cavitation and massive hemoptysis in invasive pulmonary aspergillosis. Influence of bone marrow recovery in patients with acute leukemia. *Am Rev Respir Dis* 1985;**131**:115–20.

41. De Pauw B, Walsh TJ, Donnelly JP, Stevens DA, Edwards JE, Calandra T, et al. Revised Definitions of Invasive Fungal Disease from the European Organization for Research and Treatment of Cancer/Invasive

Fungal Infections Cooperative Group and the National Institute of Allergy and Infectious Diseases Mycoses Study Group (EORTC/MSG) Consensus Group. *Clin Infect Dis* 2008;**46**:1813–21.

42. Ruhnke M, Bohme A, Buchheidt D, Donhuijsen K, Einsele H, Enzensberger R, et al. Diagnosis of invasive fungal infections in hematology and oncology–guidelines of the Infectious Diseases Working Party (AGIHO) of the German Society of Hematology and Oncology (DGHO). *Ann Hematol* 2003;**82**:S141–8.

43. Chamilos G, Luna M, Lewis RE, Bodey GP, Chemaly R, Tarrand JJ, et al. Invasive fungal infections in patients with hematologic malignancies in a tertiary care cancer center: an autopsy study over a 15-year period (1989-2003). *Haematologica* 2006;**91**:986–9.

44. Caillot D, Casasnovas O, Bernard A, Couaillier JF, Durand C, Cuisenier B, et al. Improved management of invasive pulmonary aspergillosis in neutropenic patients using early thoracic computed tomographic scan and surgery. *J Clin Oncol* 1997;**15**:139–47.

45. Caillot D, Mannone L, Cuisenier B, Couaillier JF. Role of early diagnosis and aggressive surgery in the management of invasive pulmonary aspergillosis in neutropenic patients. *Clin Microbiol Infect* 2001;**7**:54–61.

46. Greene RE, Schlamm HT, Oestmann JW, Stark P, Durand C, Lortholary O, et al. Imaging findings in acute invasive pulmonary aspergillosis: clinical significance of the halo sign. *Clin Infect Dis* 2007;**44**:373–9.

47. Soubani AO, Qureshi MA. Invasive pulmonary aspergillosis following bone marrow transplantation: risk factors and diagnostic aspect. *Haematologia (Budap)* 2002;**32**:427–37.

48. Horvath JA, Dummer S. The use of respiratory-tract cultures in the diagnosis of invasive pulmonary aspergillosis. *Am J Med* 1996;**100**:171–8.

49. Yu VL, Muder RR, Poorsattar A. Significance of isolation of Aspergillus from the respiratory tract in diagnosis of invasive pulmonary aspergillosis. Results from a three-year prospective study. *Am J Med* 1986;**81**:249–54.

50. Tang CM, Cohen J. Diagnosing fungal infections in immunocompromised hosts. *J Clin Pathol* 1992;**45**:1–5.

51. Maschmeyer G, Beinert T, Buchheidt D, Einsele H, Heussel CP, Kiehl M, et al. Diagnosis and antimicrobial therapy of pulmonary infiltrates in febrile neutropenic patients–guidelines of the Infectious Diseases Working Party (AGIHO) of the German Society of Hematology and Oncology (DGHO). *Ann Hematol* 2003;**82**:S118–26.

52. Walsh TJ, Groll AH, Overview:. non-fumigatus species of Aspergillus: perspectives on emerging pathogens in immunocompromised hosts. *Curr Opin Investig Drugs* 2001;**2**:1366–7.

53. Hachem RY, Kontoyiannis DP, Boktour MR, Afif C, Cooksley C, Bodey GP, et al. Aspergillus terreus: an emerging Amphotericin B-resistant opportunistic mold in patients with hematologic malignancies. *Cancer* 2004;**101**:1594–600.

54. Marr KA, Balajee SA, McLaughlin L, Tabouret M, Bentsen C, Walsh TJ. Detection of galactomannan antigenemia by enzyme immunoassay for the diagnosis of invasive aspergillosis: variables that affect performance. *J Infect Dis* 2004;**190**:641–9.

55. Pfeiffer CD, Fine JP, Safdar N. Diagnosis of invasive aspergillosis using a galactomannan assay: a meta-analysis. *Clin Infect Dis* 2006;**42**:1417–727.

56. Penack O, Rempf P, Graf B, Blau IW, Thiel E. Aspergillus galactomannan testing in patients with long-term neutropenia: implications for clinical management. *Ann Oncol* 2008;**19**:984–9.

57. Musher B, Fredricks D, Leisenring W, Balajee SA, Smith C, Marr KA. Aspergillus galactomannan enzyme immunoassay and quantitative PCR for diagnosis of invasive aspergillosis with bronchoalveolar lavage fluid. *J Clin Microbiol* 2004;**42**:5517–22.

58. Racil Z, Kocmanova I, Toskova M, Buresova L, Weinbergerova B, Lengerova M, et al. Galactomannan detection in bronchoalveolar lavage fluid for the diagnosis of invasive aspergillosis in patients with hematological diseases—the role of factors affecting assay performance. *Int J Infectious Diseases* 2011;**15**:e874–81.

59. Hizel K, Kokturk N, Kalkanci A, Ozturk C, Kustimur S, Tufan M. Polymerase chain reaction in the diagnosis of invasive aspergillosis. *Mycoses* 2004;**47**:338–42.

60. Reinwald M, Spiess B, Heinz WJ, Vehreschild JJ, Lass-Flörl C, Kiehl M, et al. Diagnosing pulmonary aspergillosis in patients with hematological malignancies: a multicenter prospective evaluation of an

Aspergillus PCR assay and a galactomannan ELISA in bronchoalveolar lavage samples. *Eur J Haematol* 2012;**89**:120–7.

61. Lu Y, Chen YQ, Guo YL, Qin SM, Wu C, Wang K. Diagnosis of invasive fungal disease using serum (1→3)-β -D-glucan: a bivariate meta-analysis. *Intern Med* 2011;**50**:2783–91.

62. Herbrecht R, Denning DW, Patterson TF, Bennett JE, Greene RE, Oestmann JW, et al. Voriconazole versus Amphotericin B for primary therapy of invasive aspergillosis. *N Engl J Med* 2002;**34**:408–15.

63. Segal BH, Barnhart LA, Anderson VL, Walsh TJ, Malech HL, Holland SM. Posaconazole as salvage therapy in patients with chronic granulomatous disease and invasive filamentous fungal infection. *Clin Infect Dis* 2005;**40**:1684–8.

64. Maertens JA, Raad II, Marr KA, Patterson TF, Kontoyiannis DP, Cornely OA, et al. Isavuconazole versus Voriconazole for primary treatment of invasive mould disease caused by Aspergillus and other filamentous fungi (SECURE): a phase 3, randomised-controlled, non-inferiority trial. *Lancet* 2015;**387**:760–9.

65. Spanakis EK, Aperis G, Mylonakis E. New agents for the treatment of fungal infections: clinical efficacy and gaps in coverage. *Clin Infect Dis* 2006;**43**:1060–8.

66. Walsh TJ, Anaissie EJ, Denning DW, Herbrecht R, Kontoyiannis DP, Marr KA, et al. Treatment of aspergillosis: clinical practice guidelines of the Infectious Diseases Society of America. *Clin Infect Dis* 2008;**46**:327–60.

67. Rowe JM, Andersen JW, Mazza JJ, Bennett JM, Paietta E, Hayes FA, et al. A randomized placebo-controlled phase III study of granulocyte-macrophage colony-stimulating factor in adult patients (> 55 to 70 years of age) with acute myelogenous leukemia: a study of the Eastern Cooperative Oncology Group (E1490). *Blood* 1995;**86**:457–62.

68. Mattiuzzi GN, Kantarjian H, O'Brien S, Kontoyiannis DP, Giles F, Zhou X, et al. Intravenous itraconazole for prophylaxis of systemic fungal infections in patients with acute myelogenous leukemia and high-risk myelodysplastic syndrome undergoing induction chemotherapy. *Cancer* 2004;**100**:568–73.

69. Costa SF, Alexander BD. Non-Aspergillus fungal pneumonia in transplant recipients. *Clin Chest Med* 2005;**26**:675–90.

70. Marom EM, Holmes AM, Bruzzi JF, Truong MT, O'Sullivan PJ, Kontoyiannis DP. Imaging of pulmonary fusariosis in patients with hematologic malignancies. *AJR Am J Roentgenol* 2008;**190**:1605–9.

71. Husain S, Munoz P, Forrest G, Alexander BD, Somani J, Brennan K, et al. Infections due to Scedosporium apiospermum and Scedosporium prolificans in transplant recipients: clinical characteristics and impact of antifungal agent therapy on outcome. *Clin Infect Dis* 2005;**40**:89–99.

72. Maertens J, Demuynck H, Verbeken EK, Zachée P, Verhoef GE, Vandenberghe P, et al. Mucormycosis in allogeneic bone marrow transplant recipients: report of five cases and review of the role of iron overload in the pathogenesis. *Bone Marrow Transplant* 1999;**24**:307–12.

73. Saag MS, Graybill RJ, Larsen RA, Pappas PG, Perfect JR, Powderly WG, et al. Practice guidelines for the management of cryptococcal disease. Infectious Diseases Society of America. *Clin Infect Dis* 2000;**30**:710–8.

74. Goodrich JM, Reed EC, Mori M, Fisher LD, Skerrett S, Dandliker PS, et al. Clinical features and analysis of risk factors for invasive candidal infection after marrow transplantation. *J Infect Dis* 1991;**164**:731–40.

75. Kume H, Yamazaki T, Abe M, Tanuma H, Okudaira M, Okayasu I. Increase in aspergillosis and severe mycotic infection in patients with leukemia and MDS: comparison of the data from the Annual of the Pathological Autopsy Cases in Japan in 1989. 1993 and 1997. *Pathol Int* 2003;**53**:744–50.

76. Hughes WT, Feldman S, Aur RJ, Verzosa MS, Hustu HO, Simone JV. Intensity of immunosuppressive therapy and the incidence of Pneumocystis carinii pneumonitis. *Cancer* 1975;**36**:2004–9.

77. Varthalitis I, Aoun M, Daneau D, Meunier F. Pneumocystis carinii pneumonia in patients with cancer. An increasing incidence. *Cancer* 1993;**71**:481–5.

78. Arend SM, Kroon FP, van't Wout JW. Pneumocystis carinii pneumonia in patients without AIDS, 1980 through 1993. An analysis of 78 cases. *Arch Intern Med* 1995;**155**:2436–41.

79. Pagano L, Fianchi L, Mele L, Girmenia C, Offidani M, Ricci P, et al. Pneumocystis carinii pneumonia in patients with malignant haematological diseases: 10 years' experience of infection in GIMEMA centres. *Br J Haematol* 2002;**117**:379–86.

80. Naimey GL, Wuerker RB. Comparison of histologic stains in the diagnosis of Pneumocystis carinii. *Acta Cytol* 1995;**39**:1124–7.

81. Kovacs JA, Ng VL, Masur H, Leoung G, Hadley WK, Evans G, et al. Diagnosis of Pneumocystis carinii pneumonia: improved detection in sputum with use of monoclonal antibodies. *N Engl J Med* 1988;**318**:589–93.

82. Bollee G, Sarfati C, Thiery G, Bergeron A, de Miranda S, Menotti J, et al. Clinical picture of Pneumocystis jiroveci pneumonia in cancer patients. *Chest* 2007;**132**:1305–10.

83. Glenn TJ, Blair JE, Adams RH. Coccidioidomycosis in hematopoietic stem cell transplant recipients. *Med Mycol* 2005;**43**:705–10.

84. Torres HA, Aguilera E, Safdar A, Rohatgi N, Raad II, Sepulveda C, et al. Fatal cytomegalovirus pneumonia in patients with haematological malignancies: an autopsy-based case-control study. *Clin Microbiol Infect* 2008;**14**:1160–6.

85. Ljungman P, Aschan J, Lewensohn-Fuchs I, Carlens S, Larsson K, Lönnqvist B, et al. Results of different strategies for reducing cytomegalovirus-associated mortality in allogeneic stem cell transplant recipients. *Transplantation* 1998;**66**:1330–4.

86. Franquet T, Lee KS, Muller NL. Thin-section CT findings in 32 immunocompromised patients with cytomegalovirus pneumonia who do not have AIDS. *AJR Am J Roentgenol* 2003;**181**:1059–63.

87. Tamm M, Traenkle P, Grilli B, Solèr M, Bolliger CT, Dalquen P, et al. Pulmonary cytomegalovirus infection in immunocompromised patients. *Chest* 2001;**119**:838–43.

88. Humar A, Lipton J, Welsh S, Moussa G, Messner H, Mazzulli T. A randomised trial comparing cytomegalovirus antigenemia assay vs screening bronchoscopy for the early detection and prevention of disease in allogeneic bone marrow and peripheral blood stem cell transplant recipients. *Bone Marrow Transplant* 2001;**28**:485–90.

89. Zaia JA. Prevention and treatment of cytomegalovirus pneumonia in transplant recipients. *Clin Infect Dis* 1993;**17**:S392–9.

90. Chemaly RF, Ghosh S, Bodey GP, Rohatgi N, Safdar A, Keating MJ, et al. Respiratory viral infections in adults with hematologic malignancies and human stem cell transplantation recipients: a retrospective study at a major cancer center. *Medicine (Baltimore)* 2006;**85**:278–87.

91. Whimbey E, Englund JA, Couch RB. Community respiratory virus infections in immunocompromised patients with cancer. *Am J Med* 1997;**102**:10–8.

METALLO-BETA-LACTAMASE PRODUCER *PSEUDOMONAS AERUGINOSA*: AN OPPORTUNISTIC PATHOGEN IN LUNGS

S.U. Picoli*, A.L.S. Gonçalves**

*Universidade Feevale, Novo Hamburgo, Rio Grande do Sul, Brazil; **MSc at Universidade Federal do Rio Grande do Sul, Porto Alegre, Rio Grande do Sul, Brazil

1 INTRODUCTION

Pseudomonas aeruginosa is an opportunistic human pathogen that is frequently isolated in hospital infections, especially in nosocomial pneumonia.[1] Species of this genus are characterized by their inability to ferment sugars, although most of the strains oxidatively degrade them. However, they are nutritionally versatile bacteria that are able to use a wide variety of substrates, whose optimum growth temperature varies between 30 and 37°C, although they multiply in lower temperatures.[2] Furthermore, *P. aeruginosa* has phenotypic and biochemical characteristics that allow its easy laboratorial identification; for instance, the production of water-soluble pigments named pyocyanin (white to blue-green fluorescence under ultraviolet light) and pyoverdine, in addition to producing a characteristic grape-like odor.[3]

Given that it is considered to be an opportunistic pathogen, it is generally not a high risk for healthy individuals, although it has been reported as a recurrent community-acquired pneumonia in a young patient who had no risk factors.[4] Adversely, due to the wide distribution of these bacteria in the environment, it may be problematic, especially in hospitals, since it remains on artificial respiratory systems used in Intensive Care Unit (ICU) patients, as well as in aqueous solutions.[2] Besides these features, *P. aeruginosa* is capable of producing several virulence factors, which are strongly related to this bacteria potential to cause disease.[5] This ability is mediated by the presence of genes that express different virulence factors, including *exoS* (exoenzyme S), *exoU* (exoenzyme U), *toxA* (exotoxin A), *LasB* (elastase LasB), *nan1* (neuraminidase), *plcN* (nonhemolytic phospholipase C), *plcH* (hemolytic phospholipase C), *pilB* (type IV fimbrial biogenesis protein pilB), *algD* (alginato), among others. The latter gene is responsible for the production of a mucopolysaccharide capsule that gives greater adherence of the bacteria to the respiratory epithelial cells, forming biofilms, which hamper the action of antimicrobial agents and immune system. Regarding the pathogenic potential of *P. aeruginosa,* another important characteristic is the simultaneous presence of several of this described virulence genes and enzymes, such as, the metallo-beta-lactamases (MBL). This association has a significant correlation between the virulence factors and antibiotic resistance patterns in *P. aeruginosa* multidrug resistant (MDR-PA) isolates.[6] Indeed, plasmids are one of the most important virulence mechanisms in multidrug resistant bacteria. They transport antibiotic resistance determinants contained in transposons (mobile genetic elements), which results in constant modifications to DNA.[7]

Nevertheless, patients hospitalized in ICUs with severe underlying diseases, previous antimicrobial treatment, undergoing invasive procedures, as well as extended periods of hospitalization, are of particular risk of acquiring nosocomial infections by *P. aeruginosa*. These conditions contribute to the association of the bacteria with high rates of morbidity and mortality in critically ill patients in ICUs, particularly in those with mechanical ventilator associated pneumonia.[8] In addition, exposure to several antimicrobial agents in the hospital environment may create conditions for resistance selection among the host microbiota or for pathogen transmission. Consequently, transfer of patients between admission units in the hospital may provide the introduction of new and often highly resistant clones into the ICU.[9]

2 *P. AERUGINOSA* RESISTANT TO BETA-LACTAM ANTIBIOTICS

Usually, both intrinsic and acquired wide resistance to numerous antibiotics used in medical practice contribute to *P. aeruginosa*'s pathogenicity, since bacteria from this genus may demonstrate high rates of resistance to most of the antimicrobial agents normally used in hospital environment. Therapeutic options include aminoglycosides, fluoroquinolones, broad-spectrum penicillins, monobactams, ceftazidime, fourth generation cephalosporins and carbapenems. One of the recommendations for antibiotic therapy in treatment of infections caused by *P. aeruginosa* is the use of association including a beta-lactam and another antibiotic, usually an aminoglycoside or quinolone, being the beta-lactams the basis of the chosen association.[10]

However, *P. aeruginosa* resistance to beta-lactam antibiotics may occur due to enzymatic mechanisms, for instance, the beta-lactamases production via the activation of outer membrane impermeability systems, which are related to porins[11] and also via antimicrobial agents expulsion of the bacteria cell, called the efflux pump.[10,11] Eventually, more than one mechanism may be present, resulting in a wide resistant pattern to all beta-lactams, being the beta-lactamases production the most studied mechanism over the last few decades.

MBL is an enzyme classified in the 3a group of Bush–Jacoby, in which all the biggest families of MBLs codified by plasmids are included, such as, the IMP (**IM**i**P**enemase) and the VIM (**V**erona **I**ntegron-encoded **M**etallo-beta-lactamase). They have high hydrolytic potential on penicillins, cephalosporins, carbapenems, but not on monobactams. Due to zinc metal (Zn^{2+}) present on the enzyme active site, it can be inhibit by metal chelating agents, such as, EDTA (ethylenediaminetetraacetic acid), but not by beta-lactamases inhibitors clinical available, like clavulanic acid or tazobactam.[12]

The emergence of MBL producing microorganisms has great clinical importance, as they show resistance to many antibiotics used, with sensitivity only to colistin and polymyxin B, limiting treatment options. Therefore, an appropriate screening system should be established, especially for all *P. aeruginosa* imipenem-resistant, enabling early detection of carbapenemase strains, and hence, preventing the spread of these multidrug resistant strains.[13]

MBLs are categorized in types according to its amino acid's sequence. More than 10 MBL variants were described and most of them have been already found in *P. aeruginosa:* IMP (**IMiP**enemase), VIM [**V**erona **I**ntegron-encoded **M**etallo-beta-lactamase, SPM (**Sã**o **P**aulo **M**etallo-beta-lactamase)], GIM (**G**ermany **IM**ipenemase) NDM (**N**ew **D**elhi **M**etallo-beta-lactamase), FIM (**F**lorence **IM**ipenemase).[14–18] The distribution of these enzymes is worldwide and they have been described in different countries of America, Europe, Asia, Africa and Oceania, with mean of variation between 10% and 50%.[19] However, its constant local monitoring is recommended, since epidemiology of these carbapenemase not only varies among different countries but also within the same country.

Thus, this monitoring becomes more relevant knowing that metallo-beta-lactamase producing *P. aeruginosa* (MBL-PA) isolates are associated to infection's rapid progression and evolution to death. In addition, MBL-PA are more resistant than non-MBL-PA strains and high clonal dissemination of MBL-PA strains suggest the cross transmission as an important mechanism of spreading.[20] Considering that resistance rates to polymyxin B tend to increase after generalized use of this drug, strict infection control measures should be urgently adopted, otherwise we may be facing untreatable *P. aeruginosa* nosocomial infections.[21]

Not all patients on mechanical ventilation develop lung tissue infection, since it depends on other factors such as host defenses and the *P. aeruginosa* virulence colonizing the lower respiratory tract. Nevertheless, the ventilator associated pneumonia (VAP) is an important respiratory infection seen in ICU patients and it occurs in 7–20% of those who were mechanically ventilated for periods greater than or equal to 48 h.[22] VAP is initiated by an inflammation of the lung parenchyma caused by aspirated microorganisms after mechanical ventilation and results in increased hospitalization time, higher treatment cost and higher mortality.[23] Additionally, when VAP is caused by *P. aeruginosa* the infection becomes invasive with rapid progression. It is characterized by acute leukocytosis and fever, which requires increased ventilator support besides raising significantly the mortality rates.[24,25] Production of MBL might be associated to higher mortality due to inadequacy of antimicrobial therapy in infections caused by MBL-PA, suggesting that institutions with high prevalence of MBL should review their therapeutic approaches.[24]

Commonly, nosocomial *P. aeruginosa* isolates obtained from respiratory material as bronchoalveolar lavage or endotracheal aspirate show antimicrobial resistance mediated mainly by MBLs. Additionally, biofilm formation on endotracheal tube, with subsequent embolization to distal respiratory tract, might be important in the pathogenesis of VAP.[26,27]

Microbiological criteria for VAP are quantitative and they vary according to the type of clinical material under study, where in findings of $\geq 10^6$ colony forming units (CFU)/mL in endotracheal aspirate and $\geq 10^4$ CFU/mL in bronchoalveolar lavage are considered.[25] Also, according to the American Toracic Society, all patients with suspected VAP should have blood cultures collected and a positive result can indicate the presence of pneumonia or extrapulmonary infection.[28] However, hospital-acquired pneumonia index seen in nonventilated patients is also high (>70%) and the etiological agents are similar to those found in patients with VAP, including *P. aeruginosa* multidrug resistant (MDR-PA).[8,29] Higher survival rate is expected in these patients when polymyxin is used by inhalation in addition to antimicrobial therapy,[30] noting that a study demonstrated that only this antibiotic showed overall coverage equal to or superior than 90% in patients with VAP.[31]

The result of multivariate analysis, conducted by Zavascki et al.,[32] showed that recent use of a beta-lactam antibiotic (OR 3.21; 95% CI 1.74–5.93) or a quinolone (OR 3:50; 95% CI 1:46–8:37) was a significant risk factor for infection caused by MBL-PA, even though patient to patient transmission has a major role in the dissemination of isolates. Lung was shown to be the most frequent nosocomial infection site in the patients who were evaluated (50.3%).

Regarding the chronic infections that affect lungs, cystic fibrosis (CF) is presented as the most common among Caucasians. This disease is characterized by airway obstruction due to the production of thick and viscous secretions, which increases the individual's susceptibility to pulmonary infections, being *P. aeruginosa* one of the main bacteria involved in these infections.[33,34]

Conversely, it is not entirely clear why infections in CF are chronic, noninvasive and highly resistant to eradication. Production of biofilms has been regarded as the main mechanism, but a study showed that *P. aeruginosa* might grow in bacterial aggregates, which are greatly resistant to host defenses and to antibiotics independently of the bacterial biofilm production.[35] Moreover, MBLs genes (IMP and SPM) were found concomitantly to biofilm production in *P. aeruginosa* CF isolates, suggesting that overlapping antibiotic resistance mechanisms represents a therapeutic challenge even greater than the one faced in isolates that only produce biofilms.[36]

However, the majority of CF patients acquire chronic *P. aeruginosa* infections in early life, which is responsible for much of the morbidity and mortality in individuals with the disease. Furthermore, hypermutable strains of *P. aeruginosa* were isolated from an elevated percentage of CF patients. *P. aeruginosa* follows a characteristic evolution pattern defined by selection of hypermutable strains and increase of antibiotic resistance as a consequence of long term infection in CF-affected lung. Besides affecting CF patients, *P. aeruginosa* is also recognized as a relevant pathogen in chronic obstructive pulmonary disease (COPD). These individual often have acute exacerbations, which cause great impact on these patients' quality of life. These exacerbations are the main cause of mortality among patients affected by this disease.[37]

3 DETECTION TESTS OF MBLS IN *P. AERUGINOSA*

P. aeruginosa isolates with phenotype of nonsusceptibility to antipseudomonal broad-spectrum antibiotics such as carbapenems (imipenem and meropenem) may be carriers of some MBL gene. In this context, it is essential that the microbiology laboratory promptly detect the metallo-enzyme, enabling the appropriateness of antimicrobial therapy and contributing to making infection control measures.

After observing the carbapenem-resistance phenotype, either by agar diffusion test or by automation system, it is important to confirm the MBL presence. Different methods can be used, ranging from phenotypic tests that are easy to perform and inexpensive, to even the most sensitive and specific ones like the molecular tests (Polimerase Chain Reaction-PCR, DNA probes, cloning, sequencing)[38] and MALDI-TOF MS (Matrix-Assisted Laser Desorption-Ionization Time-Of-Flight Mass Spectrometry).[39] Unfortunately, the majority of these last resources are not available in most of the clinical laboratories.

The following tests may be produced in-house in clinical laboratories, and choice of any of them for routine investigation of MBL should consider the local MBLs epidemiology. Nevertheless, these tests are prone to error due to the great diversity of the enzymes types present in this group (MBLs in *P. aeruginosa*: VIM, IMP, SPM, NDM, GIM, and FIM).

Other factors that may contribute to disagreeing results in MBL detection are: trustworthiness of the inputs used, the concentration of enzyme inhibitor (EDTA) on the carbapenem disk and also, operational errors inherent to test preparation.[40] One of the major problems faced on phenotypic MBL screening is the lack of standardization of a single test.

3.1 COMBINED DISK (CD) TEST

Variants of this test (Table 10.1) have in common the inoculum preparation equivalent to 0.5 McFarland of the microorganism under test. The inoculum is seeded on Mueller Hinton agar surface (MHA); two disks of the same carbapenem are placed on the agar and enzyme inhibitor (EDTA) is added to one of them (Fig. 10.1). After suitable incubation, typically for 24 h at 35°C, the inhibition diameters difference between the carbapenem disk with EDTA and the carbapenem alone are measured, which leads to a positive or negative result for MBL.

3.2 CARBAPENEM INACTIVATION METHOD

The Carbapenem Inactivation Method (CIM) is a new phenotypic test of low cost, which aims to detect the activity of different carbapenemases in Gram-negative bacilli, including MBLs of IMP and VIM types in *P. aeruginosa*.

The test consists in suspending a dense inoculum of the bacteria (a full 10 μL inoculation loop) in 400 μL of water. A meropenem (MEM) 10 μg disk is added in this suspension and it is incubated for at

Table 10.1 Phenotypic Tests for Metallo-Beta-Lactamase (MBL) Detection in *Pseudomonas aeruginosa* Isolates Resistant to Carbapenems

	Song et al.[41]	Sheikh et al.[40]	Qu et al.[42]	Pitout et al.[43]
2 carbapanem disks	IMP[a] 10 μg	DOR[b] 10 μg	IMP[a] 10 μg	MEM[c] 10 μg
EDTA[e]	10 μL	10 μL 0.5 M[d]	10 μL 0.5 M[d]	930 μg
aqueous solution	(30 mg/mL)	(750 μg)	(750 μg)	
MBL[f]	IMP[a]/EDTA[e] - IMP[a]	DOR[b]/EDTA[e] - DOR[b]	IMP[a]/EDTA[e] - IMP[a]	MEM[c]/EDTA[e] - MEM[c]
presence	≥5 mm	≥7 mm	≥6 mm	≥7 mm
Sensibility	100%	100%	100%	100%
Specificity	100%	64%	100%	97%
Types of MBLs[f]	IMP[g]-6	IMP[g]	IMP[g]-1	IMP[g]
detected	IMP[g]-26	VIM[h]	IMP[g]-9	VIM[h]
	VIM[h]-2		VIM[h]-2	

[a]*IMP: Imipenem.*
[b]*DOR: Doripenem.*
[c]*MEM: Meropenem.*
[d]*M: Molar.*
[e]*EDTA: Ethylenediaminetetraacetic acid.*
[f]*MBL: Metallo-beta-lactamase.*
[g]*IMP: Imipenemase.*
[h]*VIM: Verona Integron-encoded Metallo-beta-lactamase.*

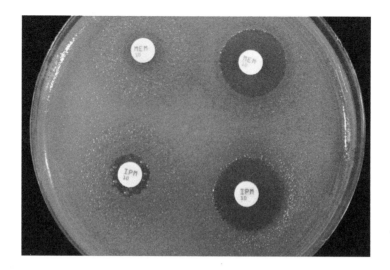

FIGURE 10.1

Phenotypic Detection of Metallo-Beta-Lactamase in *Pseudomonas aeruginosa* by the Combined Disk Test (CD Test)
Above/Upper: two meropenem disks (MEM 10 μg), plus 930 μg EDTA[43] on the right disk; MEM/EDTA—
MEM = 10.2 mm (MBL positive if ≥7 mm).
Bellow/Bottom: two imipenem disks (IPM 10 μg), plus 750 μg EDTA[42] on the right disk; IPM/EDTA—
IPM = 8.4 mm (MBL positive if ≥6 mm).

least 2 h at 35°C. The disk is removed from the suspension and it is placed on Mueller Hinton agar plate
that was previously inoculated with *Escherichia coli* strain (ATCC 25922) adjusted to 0.5 McFarland.
The plate is again incubated at 35°C for at least 6 h, although overnight incubation would be ideal. For
interpretation, it is considered carbapenemase producing bacteria in the absence of any inhibition zone
around the MEM disk because bacteria present in the suspension inactivates this antibiotic.

CIM results had high concordance (98.8%) when compared to the results of molecular tests (PCR)
to detect specific genes involved in the resistance. Similarly, positive and negative predictive values
were also high, determined at 96.3 and 99.4%, respectively.[44]

Alternatively, some quick tests for the detection of carbapenemase production may be performed.
Among them is the Carba NP (homemade) that is based on the detection of beta-lactam ring hydrolysis
of imipenem.[45] This assay has been extensively validated for carbapenemase detection, both in *Enterobacteriaceae* and in *Pseudomonas* spp.[46–48] Reading is performed within 2 h, showing high sensitivity
(94.4 %) and specificity (100%).[47] In the commercial version of this test, named Rapidec Carba NP, the
class B carbapenemase producing samples, the MBL, show a positive colorimetric response in less
than 10 min.

3.3 NONPHENOTYPIC TESTS

MALDI-TOF MS is one of the most promising and attractive carbapenemase detection techniques in
clinical isolates. It has been introduced in the routines of different clinical laboratories because results
can be obtained in shorter time when compared to other methods. Despite requiring a high initial
investment, this technique allows early identification of a resistance mechanism, making it extremely

useful for the initiation of antimicrobial therapy and it assists in controlling the transmission of this bacteria.

MALDI-TOF MS detects the activity of different carbapenemases with high sensitivity (95–99%), including the MBLs, via carbapenem hydrolysis. Briefly, the technique consists in preparing a very dense inoculum (equivalent to 3 McFarland) of the bacteria in a buffer and centrifuge. The pellet is suspended in a buffer containing the carbapenem molecule and it is incubated at 35°C for 4 h. Afterwards, it is centrifuged, a proper matrix (dihydroxybenzoic acid in 50% ethanol) is mixed to the supernatant, and measurement is performed by MALDI-TOF MS. Subsequently, spectra containing peaks representing the carbapenem molecule, its salts or its degradation products are analyzed.[49]

4 CONCLUSIONS AND FUTURE PERSPECTIVES

Respiratory infections associated with opportunistic bacteria such as metallo-beta-lactamase producing *P. aeruginosa* (MBL-PA) progresses rapidly, resulting in high morbi-mortality rates. The overall increase of this resistance has been reported in several countries and it is in evidence among the scientific community. Although several studies have been performed in an effort to establish a standard test for quick research of this powerful resistance mechanism in *P. aeruginosa*, it has not yet been possible to elect a single reliable test. This condition can be attributed to the diversity of circulating MBLs types (VIM, IMP, SPM, GIM, NDM, and FIM), reinforcing the need for local molecular epidemiological studies of these enzymes. It is presumed from prior knowledge of this data that it will be possible to establish a quick test, preferably with low cost, reliable, and feasible to be used in clinical laboratories covered by specific geographical location. However, it is evident that strict infection control measures, including restrictive antibiotic use policies in the hospital environment, isolation of patients with MBL-PA, especially ICU patients, should be applied whenever necessary, aiming to avoid the increase of nosocomial infections associated with MBL-PA.

REFERENCES

1. Linch JP. Hospital-acquired infections: realities of risks and resistance. *Chest* 2001;**119**:373S–84S.
2. Kiska DL, Gilligan PH. *Pseudomonas*. In: Murray PR, Baron EJ, Pfaller MA, Tenover FC, Yolken RH, editors. *Manual of Clinical Microbiology*. 7th ed. Washington: American Society for Microbiology; 1999. p. 517–25.
3. Koneman EW, Allen SD, Janda WM, Schereckenberger PC, Winn WCJ. Bacilos Gram negativos não-fermentadores. *Diagnóstico microbiológico: texto e atlas colorido*. Rio de Janeiro, Brasil: Medsi; 2001. p. 263–319.
4. Fujii A, Seki M, Higashiguchi M, Tachibana I, Kumanogoh A, Tomono K. Community-acquired, hospital-acquired, and healthcare-associated pneumonia caused by *Pseudomonas aeruginosa. Respir Med Case Rep* 2014;**12**:30–3.
5. Pitt TL, Barth AL. *Pseudomonas aeruginosa* and other medically important Pseudomonads. In: Emmerson AM, Hawkey PM, Gillespie SH, editors. *Principles and Practice of Clinical Bacteriology*. United Kingdom: Wiley; 1997. p. 493–517.
6. Sonbol FI, Khalil MAEF, Mohamed AB, Ali SS. Correlation between antibiotic resistance and virulence of *Pseudomonas aeruginosa* clinical isolates. *Turk J Med Sci* 2015;**45**:568–77.

7. Woodford N, Turton JF, Livermore DM. Multiresistant Gram-negative bacteria: the role of high-risk clones in the dissemination of antibiotic resistance. *FEMS Microbiol Rev* 2011;**35**:736–55.
8. American Toracic Society Documents. Guidelines for the management of adults with hospital-acquired, ventilator-associated, and healthcare-associated pneumonia. *Am J Respir Crit Care Med* 2005;**171**:388–416.
9. Streit JM, Jones RN, Sader HS, Fritsche TR. Assessment of pathogen occurrences and resistance profiles among infected patients in the intensive care unit: report from the SENTRY Antimicrobial Surveillance Program (North America, 2001). *Int J Antimicrob Agents* 2004;**24**:111–8.
10. Nordmann P. Mechanism of resistance to betalactam antibiotics in *Pseudomonas aeruginosa*. *Ann Fr Anesth Reanim* 2003;**22**(6):527–30.
11. Livermore DM. Of *Pseudomonas*, porins, pumps and carbapenems. *J Antimicrob Chemother* 2001;**47**:247–50.
12. Bush K, Jacoby GA. Updated functional classification of beta-lactamases. *Antimicrob Agents Chemother* 2010;**54**(3):969–76.
13. EL-Mosallamy WAES, Osman AS, Tabl HAEM, Tabbakh ASM. Phenotypic and genotypic methods for detection of metallo-beta-lactamase (MBL) producing *Pseudomonas aeruginosa*. *Egypt J Med Microbiol* 2015;**24**(3):27–35.
14. Martins AF, Zavascki AP, Gaspareto PB, Barth AL. Dissemination of *Pseudomonas aeruginosa* producing SPM-1-like and IMP-1-like metallo-beta-lactamases in hospitals from southern Brazil. *Infection* 2007;**35**:457–60.
15. Lee K, Lim JB, Yum JH, Yong D, Chong Y, Kim JM, et al. *bla*VIM-2 cassette-containing novel integrons in metallo-beta-lactamase-producing *Pseudomonas aeruginosa* and *Pseudomonas putida* isolates disseminated in a Korean hospital. *Antimicrob Agents Chemother* 2002;**46**:1053–8.
16. Castanheira M, Toleman MA, Jones RN, Schmidt FJ, Walsh TR. Molecular characterization of a beta-lactamase gene *blaGIM*-1 encoding a new subclass of metallo-beta-lactamase. *Antimicrob Agents Chemother* 2004;**48**(12):4654–61.
17. Jovcic B, Lepsanovic Z, Suljagic V, Rackov G, Begovic J, Topisirovic L, et al. Emergence of NDM-1 metallo-beta-lactamase in *Pseudomonas aeruginosa* clinical isolates from Serbia. *Antimicrob Agents Chemother* 2011;**55**:3929–31.
18. Pollini S, Maradei S, Pecile P, Olivo G, Luzzaro F, Docquier JD, et al. FIM-1, a new acquired metallo-beta-lactamase from a *Pseudomonas aeruginosa* clinical isolate from Italy. *Antimicrob Agents Chemother* 2013;**57**:410–6.
19. Hong DJ, Bae IK, Jang I, Jeong SH, Kang H, Lee K. Epidemiology and characteristics of metallo-beta-lactamase-producing *Pseudomonas aeruginosa*. *Infect Chemother* 2015;**47**(2):81–97.
20. Lucena A, Dalla Costa LM, Nogueira KS, Matos AP, Gales AC, Paganini MC, et al. Nosocomial infections with metallo-beta-lactamase-producing *Pseudomonas aeruginosa*: molecular epidemiology, risk factors, clinical features and outcomes. *J Hosp Infect* 2014;**87**(4):234–40.
21. Zavascki AP, Goldani LZ, Gaspareto PB, Gonçalves ALS, Martins AF, Barth AL. High prevalence of metallo-beta-lactamase-mediated resistance challenging antimicrobial therapy against *Pseudomonas aeruginosa* in a Brazilian teaching hospital. *Epidemiol Infect* 2007;**135**:343–5.
22. Xiao H, Ye X, Liu Q, Li L. Antibiotic susceptibility and genotype patterns of *Pseudomonas aeruginosa* from mechanical ventilation-associated pneumonia in intensive care units. *Biomed Rep* 2013;**1**(4):589–93.
23. Alp E, Güven M, Yıldız O, Aygen B, Voss A, Doganay M. Incidence, risk factors and mortality of nosocomial pneumonia in Intensive Care Units: a prospective study. *Ann Clin Microbiol Antimicrob* 2004;**3**:17.
24. Zavascki AP, Barth AL, Fernandes JF, Moro ALD, Gonçalves ALS, Goldani LZ. Reappraisal of *Pseudomonas aeruginosa* hospital-acquired pneumonia mortality in the era of metallo-beta-lactamase-mediated multidrug resistance: a prospective observational study. *Crit Care* 2006;**10**:R114.
25. Fricks-Lima J, Hendrickson CM, Allgaier M, Zhuo H, Wiener-Kronish JP, Lynch SV, et al. Differences in biofilm formation and antimicrobial resistance of *Pseudomonas aeruginosa* isolated from airways of mechanically ventilated patients and cystic fibrosis patients. *Int J Antimicrob Agents* 2011;**37**(4):309–15.

26. Inglis TJ, Millar MR, Jones JG, Robinson DA. Tracheal tube biofilm as a source of bacterial colonization of the lung. *J Clin Microbiol* 1989;**27**:2014–8.

27. Adair CG, Gorman SP, Feron BM, Byers LM, Jones DS, Goldsmith CE, et al. Implications of endotracheal tube biofilm for ventilator-associated pneumonia. *Intens Care Med* 1999;**25**:1072–6.

28. Luna CM, Videla A, Mattera J, Vay C, Famiglietti A, Vujacich P, et al. Blood cultures have limited value in predicting severity of illness and as a diagnostic tool in ventilator-associated pneumonia. *Chest* 1999;**116**:1075–84.

29. Van Eldere J. Multicentre surveillance of *Pseudomonas aeruginosa* susceptibility patterns in nosocomial infections. *J Antimicrob Chemother* 2003;**51**:347–52.

30. Hamer DH. Treatment of nosocomial pneumonia and tracheobronchitis caused by multidrug-resistant *Pseudomonas aeruginosa* with aerosolized colistin. *Am J Respir Crit Care Med* 2000;**162**:328–3230.

31. Bassetti M, Taramasso L, Giacobbe DR, Pelosi P. Management of ventilator-associated pneumonia: epidemiology, diagnosis and antimicrobial therapy. *Expert Rev Anti Infect Ther* 2012;**10**(5):585–96.

32. Zavascki AP, Barth LA, Gaspareto PB, Gonçalves ALS, Moro ALD, Fernandes JF, et al. Risk factors for nosocomial infections due to *Pseudomonas aeruginosa* producing metallo-beta-lactamase in two tertiary-care teaching hospitals. *J Antimicrob Chemother* 2006;**58**(4):882–5.

33. Mogayzel PT, Naureckas Jr ET, Robinson KA, Mueller G, Hadjiliadis D, Hoag JB, et al. Cystic fibrosis pulmonary guidelines chronic medications for maintenance of lung health. *Am J Respir Crit Care Med* 2013;**187**(7):680–9.

34. Hauser AR, Jain M, Bar-Meir M, McColley SA. Clinical significance of microbial infection and adaptation in cystic fibrosis. *Clin Microbiol Rev* 2011;**24**(1):29–70.

35. Staudinger BJ, Muller JF, Halldórsson S, Boles B, Angermeyer A, Nguyen D, et al. Conditions associated with the cystic fibrosis defect promote chronic *Pseudomonas aeruginosa* infection. *Am J Respir Crit Care Med* 2014;**189**(7):812–24.

36. Perez LRR, Antunes ALS, Freitas ALP, Barth AL. When the resistance gets clingy: *Pseudomonas aeruginosa* harboring metallo-beta-lactamase gene shows high ability to produce biofilm. *Eur J Clin Microbiol Infect Dis* 2012;**31**:711–4.

37. Martínez-Solano L, Macia MD, Fajardo A, Oliver A, Martinez JL. Chronic *Pseudomonas aeruginosa* infection in chronic obstructive pulmonary disease. *Clin Infect Dis* 2008;**47**(12):1526–33.

38. Wirth FW, Picoli SU, Cantarelli VV, Gonçalves ALS, Brust FR, Santos LMO, et al. Metallo-beta-lactamase-producing *Pseudomonas aeruginosa* in two hospitals from southern Brazil. *Braz J Infect Dis* 2009;**13**(3):170–2.

39. Chong PM, McCorrister SJ, Unger MS, Boyd DA, Mulvey MR, Westmacott GR. MALDI-TOF MS detection of carbapenemase activity in clinical isolates of Enterobacteriaceae spp., *Pseudomonas aeruginosa* and *Acinetobacter baumannii* compared against the Carba-NP assay. *J Microbiol Methods* 2015;**111**:21–3.

40. Sheikh AF, Rostami S, Jolodar A, Tabatabaiefar MA, Khorvash F, Saki A, et al. Detection of metallo-beta-lactamases among carbapenem-resistant *Pseudomonas aeruginosa*. *Jundishapur J Microbiol* 2014;**7**(11):e12289.

41. Song W, Hong SG, Yong D, Jeong SH, Kim HS, Kim H-S, et al. Combined use of the modified Hodge test and carbapenemase inhibition test for detection of carbapenemase producing *Enterobacteriaceae* and metallo-beta-lactamase producing *Pseudomonas* spp. *Ann Lab Med* 2015;**35**(2):212–9.

42. Qu T, Zhang J, Wang J, Tao J, Yu Y, Chen Y, et al. Evaluation of phenotypic tests for detection of metallo-beta-lactamase producing *Pseudomonas aeruginosa* strains in China. *J Clin Microbiol* 2009;**47**(4):1136–42.

43. Pitout JDD, Gregson DB, Poirel L, McClure J-A, Le P, Church DL. Detection of *Pseudomonas aeruginosa* producing metallo-beta-lactamases in a large centralized laboratory. *J Clin Microbiol* 2005;**43**(7):3129–35.

44. Van der Zwaluw K, de Haan A, Pluister GN, Bootsma HJ, de Neeling AJ, Schouls LM. The carbapenem inactivation method (CIM), a simple and low-cost alternative for the Carba NP test to assess phenotypic carbapenemase activity in gram negative rods. *PLoS ONE* 2015;**10**(3):e0123690.

45. Poirel L, Nordmann P. Rapidec Carba NP test for rapid detection of carbapenemase producers. *J Clin Microbiol* 2015;**53**(9):3003–8.
46. Nordmann P, Poirel L, Dortet L. Rapid detection of carbapenamase-producing *Enterobacteriaceae. Emerg Infect Dis* 2012;**18**:1503–7.
47. Dordet L, Poirel L, Nordmann P. Rapid detection of carbapenemase producing *Pseudomonas* spp. *J Clin Microbiol* 2012;**50**(11):3773–6.
48. Tai AS, Sidjabat HE, Kidd TJ, Whiley DM, Paterson DL, Bell SC. Evaluation of phenotypic screening tests for carbapenemase production in *Pseudomonas aeruginosa* from patients with cystic fibrosis. *J Microbiol Methods* 2015;**111C**:105–7.
49. Hrabák J, Chudáčková E, Walková R. Matrix-assisted laser desorption ionization–time of flight (MALDI-TOF) mass spectrometry for detection of antibiotic resistance mechanisms: from research to routine diagnosis. *Clin Microbiol Rev* 2013;**26**(1):103–14.

MYCOBACTERIUM TUBERCULOSIS: CLINICAL AND MICROBIOLOGICAL ASPECTS

11

R.Y. Ramírez-Rueda

*Pedagogical and Technological University of Colombia, Faculty of Health Sciences,
School of Nursing, Tunja, Colombia*

1 INTRODUCTION

Tuberculosis (TB) is a chronic infectious disease that usually affects the lungs. It is mainly caused by *Mycobacterium tuberculosis* (Mt). This infection is transmitted directly without contact, through coughing by micro-droplets of saliva (flügge) that transport the bacterium, which are found inside the airways of people with active tuberculosis.

The bacterial wall of Mt contains arabinogalactan, esterified with fatty acids of heavy molecular weight termed mycolic acids, which turns it into in an acid-fast bacillus (AFB), a particular feature of *Mycobacterium* genus. Virulence factors of Mt are responsible for its survival inside of the human body and also for many of the deleterious effects on the host. Such effects are especially enhanced in immunosuppressed hosts. However, *M. tuberculosis* infection is usually asymptomatic in healthy people because their immune system acts to develop a barrier (tuberculous granuloma) around the bacterium. Ordinarily, TB appears as a pulmonary syndrome called pulmonary TB. After beating the immune system, Mt replicates slowly but continuously in the lungs until the occurrence of the first symptoms, which marks the beginning of active tuberculosis.[1]

The oldest evidence of a TB case dates back to 9000 BC. This proof of coevolution with the bacterium was found through studies of paleopathology in the remains of an approximately 1-year-old child found near the Mediterranean Sea.[2] Numerous studies have been reported demonstrating the existence of TB through human evolution and its cosmopolitan distribution. Another archeological evidence for the presence of the disease is the demonstration of spinal tuberculosis, or Pott's disease, in ancient Egyptian mummies over 5500 years ago.[3]

According to the World Health Organization, TB currently affects more than 9 million people worldwide and causes around 1.5 million deaths per year.[4] Morbidity and mortality in TB are associated with coinfection of primary syndromes which affects the immune system, and acquired immune deficiency syndrome (AIDS), is the principal associated cause of TB. AIDS infected people carry a greater risk of developing severe diseases because it accelerates the course of HIV disease by activating viral replication and increased depletion rate of CD4+ T cells. Progress in scaling up interventions to deal with the AIDS/TB coepidemic continues. However, the increase of morbidity is greater than the

The Microbiology of Respiratory System Infections. http://dx.doi.org/10.1016/B978-0-12-804543-5.00011-7

efforts made at this level.[5] These efforts, and those made in the treatment of TB, consist of improving and introducing new diagnostic techniques, drugs, vaccine research, and surveillance, not only in TB but in the diseases associated with this disease.

2 DESCRIPTION OF CAUSAL MICROORGANISM

M. tuberculosis is an AFB, a characteristic conferred to the bacterium by the structure of its bacterial wall, which in addition to the peptidoglycan (common component of bacterial cell wall), has arabinogalactan. The distal ends of the arabinogalactan are esterified with fatty acids of heavy molecular weight, termed mycolic acids, which have a size and a unique structure for mycobacteria. The disease caused by the bacilli is determined by the immune status of the host and factors of virulence within the bacteria. Upon entering the body, Mt is able to survive the attack of macrophages, blocking the acidification of the phagosome (through the production of ammonia), which is essential in the formation of the phagolysosome and the activation of the bactericidal factors released during the merger. An alternative bacterial mechanism to avoid its destruction within the phagolysosome consists of blocking the oxidative phosphorylation process and suppressing the production of reactive Oxygen and Nitrogen species by means of a sulfated glycolipid from the bacterial wall called sulfolipid 1.[6]

There is no doubt that much of the virulence of the Mt is in its envelope (wall and cell membrane), so that apart from the lipids (present in large number and variety) there are also important proteins such as the Exported repetitive protein (Erp). The absence of Erp in mutant strains decreases the replication of the bacteria within macrophages.[7] The mycocerosic acid synthase (mas) is essential for the synthesis of mycocerosic acids and their derivatives. When there is deficiency of one of these, such as phtioceroldimycocerasic acid, strains with attenuated growth in macrophages in vitro and murine models are observed.[8] Lipids such as phtioceroldimycocerosate (PDIM) have been also nominated as virulent factors, showing growth reduction in vivo and in murine models when Mt strains defective in PDIM or in fad26/28 (fatty-acid-Coa synthase) were inoculated.[9] In addition to the production of PDIM, its transport and localization in the cell membrane of Mt by transporting proteins such as mmpl7 are essential for the virulence. For this reason, in Mt mutant strains for the gene encoding this protein, similar effects were observed as those described above. The importance of Mt growth inside phagocytes is highlighted by the enzyme fibronectin binding protein mycolytransferase and its essential part in the synthesis of mycolic acid, as attenuated growth phenotypes were observed in human monocytes and murine macrophages when inoculated with gene defective strains coding for this enzyme.[10] Other enzymes, such as methoxymycolic acid synthase 4 (mmaA4) and mycolic acid cyclopropane synthase (pcaA), are important in the cell permeability and their absence produces phenotypes with decreased growth in vivo and a reduction of mortality in mice.[11,12]

Mt also has virulent factors such as lipoarabinomannane (LAM) which is able to inmunomodulate the host response decreasing the production of interferon-γ, scavenging the Oxygen free radicals and blocking the production of protein-kinase C.[13] Moreover, extrapulmonary dissemination could be partially due to the presence of proteins such as heparin-binding hemaglutin (HbhA) which decreases the bacterial phagocytosis in the lung.[14]

The bacterial metabolism plays an important role in Mt virulence and adaptation to different substrates within the host. This microorganism has several enzymes as isozitratelyase (Icl),[15] lipase/esterase LipF, polyketide synthase (FasD33),[16] a great number of phospholipases C (Rv2351c, Rv2350c,

Rv2349c, Rv1755c, plcA, plcB, plcC, plcD),[17] pantothenatesynthetases and aspartate-1-decarboxylase (Rv3602c, Rv3601c, pan C, panD)[18]; which help metabolize several lipids required for its development and replication, unlike mutant strains without them.

The biosynthesis of aminoacids and purines is essential for the development and expression of virulence in Mt, just like lipid synthesis. The bacterium uses enzymes such as isopropylmalate isomerase (LeuD), anthranilate phosphoribosiltransferase (TrpD), pyrroline-5-carboxlate reductase (ProC) and 1-phosphoribosylaminoimidazole-succinocarboxamide synthase (PurC) for the biosynthesis of amino acids and purines.[19–21] Magnesium and Iron are essential for the life of various pathogens (including Mt), this is the reason why some transporters and acceptors are required of these metals. In Mt the presence of Mg^{2+} transport P-type ATPase, ABC transporters such as Iron ABC transporter (coded by mbtB) and Iron-dependent regulatory proteins (IrtAB and ideR) are important for the synthesis of siderophores, the micobactina and the carboximicobactina are the most important to Mt. These are essential in the biosynthesis of cytochromes and hemoproteins, acting also as cofactors of proteins involved in the synthesis of amino acids, pyrimidines and bacterial DNA.[22] These siderophores are developed by the bacterium in response to the low concentration of Iron inside the macrophages of mammals (approximately 1000 times less than the concentration required for normal bacterium growth) and, in their absence, Mt has an attenuated growth in macrophages and lungs of murine models.[23]

Adaptation to different Oxygen concentrations inside the host, in different stages of the disease, is also a significant feature in bacterial virulence. Late stages of TB such as granulomas, cause microaerophilic microenvironments with the concomitant production of Oxygen free radicals (oxidative stress). Enzymes such as alkyl hydroperoxidereductase and catalase-peroxidase enzyme, which catabolize hydrogen peroxide and other organic peroxides, are important in the elimination of Oxygen free radicals. In the absence of such enzymes, Mt and *Mycobacterium bovis* show a growth decrease in animal models.[24] With the same purpose, Mt also exhibits enzymes superoxide dismutase-type which detoxify the bacterium of radicals of Oxygen; also proposing that such enzymes could inhibit the redox signaling necessary for the initiation of the immune response generated by the macrophage.[25]

Bacterial adaptability to express certain genes in response to external stimuli is also considered as a virulence factor. It determines bacterial survival when subjected to microenvironments such as inside the human body. Sigma type transcriptional factors found in Mt are A, C, D, E, F, G, H, and L; which are responsible in allowing the expression of a diverse repertoire of virulence factors. These serve, among other things, in the transcription of housekeeping genes to activate the latency, withstand increased temperatures, and act as a buffer in the treatment with detergents such as sodium dodecyl sulfate (SDS) or the action of oxidizing agents such as diamides.[26–28]

Another way bacteria respond to environmental changes is the expression of regulatory factors such as the two-component systems, the regulators of response, and kinases sensors. In Mt, 11 two-component systems have been described until now, without which the bacterium usually loses or minimizes its ability to replicate intracellularly in vitro and in animal models. One of the most important two-component system in Mt is the PhoP-Phor system, which is related to the regulation of lipid metabolism and cellular respiration processes.[29] Other outstanding two-component systems in Mt are SenX3-RegX3 (which regulates 100 genes involved in the maintenance of the cell envelope, some regulatory functions, and energy metabolism),[30] DosR/S/T (responsible of the latent stage)[31], and MprA/AprB (which responds to the damage in the cell envelop caused by alkaline pH, detergents, and antibiotics).[32]

To control the expression of large groups of genes, Mt uses other transcriptional regulators such as hspR, which act as transcriptional repressors of the *hsp70* heat-shock genes, whose repression starts at 37°C and is suppressed at 45°C. The absence of these transcriptional repressors is reflected in a colony forming units (CFU) reduction in animal model organs.[33] Sporulation is an important phenomenon in the survival of bacteria in unfavorable environments for their growth. Mt is unable of sporulation, but it remains latent; for this reason ortholog genes of the WhiB family described originally in *S. coelicolor* have been studied; as well as genes coding for the WhiD protein, which interrupts the sporulation and septation in this bacterium.[34] Studies conducted in a homologous gene to WhiB2 (WhmD) in *M. smegmatis* have demonstrated that this gene is required for cell division[35]; whereas in Mt suppression of WhiB2 does not affect the growth of the bacterium in murine models and Guinea pigs; however, it affects its survival. The half-life of Mt was found to be increased in the mutant strains for the gene WhiB3 (350 days), compared with mice infected with wild strains (225 days). Recently, it has been shown that the function of whiB2 in Mt is the control of cell division (Rv3260c gene) as well as in *M. smegmatis*.[36]

3 RESPIRATORY DISEASE CAUSED BY *M. TUBERCULOSIS*

Exposure to *M. tuberculosis* often results in the development of latent tuberculosis (LTB) with a risk of 5–10% for progression to active tuberculosis (ATB). Most TB cases occur in the first 2 years after the establishment of the bacteria in the lungs.[37]

3.1 LATENT TUBERCULOSIS

Mt primary infection begins when the bacilli are able to reach the alveoli, where they are phagocytized by alveolar macrophages that transport them across the lymphatic system to regional lymph nodes (hilar, mediastinal and sometimes supraclavicular or retroperitoneal), producing a nonspecific inflammatory response which is usually asymptomatic. This immune response results in LTB, classically defined as an immunological sensitization to bacteria in the absence of transmission and manifestations of active disease (fever, chills, night sweats, weight lost, cough, hemoptysis, or opacities in the chest X-ray) due to the absence of bacterial replication. A granuloma is formed at the site of inoculation of Mt, which becomes necrotic and finally calcifies. In most cases (in immunocompetent hosts) the bacilli are destroyed, and then the only evidence of infection is a positive tuberculin hypersensitivity skin test; but sometimes bacilli can survive, leading to the LTB.[38]

The immunological reaction to Mt is evaluated using the tuberculin skin test (TST), also called Mantoux test or through Interferon gamma release assays (IGRAs). A positive result for any of the methods mentioned above is indicative of LTB, whereas microbiological test such as smear microscopy and culture for Mt should be negative, as well as the individual's ability to transmit the infection.[39] The homogeneity in the definition of LTB is currently being discussed because it does not consider how long the focus has been present, which could vary depending on the virulence of the Mt strain and the individual host susceptibility. A fact that reinforces the heterogeneity theory in LTB is the evidence of replication of Mt inside the granuloma of patients suffering LTB; supported by decreased risk of progression to ATB when treatment is performed with isoniazid, an antibiotic that inhibits bacterial cell wall synthesis (a process that occurs only when bacterial replication is active).[40]

3.2 ACTIVE TUBERCULOSIS

About 2–10 weeks after the inoculation of Mt, an injury appears demonstrable by X-ray in cases of ATB, which is caused by the necrosis of the primary focus after the accumulation of antigens in that focus. The accumulation of these antigens in sufficient quantities causes cell-mediated hypersensitivity, resulting in necrosis with calcification radiologically visible that could be detected through a tuberculin test or purified protein derivative (PPD). The combination of a peripheral lung injury and a calcified parahilar node is known as Ghon complex. TB disease or ATB develops when this immune response fails and the bacterium is not contained. TB disease or ATB is caused either by reactivation of the primary focus or by progression of a primary infection. The development of ATB occurs (on an average) within 2 years since the Mt primary infection has occurred. ATB is also produced by reinfection, which means a new infection that goes beyond the holding capacity of the immune system. In these cases the host's immune reaction results in a pathological lesion typically localized, and it often presents extensive tissue destruction and cavitation.[41] The risk of ATB is high in immunocompromised situations, such as extremes of age or people with debilitating diseases such as cancer or diabetes.[42] When the infection is not contained at that level, bacilli may reach to the bloodstream and spread. Most of the pulmonary lesions or those disseminated lesions scar, turning into foci of potential future reactivation for TB. If such dissemination occurs, it may result in meningeal or miliary tuberculosis, which is particularly fatal in immunocompromised individuals.

4 DIAGNOSIS OF PULMONARY TUBERCULOSIS

Pulmonary TB presents a broad spectrum of signs and symptoms in patients, which forces clinicians to establish a differential diagnosis with other lung infectious diseases such as bacterial or viral pneumonia, pneumocystosis, histoplasmosis, acquired bronchiectasis, or lung abscess; or with noninfectious conditions such as asthma, congestive heart failure, nonacquired bronchiectasis, lung cancer, or chronic obstructive pulmonary disease. Accurate diagnosis of TB must combine clinical and laboratory elements in order to confirm the clinical suspicion.

4.1 CLINICAL DIAGNOSIS OF PULMONARY TUBERCULOSIS

Pulmonary tuberculosis has nonspecific clinical manifestations. Usually, the affected individual has clinical manifestations over several weeks with cough, expectoration (sometimes hemoptoic), chest pain and nonspecific symptoms (fever or low grade fever, sweating, asthenia, anorexia, and weight loss). The results after the chest examination may be unremarkable, but localized rales or adventitious sounds with asymmetry in breathing sounds can be auscultated. It is essential to suspect the presence of disease in all patients with respiratory symptoms lasting more than 2–3 weeks and in all patients with hemoptysis, independent of the duration of respiratory symptoms. In addition to the clinical diagnosis, the epidemiological suspicion can be added when patients have an underlying disease causing immunosuppression (AIDS, cancer, etc.), are under 5 or over 60 years old, are smokers, alcoholics or drug addicts; and patients who had TB in the past or maintain an intense and prolonged contact with people suffering ATB.[43]

4.2 LABORATORY DIAGNOSTIC OF PULMONARY TUBERCULOSIS

Clinical suspicion of TB requires confirmation by laboratory techniques, which can range from simple microscopic observations to determinations requiring molecular biology and other cutting edge techniques.

4.2.1 Microbiological tests

Microbiological testing has been and remains the mainstay of the diagnosis of TB. The parameters that contribute to the diagnosis of pulmonary TB can range from simple observation to the confirmation of bacillus species by molecular biological methods. These tests are explained in the following sections.

Sputum smear

The sputum smear test consists of the microscopic examination of AFB in an extended lung secretion from patients with presumptive symptoms of TB. The smear is stained with Ziehl Neelsen stain (ZN), in which carbolfuchsin is retained by the mycobacterium (due to the presence of mycolic acids and waxes in its cell wall) after discoloration process with acid alcohol. The mycobacterium is observed as red bacilli on a blue background provided by the methylene blue, which acts as contrast dye. The ZN stain sputum allows the detection of all members of *Mycobacterium* genus, but the presence of AFB is not a sufficient parameter for confirm the presence of Mt in the samples evaluated by this technique.[44] Other limitations of this testing include the requirement of 5.000 to 10.000 AFB/mL in the sample (optimally 100.000) in order to be detected. This, along with the fact that approximately a 30–50% of TB patients are smear sputum negative, decreases the sensitivity of smear microscopy, and makes it impossible to consider a negative result as a unique parameter to rule out the disease. The diagnostic sensitivity of smear microscopy is estimated between 22–43% in a single sample and 50–70% with two to three samples examined in different days. Therefore, the recommendation in the diagnosis of pulmonary TB is to examine three sputum samples through ZN staining on three different days, preferably in the morning samples.[45] The specificity of the smear sputum microscopy depends on several factors, including the quality of coloration and the expertise of the examiner in order to distinguish AFB from acid-fast elements such as food particles, precipitated dye, environmental AFB, normal microbiota, atypical mycobacteria, or *Nocardia* species; fibers (cotton, wool, etc.), pollen granules; and even defects on the glass slide where the smear was made. It is also important to consider that false negatives could occur due to deficiencies in the dying of the sample, reading of the test, sputum collection (saliva collection or nasopharyngeal secretions), and the selection of the portion of the sample used for the test.[45–47]

Fluorescent dyes

The fluorescent stains for Mt as auramine O, rhodamine, or a combination of these, represent an alternative to the ZN staining. With these stains the AFB appear fluorescent. With auramine yellow to orange, with rhodamine red, and with auramine-rhodamine are reddish yellow on a dark background when using potassium permanganate as contrast dye.[48] An advantage of using fluorochromes in the observation of AFB is that these allow its detection in a shorter time, compared to regular ZN staining. Thus, the bacilli can be observed with low power lenses, and then confirming its morphology with the oil immersion technique. Another feature of the technique is that it allows subsequent staining of smears with ZN in order to study the morphology of the bacilli and its acid-fast properties. The sensitivity of fluorescent stains in the diagnosis of pulmonary TB is greater than in ZN, such sensitivity increases in samples with low quantity of bacilli. The specificity of these two techniques is similar, and some studies suggest that the difference in sensitivity and specificity using different fluorescent compounds is not significant.[49] It is important to mention that this technique, as well as the ZN stain, could lead to false positive results due to the presence of artifacts in the preparation[50]; or false negatives due to poor sampling preparation or poor staining technique, particularly considering the fact that fluorescence is stable for only 3 days even when the glass slide is protected from light.[51]

Culture

This technique is considered to be the gold standard in the diagnosis of TB because it confirms the positive results of the microscopic test. The culture has much a greater sensitivity than the sputum smear staining, being able to detect between 10 and 100 AFB/mL of sputum; allowing the isolation of the mycobacterium for identification and determination of antibiotic susceptibility. The biggest disadvantage of cultures (using traditional methods) is the long waiting time for the results, which ranges from 4 to 8 weeks on solid media as Ogawa Kudoh (OK) or Lowestein Jensen (LJ); a fact that is determined by the metabolic characteristics of Mt.[44] Some techniques have been developed to minimize this incubation period that involve the use of liquid media incorporated into automated systems that provide results in a significantly shorter time.

Cultures in solid medium

Solid consistency of these kind of media can be provided by compounds such as egg (Ogawa Kudoh, Lowenstein Jensen and Coletsos) or agar (7H10 and 7H11 of Middlebrook). The egg-based media are enriched with the addition of an inhibitor called malachite green. These media allow the recovery of most of the mycobacterial species due to their components, inhibitory power of the contaminant flora, and their buffering capability that allows the neutralization of multiple toxic products present in the clinical samples. Furthermore their half-life is long (6 months at a temperature between 2 and 8°C), and the phenotypic characteristics of the isolated strains represent a good degree of fidelity. Among the disadvantages are the slowness with which Mt develops in the media and the difficulties of preparation (coagulation method), leading to variations between batches.[52] The agar-based media allows a rapid detection of colony growth of Mt (10–12 days), and its recovery for susceptibility testing. The Middlebrook 7H11 media containing 0.1% of casein hydrolisate favors the recovery of isoniazid-resistant Mt strains. The disadvantages with respect to the egg-based media, are the short half-life (1 month at a temperature between 2 and 8°C) and a higher rate of contamination.[53-54]

An alternative technique developed for the identification and susceptibility testing of Mt is the mycobacteriophage system. Phage-based assays for Mt detection have high specificity (ranging 0.83–1.00), but modest and variable sensitivity (ranging 0.21–0.88). This method is more useful for detecting rifampicin resistance directly from sputum smear-positive samples or indirectly from culture. The test has a manual format and the results are read visually. Results are available within 2 days after of inoculation of mycobacteriophages, and use basic microbiological equipment available in most laboratories.[55]

Cultures in liquid media

This test uses cultures of pulmonary secretions in liquid media into automated systems for finding Mt, which is the most effective recovery of this bacterium and also the fastest method. Consequently, this test is always recommended for the primary isolation of clinical samples. Liquid media systems are the basis for identification and susceptibility testing in automated incubation and reading systems. The disadvantage of these media is the inability to distinguish colonies and therefore possible contamination, as well as its high cost. However, its performance and adaptation to the new technologies of identification and systematization make them ideal for fast and effective detection of Mt.[56] These media can be classified into manual reading media and automated reading media. The manual reading culture media contains mycobacterial growth indicators without apparent presence of bacterium. Such indicators may be tetrazolium salts present in culture media as MB Redox medium or fluorescent compounds such as Ruthenium pentahydrate component of Mycobacterial Growth Indicator Tube (MGIT).[57] On the other hand, semiautomated systems such as BACTEC 460 TB use a radiometric marker (palmitic acid labeled with ^{14}C) as an indicator for the detection of mycobacteria, and can be used for the presumptive identification of Mt and susceptibility testing. Unfortunately, due to a lack of integration with an automated reading system, it may have problems of cross-contamination, and there may be a possible production of aerosols during handling and reading. The evolution of automated system makes it possible to produce systems that combine incubation and reading, such as the BACTEC MGIT 960 which uses MGIT media detecting O_2 consumption by fluorometric sensors with the ability to perform susceptibility testing for first-line antimycobacterial drugs (including pyrazinamide).[58] There are alternatives such as ESP Culture System I, whose detection system uses pressure sensors detecting the O_2 consumption of the mycobacteria. The basic medium of this system is a Middlebrook 7H9 media with enriching supplements and inhibition of contaminant agents.[59] This system is able to perform susceptibility testing for isoniazid, rifampicin, and Ethambutol.[60] There are also automated methods based on colorimetric methods such as MB/BacT ALERT 3D, which uses production of CO_2 as indicator of bacterial growth, and Middlebrook 7H9 media with growth and inhibition factors for samples that come from human compartments that are normally nonsterile. Once inoculated, cultures remain in the "incubator-reader" system without additional handling until the equipment notifies the results. Following a positive result, susceptibility testing can be performed for first-line antimycobacterial drugs (excluding pyrazinamide).[61]

4.2.2 Molecular tests

Several methods for detecting DNA of Mt directly from samples or positive cultures have been developed. These tests take a great value today due to the increasingly frequent occurrence of multi-drug resistance Mt strains. Some of the most used tests for MDR detection are the molecular line probe assays, which use multiplex polymerase chain reaction (PCR) amplification and reverse hybridization to identify Mt complex and gene mutations associated with rifampicin and isoniazid resistance. Line probe assays are highly sensitive and specific for detection of rifampicin resistance ($\geq 97\%$ and $\geq 99\%$) and isoniazid resistance ($\geq 90\%$ and $\geq 99\%$) on culture isolates and smear-positive sputum.[62] Amplification and detection of Mt DNA is one of the fastest and most sensitive ways to detect tuberculosis, and may also allow the detection of genetic mutations associated with drug resistance. Some of these methods are complicated and require specialized training, and can only be performed on material that has been subjected to processing and DNA extraction. Between the automatic systems of detection stands the GeneXpert system (Cepheid), which is a closed, self-contained, fully-integrated and automated platform that allows a relatively untrained operator to perform sample processing, DNA amplification, and detection of a large variety of microorganisms that cause infectious diseases in less than 2 h.[63] In the TB field, GeneXpert systems has developed a semi-nested RT-PCR assay for Mt that simultaneously detects rifampicin resistance called Xpert MTB/RIF, this assay has shown excellent performance in a multicentre study, where in a unique direct MTB/RIF testing detected 92.2% of cases of pulmonary tuberculosis, including 72.5% of those with smear-negative disease, which was equivalent to that reported for solid culture.[64] A striking molecular method is a loop-mediated isothermal amplification assay, which is a simple DNA amplification method that does not require a thermocycler or detection system and reportedly allows visual detection of amplification.[65]

4.2.3 Indirect tests

These are assays based on detection of organism's effects on the host. These can be classified into antibody detection test (humoral response) and tests for cellular response detection, either by direct cell production and its effect in vivo or by the production of some molecules released by leukocytes sensitized with a TB antigen.

Serodiagnostic tests

Unlike many infectious diseases whose diagnosis can be supported strongly in the application of serodiagnostic tests, it has not been possible to develop a test for TB with high sensitivity and specificity. This is most probably because the primary immune response to Mt is cell-mediated, instead of being a humoral response. Evidence suggests that the humoral response to the bacilli is heterogeneous.[66] Currently, the International Standards for TB Care discourage about the use of serological tests in routine practice and no international guidelines recommend their use; however, many commercial serological tests for TB diagnosis are offered for sale in many countries around the world. WHO is emphatic in their policy statements to recommend strongly that these commercial tests not be used for the diagnosis of pulmonary and extrapulmonary TB. Results from studies performed in several countries of the most widely used tests showed sensitivity at 76% and 59%, and specificity at 92% and 91% in smear-positive and smear-negative patients, respectively. In conclusion, currently available commercial serodiagnostic tests provide inconsistent and imprecise results.[67] Independent studies, such as one made in India where 1.259 subjects were included in the study, using three different enzyme immunoassays, also conclude that the commercial serological test evaluated showed poor sensitivity and specificity, and suggests no utility for the detection of pulmonary tuberculosis.[68]

Immunological tests

The TST is a traditional method used as a screening method for the diagnosis of TB, which is also called PPD or Mantoux test. Tuberculin is a method based on a delayed-type hypersensitivity reaction, developed in the skin, after intradermal inoculation with PPD. The Mantoux test is typically used to determine immunity to TB in humans and positive reactions develop not only in individuals previously exposed to Mt but also in those previously immunized with the Bacillus of Calmette and Guérin (BCG) vaccine. After 24–72 h later the tuberculin (PPD-RT23) is inoculated in the forearm, in a dose of 0.1 mL; an induration that is larger in diameter than a certain size (ranging from 5 to 15 mm) is considered a positive result and indicates infection with Mt.[69] Despite its importance, the Mantoux test is falsely negative in 10–47% of patients with active disease.[70] The PPD test has lower sensitivity in populations of immunocompromised patients, recently infected individuals, and very young children. Specificity is low because the PPD contains various antigens widely shared among different species of mycobacteria.[71] In the presence of coinfection with HIV, a far higher proportion of patients with active disease will have a false-negative test result. The degree of immunosuppression determines the rate of false negatives, being 30% for patients with a CD4 T-lymphocyte cell count of more than 500/mL, compared with almost a 100% of patients with a CD4 T-lymphocyte cell count of less than 200/mL.[72] The Mantoux test will be replaced in the future by specific antigens only found in Mt. Recombinant dimer ESAT-6 (rdESAT-6) has been successfully tested for this purpose. Improvements of this new skin test continue to be developed using other antigens such as CFP-10.[73]

To improve the performance of Mantoux test, new tests such as T-cell-based IGRA have been designed. IGRAs are in vitro blood tests based on interferon-gamma release after stimulation by antigens such as Early Secreted Antigenic Target-6 (ESAT-6) and Culture Filtrate Protein-10 (CFP-10). These antigens are not shared with any of BCG vaccine strains or certain species of nontuberculosis mycobacteria. Indeterminate results are likely in immunocompromised individuals with low CD4+ cell counts.[74] There are currently two commercial IGRAs: the QuantiFERON-TB Gold In-Tube (QFT) and the two-stage T-SPOT. TB assay. There is strong evidence that IGRAs, especially QFT, have excellent specificity unaffected by BCG vaccination and appear to correlate well with markers of TB exposure. Regarding the sensitivity, T-SPOT.TB appears to be more sensitive than QFT. This could be partly because the cut-off for T-SPOT.TB is designed to maximize sensitivity, while the cut-off for QFT is designed to maximize specificity. Resuming, the estimated sensitivity of IGRAs in patients with active TB is between 75% and 95%, decreasing in immunocompromised individuals (60–80%), and estimated specificity in healthy persons with no TB disease or exposure is between 95% and 100%.[75]

4.3 NEW PERSPECTIVES IN TB DIAGNOSTIC

In recent years, new alternatives for TB diagnostic have been developed as a solution to the problems that occur with current techniques. Tests such as breathalyser screening test, which use an instrument for detection of volatile organic compounds of Mt, is performed and a readout is obtained in under 10 min. The instrument is fully portable and runs off rechargeable AA batteries.[76] The detection of Mt molecules in urine is a test that is based on detection of lipoarabinomannan (LAM) excreted in the urine of TB patients. Urinary antigen detection may be of particular value in diagnosing TB in HIV-coinfected patients. It may prove valuable for rapid and simple diagnosis of TB, particularly in developing countries at peripheral levels.[77] An increase of the sensitivity of sputum samples is being searched for through the use of techniques of concentrating sputum, such as sputum filtration. In this technique, sputum is liquefied and passed through a filter, which is then stained or cultured by standard techniques. Filtration considerably concentrates mycobacteria, increasing sensitivity. Another advantage of using concentrating sputum is the reduction in time spent on sputum examination.[78] TB diagnosis using immunodominant secreted antigens of Mt is being developed. In this method, skin patches delivering MPT64, ESAT6, or CFP-10 (specific proteins of Mt) are placed on the patient's skin and in those with active infectious TB a localized immune response consisting of erythema and/or vesiculation appears 3–4 days after application to the skin.[79]

5 CONCLUSION AND FUTURE PERSPECTIVES

Infectious diseases have accompanied people since the beginning of the time. Likewise, because humans were affected by disease, they have always looked for cures. In the case of TB the story is the same and, as in most infectious diseases, despite the significant progress made to fight against it the disease succeeds to the present day. However, at present, the health situation is much better than it was. This is caused in part by the actions implemented against infectious diseases. The development of antibiotics against Mt, implementation in diagnostic methods, prophylaxis through vaccination, and improving the conditions of poverty, have been masterpieces in reducing TB.

The development of molecular biology and the other genomic technologies has aided the diagnosis of TB. For example, it has allowed the fast detection of TB. In comparison with traditional techniques, current methods provide much more information in a shorter time.

A more effective TB vaccine is expected to be introduced over the next few decades. In addition, we also expect the introduction of antibiotics that can be used against MDR-Mt, and fast and accurate diagnostic techniques that are more affordable for developing countries. The ultimate goal is the eradication of TB. This should be parallel to the reduction in cases of AIDS, which is one of the main risk factors for acquiring TB.

Fundamental and applied research on TB and the maintenance and proper implementation of public health policies should be the cornerstone on which must be founded the fight against one of the most important and ancient infectious diseases in the history of humanity.

REFERENCES

1. Hopewell PC, Kato-Maeda M, Ernst JD. 35 - Tuberculosis. In: Murray JA, Nadel AS, editors. *Murray and Nadel's textbook of respiratory medicine*. 6th ed. Philadelphia: W.B. Saunders; 2016 p. 593.e20–628.e20.
2. Holloway KL, Henneberg RJ, de Barros Lopes M, Henneberg M. Evolution of human tuberculosis: a systematic review and meta-analysis of paleopathological evidence. *HOMO* 2011;**62**(6):402–58.
3. Daniel TM. The history of tuberculosis. *Resp Med* 2006;**100**(11):1862–70.
4. World Health Organization. Global tuberculosis report 2014; 2014. Available from: http://apps.who.int/iris/bitstream/10665/137094/1/9789241564809_eng.pdf?ua=1.
5. Pawlowski A, Jansson M, Sköld M, Rottenberg ME, Källenius G. Tuberculosis and HIV co-infection. *PLoS Pathog* 2012;**8**(2):e1002464.
6. Zhang L, Goren M, Holzer T, Andersen B. Effect of Mycobacterium tuberculosis-derived sulfolipid I on human phagocytic cells. *Infect Immun* 1988;**56**:2876–83.
7. Berthet F, Lagranderie M, Gounon P, Laurent-Winter C, Ensergueix D, Chavarot P, et al. Attenuation of virulence by disruption of the *Mycobacterium tuberculosis* erp gene. *Science* 1998;**282**:759–62.
8. Sirakova T, Dubey V, Cynamon M, Kolattukudy P. Attenuation of Mycobacterium tuberculosis by disruption of a mas-like gene or a chalcone synthaselike gene, which causes deficiency in dimycocerosyl phthiocerol synthesis. *J Bacteriol* 2003;**185**:2999–3008.
9. Camacho L, Ensergueix D, Perez E, Gicquel B, Guilhot C. Identification of a virulence gene cluster of *Mycobacterium tuberculosis* by signature-tagged transposon mutagenesis. *Mol Microbiol* 1999;**34**:257–67.
10. Armitige LY, Jagannath C, Wanger AR, Norris SJ. Disruption of the genes encoding antigen 85A and antigen 85B of Mycobacterium tuberculosis H37Rv: effect on growth in culture and in macrophages. *Infect Immun* 2000;**68**:767–78.
11. Dubnau EJ, Chan C, Raynaud VP, Mohan MA, Laneelle K, et al. Oxygenated mycolic acids are necessary for virulence of Mycobacterium tuberculosis in mice. *Mol Microbiol* 2000;**36**:630–7.
12. Glickman MS, Cox J, Jacobs W. A novel mycolic acid cyclopropane synthetase is required for cording, persistence, and virulence of *Mycobacterium tuberculosis*. *Mol Cell* 2000;**5**:717–27.
13. Chan J, Ran X, Hunter S, Brennan P, Bloom B. Lipoarabinomannan, a possible virulence factor involved in persistence of *Mycobacterium tuberculosis* within macrophages. *Infect Immun* 1991;**59**:1755–61.
14. Pethe K, Alonso F, Biet G, Delogu MJ, Brennan C, Locht C, et al. The heparin-binding haemagglutinin of M. tuberculosis is required for extrapulmonary dissemination. *Nature* 2001;**412**:190–4.
15. Dubnau EP, Fontan R, Manganelli S, Soares-Appel S, Smith I. Mycobacterium tuberculosis genes induced during infection of human macrophages. *Infect Immun* 2002;**70**:2787–95.

16. Rindi L, Fattorini L, Bonanni D, Iona E, Freer G, Tan D, et al. Involvement of the fadD33 gene in the growth of *Mycobacterium tuberculosis* in the liver of BALB/c mice. *Microbiology* 2002;**148**:3873–80.

17. Raynaud C, Guilhot C, Rauzier J, Bordat Y, Pelicic V, Manganelli R, et al. Phospholipases C are involved in the virulence of *Mycobacterium tuberculosis*. *Mol Microbiol* 2002;**45**:203–17.

18. Sambandamurthy VK, Wang X, Chen B, Russell R, Derrick S, Collins F, et al. A pantothenate auxotroph of *Mycobacterium tuberculosis* is highly attenuated and protects mice against tuberculosis. *Nat Med* 2002;**8**:1171–4.

19. Hondalus MK, Bardarov S, Russell R, Chan J, Jacobs WJ, Bloom B. Attenuation of and protection induced by a leucine auxotroph of *Mycobacterium tuberculosis*. *Infect Immun* 2000;**68**:2888–98.

20. Smith DA, Parish T, Stoker NG, Bancroft GB. Characterization of auxotrophic mutants of *Mycobacterium tuberculosis* and their potential as vaccine candidates. *Infect Immun* 2001;**69**:1142–50.

21. Jackson MS, Phalen W, Lagranderie M, Ensergueix D, Chavarot P, Marchal G, et al. Persistence and protective efficacy of a *Mycobacterium tuberculosis* auxotroph vaccine. *Infect Immun* 1999;**67**:2867–73.

22. Quadri L. Iron uptake in Mycobacteria. In: Daffé M, Reyrat J, editors. *The mycobacterial cell envelope*. Washington: ASM Press; 2008. p. 167–84.

23. Litwin CM, B CS. Role of iron in regulation of virulence genes. *Clin Microbiol Rev* 1993;**6**:137–49.

24. Wilson TM, De Lisle GW, Collins DM. Effect of inhA and katG on isoniazid resistance and virulence of *Mycobacterium bovis*. *Mol Microbiol* 1995;**15**:1009–15.

25. Edwards KM, Cynamon MH, Voladri RK, Hager CC, DeStefano MS, Tham KT, et al. Iron-cofactored superoxide dismutase inhibits host responses to *Mycobacterium tuberculosis*. *Am J Respir Crit Care Med* 2001;**164**:2213–9.

26. Steyn AJ, Collins DM, Hondalus MK, Jacobs Jr WR, Kawakami RP, Bloom BR. Mycobacterium tuberculosis WhiB3 interacts with RpoV to affect host survival but is dispensable for in vivo growth. *Proc Natl Acad Sci USA* 2002;**99**(5):3147–52.

27. Beaucher J, Rodriguez S, Jacques PE, Smith I, Brzezinski R, Gaudreau L. Novel Mycobacterium tuberculosis anti-sigma factor antagonists control sigma F activity by distinct mechanisms. *Mol Microbiol* 2002;**45**:1527–40.

28. Manganelli R, Dubnau E, Tyagi S, Kramer FM, Smith I. Differential expression of 10 sigma factor genes in *Mycobacterium tuberculosis*. *Mol Microbiol* 1999;**31**:715–24.

29. Gonzalo-Asensio J, Mostowy S, Harders-Westerveen J, Huygen K, Hernández-Pando R, Thole J, et al. PhoP: a missing piece in the intricate puzzle of *Mycobacterium tuberculosis* virulence. *PLoS One* 2008;**3**:e3496 p. Available from: http://dx.doi.org/10.1371/journal.pone.0003496.

30. Parish T, Smith DA, Roberts G, Betts J, N.G S. The senX3-regX3 two-component regulatory system of *Mycobacterium tuberculosis* is required for virulence. *Microbiology* 2003;**149**:1423–35.

31. Boon C, Dick T. Mycobacterium bovis BCG response regulator essential for hypoxic dormancy. *J Bacteriol* 2002;**184**:6760–7.

32. Zahrt TC, Deretic V. Mycobacterium tuberculosis signal transduction system required for persistent infections. *Proc Natl Acad Sci USA* 2001;**98**:12706–11.

33. Gomez M, Smith I. Determinants of mycobacterial gene expression. In: Hatfull GF, Jacobs Je WR, editors. *Molecular genetics of mycobacteria*. Washington, DC: American Society for Microbiology; 2000. p. 111–29.

34. Molle V, Palframan WJ, Findlay KC, Buttner M. WhiD and WhiB, homologous proteins required for different stages of sporulation in *Streptomyces coelicolor* A3(2). *J Bacteriol* 2000;**182**:1286–95.

35. Gomez JE, Bishai WR. whmD is an essential mycobacterial gene required for proper septation and cell division. *Proc Natl Acad Sci USA* 2000;**97**(15):8554–9.

36. Fu LM, Shinnick TM. Genome-wide exploration of the drug action of capreomycin on *Mycobacterium tuberculosis* using Affymetrix oligonucleotide GeneChips. *J Infection* 2007;**54**(3):277–84.

37. Centers for Disease Control and Prevention (CDC). Basic TB facts, risk factors 2012. Available from: http://www.cdc.gov/tb/topic/basics/risk.htm.

38. Hartman-Adams H, Clark K, Juckett G. Update on latent tuberculosis infection. *Am Fam Physician* 2014;**89**(11):889–96.
39. Salgame P, Geadas C, Collins L, Jones-López E, Ellner JJ. Latent tuberculosis infection—revisiting and revising concepts. *Tuberculosis* 2015;**95**(4):373–84.
40. Mack U, Migliori GB, Sester M, Rieder HL, Ehlers S, Goletti D, et al. LTBI: latent tuberculosis infection or lasting immune responses to *M. tuberculosis*? A TBNET consensus statement. *Eur Respir J* 2009;**33**(5): 956–73.
41. Russell DG, Cardona P-J, Kim M-J, Allain S, Altare F. Foamy macrophages and the progression of the human tuberculosis granuloma. *Nat Immunol* 2009;**10**(9):943–8.
42. Lönnroth K, Jaramillo E, Williams BG, Dye C, Raviglione M. Drivers of tuberculosis epidemics: the role of risk factors and social determinants. *Soc Sci Med* 2009;**68**(12):2240–6.
43. Richeldi L. An update on the diagnosis of tuberculosis infection. *Am J Respir Crit Care Med* 2006;**174**(7): 736–42.
44. Pfyffer GE. Mycobacterium: general characteristics, laboratory detection and staining procedures. In: Murray PRBE, Jorgensen JH, Landry ML, Pfaller MA, editors. *Manual of clinical microbiology*. 9th ed. Washington DC: ASM Press; 2007. p. 543–72.
45. American Thoracic Society, Centers for Disease Control and Prevention, Infectious Diseases Society of America. . Diagnostic standards and classification of tuberculosis in adults and children. *Am J Respir Crit Care Med* 2000;**161**(4):1376–95.
46. Takahashi S. *Handbook of direct smear examination of sputum for tubercle bacillus*. Tokyo, Japan: South-East Asian Medical Information Centre; 1975.
47. Smithwick R. *Laboratory manual for acid-fast microscopy*. Atlanta, GA, USA: Department of Health, Education and Welfare; 1976.
48. Bogen E. Detection of tubercle bacilli by fluorescence microscopy. *Am J Respir Crit Care Med* 1941;**44**(3): 267–71.
49. Steingart KR, Henry M, Ng V, Hopewell PC, Ramsay A, Cunningham J, et al. Fluorescence versus conventional sputum smear microscopy for tuberculosis: a systematic review. *Lancet Infec Dis* 2006;**6**(9):570–81.
50. Boyd J, Marr J. Decreasing reliability of acid-fast smear techniques for detection of tuberculosis. *Ann Intern Med* 1975;**82**:489–92.
51. Toman K. What are the advantages and disadvantages of fluorescence microscopy? In: Frieden T, editor. *Toman's tuberculosis: case detection, treatment, and monitoring—questions and answer*. 2nd ed. Geneva: World Health Organization; 2004. p. 31–4.
52. Mitchison DA, Allen BW, Carrol L, Dickinson JM, Aber VR. A selective oleic acid albumin agar medium for tubercle bacilli. *J Med Microbiol* 1972;**5**(2):165–75.
53. Cohn ML, Waggoner RF, McClatchy JK. The 7H11 medium for the cultivation of mycobacteria. *Am Rev Resp Dis* 1968;**98**(2):295–6.
54. Joloba ML, Johnson JL, Feng P-JI, Bozeman L, Goldberg SV, Morgan K, et al. What is the most reliable solid culture medium for tuberculosis treatment trials? *Tuberculosis* 2014;**94**(3):311–6.
55. Kalantri S, Pai M, Pascopella L, Riley L, Reingold A. Bacteriophage- based tests for the detection of *Mycobacterium tuberculosis* in clinical specimens: a systematic review and meta-analysis. *BMC Infect Dis* 2005;**5**(1):59.
56. Perkins MD. New diagnostic tools for tuberculosis [The Eddie O'Brien Lecture]. *Int J Tuberc Lung Dis* 2000;**4**(12):S182–8.
57. Palomino JC, Martin A, Von Groll A, Portaels F. Rapid culture-based methods for drug-resistance detection in *Mycobacterium tuberculosis*. *J Microb Meth* 2008;**75**(2):161–6.
58. Pinheiro MD, Ribeiro MM. Comparison of the Bactec 460TB system and the Bactec MGIT 960 system in recovery of mycobacteria from clinical specimens. *Clin Microbiol Infect* 2000;**6**(3):171–3.
59. Woods GL, Fish G, Plaunt M, Murphy T. Clinical evaluation of difco ESP culture system II for growth and detection of mycobacteria. *J Clin Microbiol* 1997;**35**(1):121–4.

60. Richter E, Rüsch-Gerdes S, Hillemann D. Drug-susceptibility testing in TB: current status and future prospects. *Expert Rev Respir Med* 2009;**3**(5):497–510.

61. Ängeby KAK, Werngren J, Toro JC, Hedström G, Petrini B, Hoffner SE. Evaluation of the BacT/ALERT 3D system for recovery and drug susceptibility testing of *Mycobacterium tuberculosis*. *Clin Microbiol Infec* 2003;**9**(11):1148–52.

62. Ling DI, Zwerling AA, Pai M. GenoType MTBDR assays for the diagnosis of multidrug-resistant tuberculosis: a meta-analysis. *Eur Respir J* 2008;**32**(5):1165–74.

63. Niemz A, Ferguson TM, Boyle DS. Point-of-care nucleic acid testing for infectious diseases. *Trends Biotechnol* 2011;**29**(5):240–50.

64. Boehme CC, Nabeta P, Hillemann D, Nicol MP, Shenai S, Krapp F, et al. Rapid molecular detection of tuberculosis and rifampin resistance. *New Engl J Med* 2010;**363**(11):1005–15.

65. Bi A, Nakajima C, Fukushima Y, Tamaru A, Sugawara I, Kimura A, et al. A rapid loop-mediated isothermal amplification assay targeting hspX for the detection of *Mycobacterium tuberculosis* complex. *Jpn J Infect Dis* 2012;**65**(3):247–51.

66. Lyashchenko K, Colangeli R, Houde M, Al Jahdali H, Menzies D, Gennaro M. Heterogeneous antibody responses in tuberculosis. *Infect Immun* 1998;**66**(8):3936–40.

67. World Health Organization. Commercial serodiagnostic tests for diagnosis of active tuberculosis Geneva: World Health Organization; 2011. Available from: http://www.who.int/tdr/publications/tdr-research-publications/diagnostics-evaluation-2/en/.

68. Singh S, Singh J, Kumar S, Gopinath K, Balooni V, Singh N, et al. Poor performance of serological tests in the diagnosis of pulmonary tuberculosis: evidence from a contact tracing field study. *PLoS ONE* 2012;**7**(7):e40213.

69. Vukmanovic-Stejic M, Reed JR, Lacy KE, Rustin MHA, Akbar AN. Mantoux Test as a model for a secondary immune response in humans. *Immunol Lett* 2006;**107**(2):93–101.

70. Al Zahrani K, Al Jahdali H, Menzies D. Does size matter? Utility of size of tuberculin reactions for the diagnosis of mycobacterial disease. *Am J Respir Crit Care Med* 2000;**162**:1419–22.

71. Andersen P, Munk M, Pollock J, Doherty T. Specific immune-based diagnosis of tuberculosis. *Lancet* 2000;**356**(9235):1099–104.

72. Markowitz N, Hansen NI, Wilcosky TC, Hopewell PC, Glassroth J, Kvale PA, et al. Tuberculin and anergy testing in HIV-seropositive and HIV-seronegative persons. *Ann Intern Med* 1993;**119**(3):185–93.

73. Arend SM, Franken WPJ, Aggerbeck H, Prins C, van Dissel JT, Thierry-Carstensen B, et al. Double-blind randomized Phase I study comparing rdESAT-6 to tuberculin as skin test reagent in the diagnosis of tuberculosis infection. *Tuberculosis* 2008;**88**(3):249–61.

74. Dinnes J, Deeks J, Kunst H, Gibson A, Cummins E. A systematic review of rapid diagnostic tests for the detection of tuberculosis infection. *Health Technol Asses* 2007;**11**(3):1–196.

75. Pai M, Denkinger CM, Kik S, Rangaka MX, Zwerling A, Oxlade O, et al. Gamma interferon release assays for detection of *Mycobacterium tuberculosis* infection. *Clin Microbiol Rev* 2014;**27**(1):3–20.

76. Amy S-T, Michael E, Stephen C. Validating a breath collection and analysis system for the new tuberculosis breath test. *J Breath Res* 2013;**7**(3):037108.

77. Hamasur B, Bruchfeld J, Haile M, Pawlowski A, Bjorvatn B, Källenius G, et al. Rapid diagnosis of tuberculosis by detection of mycobacterial lipoarabinomannan in urine. *J Microbiol Meth* 2001;**45**(1):41–52.

78. Ramos E, Schumacher SG, Siedner M, Herrera B, Quino W, Alvarado J, et al. Optimizing tuberculosis testing for basic laboratories. *J Trop Med Hyg* 2010;**83**(4):896–901.

79. Bekmurzayeva A, Sypabekova M, Kanayeva D. Tuberculosis diagnosis using immunodominant, secreted antigens of *Mycobacterium tuberculosis*. *Tuberculosis* 2013;**93**(4):381–8.

PULMONARY ASPERGILLOSIS: DIAGNOSIS AND TREATMENT

12

S. Quereshi*, P. Paralikar**, R. Pandit**, M. Razzaghi-Abyaneh[†], K. Kon[‡], M. Rai**

*Department of Microbiology and Biotechnology, Indira Priyadarshini College, Chhindwara, Madhya Pradesh, India; **Nanobiotechnology Laboratory, Department of Biotechnology, SGB Amravati University, Amravati, Maharashtra, India; [†]Department of Mycology, Pasteur Institute of Iran, Tehran, Iran; [‡]Department of Microbiology, Virology and Immunology, Kharkiv National Medical University, Kharkiv, Ukraine

1 INTRODUCTION

Aspergillus is a cosmopolitan fungus that usually occurs on organic debris, compost, food, stored grain, compost heaps, air vents, and airborne dust or in other decaying vegetation. It is aerobic and grows mostly on a high carbon source such as monosaccharides and polysaccharides. It belongs to Deutromycetes, which is devoid of a sexual reproductive growth phase. The species of *Aspergillus* which are associated with human illness are *Aspergillus fumigatus* and *Aspergillus niger* and less frequently, *Aspergillus flavus* and *Aspergillus clavatus*. The term "aspergillosis" indicates, an infection of the airways, which involves the respiratory tract. The cutaneous infection is caused by *Aspergillus* species.[1] Other sites of *Aspergillus* infection present in human include auditory canal, skin, nails, eyes, sinuses, meninges, and bones. The central nervous system, cardiovascular system, and other tissues may be infected. The majority of *Aspergillus* species can tolerate temperature up to 49°C. Nearly about 200 species of *Aspergillus* are known, but it is estimated that around 40 *Aspergillus* species are responsible to cause infections in humans. *Aspergillus* reproduces by means of conidia. They are approximately 2–4 µm in diameter. As conidia are airborne, they can be easily inhaled and cause infections in the lungs.[2] Gomori methanamine silver stain can be used to observe sputum or mucus hyphae.[3,4]

Aspergillosis of the head and neck region affects the nasal and paranasal sinuses. *Aspergillus* is the most common fungus responsible for the paranasal sinus involvement. Generally *A. fumigatus* is the most predominant causative agent followed by *A. flavus*. It is a type of invasive aspergillosis, that affects the patients with chronic nasal nodule and mucoid impaction of the sinuses. The immunological pathogenesis is not fully understood, but it is assumed that continuous inhalation of *Aspergillus* spores results in the colonization in the sputum plugs and leads to minute tissue damage. Spores of *Aspergillus* secret certain proteolytic enzymes, as a result bronchial wall invasion takes place and antigen absorption increases. Antigen release results in the production of IgE, IgG, and IgA. Hypersensitivity reaction occurs and results in tissue inflammation. Mucus plugs consist of Charcot-Leyden crystals, eosinophils, and hyphae of *A. fumigatus*.[1,4,5]

The most frequently affected sinus is the maxillary sinus. It is categorized into invasive and noninvasive types on the basis of the fungal invasion of the bones and the mucosal layer.[6] Chronic invasive sinonasal aspergillosis is mostly noticed in immunocompetent patients, residing in dry air climate countries such as India, Saudi Arabia, and Sudan. In Sudan area, invasive sinonasal aspergillosis occurs due to *A. flavus*. Further progress in this type of aspergillosis for few months leads to granulomatous reactions. Invasive aspergillosis is further subdivided into three types: namely acute, fulminant, chronic invasive, and granulomatous invasive. Noninvasive aspergillosis is classified into two types allergic rhinosinusitis and fungal ball rhinosinusitis. The common symptoms of noninvasive aspergillosis are the invasion of sinus mucosa and bone atrophy.[1–4]

Aspergilloma is defined as the condition in which accumulation of *Aspergillus* hyphae, cellular debris, and fibrin are present within a pulmonary cavity. Aspergilloma is a fungus ball mycetoma that develops in a preexisting cavity in the lung parenchyma. Underlying causes of this cavity may include treating tuberculosis or other necrotizing infections, sarcoidosis, CF and emphysematous bullae. The ball of fungus may translocate within the cavity, but does not invade the cavity wall. However, as a complication, hemoptysis (the coughing of blood) may occur. Patients with a previous history of such lung diseases as tuberculosis, sarcoidosis, cystic fibrosis, or other are most susceptible to an aspergilloma. Usually, there are no specific symptoms in aspergilloma, but in many patients the haemoptysis may be infrequent and in small quantity, but can be severe and require urgent medical intervention.[7]

2 CLASSIFICATION AND MICROBIOLOGY OF PULMONARY ASPERGILLOSIS

Pulmonary aspergillosis is a disease of the lungs, which is characterized by the colonization of *Aspergillus* spores on the invasive infection.[4] *Aspergillus* species, such as *A. fumigatus, A. niger, A. terreus,* and *A. flavus* have been known to exhibit many life-threatening diseases in humans, especially in the immunocompromised patients. Generally, these are present in the intrapulmonary region, nasal cavity, auditory canal and cornea. These species generate allergies, chronic and saprophytic conditions and result in different forms of aspergillosis. *A. fumigatus* is commonly found in the patients suffering from invasive aspergillosis. Other predominantly occurring types of *Aspergillus* species are *A. flavus, A. niger* and *A. terreus*.[8–11] Generally, Pulmonary aspergillosis can be subdivided into ABPA, chronic pulmonary aspergillosis (CPA), invasive pulmonary aspergillosis (IPA), and simple pulmonary aspergilloma (SPA).[12,13]

2.1 ALLERGIC BRONCHOPULMONARY ASPERGILLOSIS

ABPA is a lung mycosis that has been recorded in people who are allergic to the *Aspergillus*. In a case of ABPA, the allergic reaction takes place between the immune system of the host and *Aspergillus*, which rises into lung inflammation.[14] Common problems associated with ABPA are bronchospasm and mucus build up, which eventually lead to the breathing problem, cough, and airway obstruction.[15] Bronchiectasis is the disease characterized by damage in the airways, which affects lungs badly and has been observed in the few patients of ABPA. ABPA is a disorder which is caused by *A. fumigatus*.[16] *Aspergillus* sensitization is the first progression step that results in the advancement of ABPA.[17] From clinical findings, it was found that *A. flavus, A. niger* and *A. fumigatus* are responsible for ABPA.[18] *Aspergillus* sensitization is defined as an immediate hypersensitivity reaction against antigens of *A. fumigatus*.[19,20] Another disease called allergic bronchopulmonary mycosis ABPM is a disease which is similar to ABPA like syndrome, but it is affected by fungi other than *A. fumigatus*. Allergic brochopulomonary fungosis occurs, when *A. fumigatus* is present in association with other fungi, such as, *Helminthosporium* species, *Stemphylium lanuginosum, Fusarium vasinfectum, Dreschslera hawaiiensis, Candida* species, *Curvularia* species, *Schizophyllum commune*, and species of *Aspergillus* such as *A. nidulans, A. niger, A. flavus* and *A. oryzae*.[21]

In 1952, ABPA was first reported in England by Hinson et al.[22] and it has now emerged all over the world. Bronchial colonization of *A. fumigatus* results in a hypersensitive lung disease ABPA. This is a disease which generally affects patients of asthma and cystic fibrosis. It is estimated that 1–2% of asthma patients, 7–14% steroid-dependent patients, and 2–15% of cystic fibrosis patients are supposed to be the sufferer of ABPA.[15] This disease is characterized by asthma, eosinophilia and pulmonary infiltrates. The actual susceptibility of ABPA in asthma patient is not exactly known. It is reported that high concentration of spore exposure increases the chances of ABPA. It is estimated that defects in adaptive and innate immunity may result in the persistence of *A. fumigatus*.[19] Some researchers have stated that exposure to the spores of *A. fumigatus* may be responsible for ABPA in asthmatic patients. The environment is not considered as the reason for the development of ABPA. Although all asthma patients are exposed to the same environment, very few of them are prone to the ABPA. The outcome of inhalation of *A. fumigatus* conidia may give rise to hyphal growth. Many types of proteins are released by *A. fumigatus* particularly pro-inflammatory cytokines by the airway epithelium. Certain proteases released by *Aspergillus* are directly toxic to the pulmonary epithelial cell and causes cell detachment and cell death.[23] Depending on radiological images, ABPA has been classified into three groups, which are: ABPA Seropositive (ABPA-S), ABPA-Central Bronchiectasis (ABPA-CB) and ABPA Other Radiological Findings (ABPA-CB-ORF).[19]

According to clinical data, the following five different stages of ABPA have been observed:[7,14]

Stage 1: This is also known as acute stage. In this stage, the level of IgE and IgG specific to *A. flavus* is increased. Pulmonary infiltrate was observed in the middle and lower lobe of the lungs along with increase in the level of IgE.

Stage 2: This is called remission. In this stage, stage I condition persists for more than 6 months. Ig G level in the serum may be normal or slightly higher than the normal range. If stage 2 persists for longer duration, it may result in the development of stage 3.

Stage 3: This is known as exacerbation. In this stage, relapse of stage 2 is observed. The in-filtrates of Ig E are observed in the middle and lower lobes of lungs. At this level patient required proper medications for ABPA.

Stage 4: This is known as corticosteroid-dependent asthma. At this level, patients are supposed to suffer from constant cough and wheezing like symptoms. Ig E level is increased in the blood even on the oral corticosteroid consumption. Symptoms become more critical if corticosteroids consumption is stopped.

Stage 5: This is the last stage, which is known as end stage. In this stage, the patients whose diagnosis was missed during first stage and those patients who had taken treatment for steroids for the treatment of asthma may become prone to bronchiectasis and fibrosis.

2.2 CHRONIC PULMONARY ASPERGILLOSIS

CPA is a gradual and progressive inflammatory pulmonary syndrome. CPA is also named as semi-invasive aspergillosis or subacute invasive aspergillosis. The names chronic necrotizing pulmonary aspergillosis (CNPA) and semi-invasive aspergillosis were given by Gefter et al.[24] According to recent guidelines from the Infectious Diseases Society of America (IDSA), there are three major subtypes of chronic types of pulmonary aspergillosis: (CNPA), chronic cavitary pulmonary aspergillosis (CCPA) and aspergilloma.[25] CCPA is characterized by the occurrence of a large number of cavities with the presence or absence of fungal ball along with *Aspergillus* antibodies and increased in the number of inflammatory markers. Aspergilloma is defined as the condition, in which accumulation of *Aspergillus* hyphae, cellular debris, and fibrin are present within a pulmonary cavity.[7] CNPA progresses gradually in a few months or even in a year, which results in lung destruction, such as, progressive cavitation, fibrosis. Common symptoms associated with CNPA are weight loss, productive cough, chronic sputum, hemosputum, or hemoptysis.[26]

From the previous case reports, it was found that *A. fumigatus* is supposed to be the most common causative agent present in the CPA patients. CNPA is observed in middle-aged people and elderly patients.[25] Patients with chronic lung disease, such as, pulmonary tuberculosis, cystic fibrosis, chronic obstructive lung disease, pneumoconiosis, lung infarction and sarcoidosis are susceptible to CNPA.[27] Immunocompromised patients, such as, suffering from diabetes mellitus, chronic liver disease, malnutrition, and alcoholism are prone to CNPA.[7] The pulmonary diseases that are responsible for cavitation and which may result in the CPA are emphysematous bullae, chronic obstructive pulmonary disease, bronchiectasis, histoplasmosi, and rheumatoid nodules.[28] CNPA is histologically characterized by necrosis of lung tissue followed by acute or chronic inflammation of the cavity wall. Criteria for the detection of CNPA are divided into three parts: clinical detection, radiological detection and laboratory detection. Clinical symptoms of CNPA are weight loss, productive cough, and hemoptysis. Radiological criteria include the symptoms like cavitary pulmonary lesions, paracavitary infiltrates, new cavity formation or expansion of cavity size. Laboratory criteria include increased levels of inflammatory markers. Laboratory criteria are supposed to be helpful for the early detection and therapy of aspergillosis.[7] CNPA differential diagnosis is linked with dreadful lung diseases such as infectious actinomycosis, necrotizing pneumonias tuberculosis, and lung abscesses. Other noninvasive diseases associated with lung cavities are Wegener's granulomatosis, sarcoidosis, and lymphomatoid granulomatosis. Lung diseases with primary tumors in the lungs are also related with differential diagnosis, including lymphoma and Kaposi's sarcoma, Langerhans' cell histiocytosis, and autoimmune diseases such as ankylosing spondylitis, rheumatoid arthritis, systemic lupus erythematosus, and primary amyloidosis.[28]

2.3 TRACHEOPULMONARY ASPERGILLOSIS

Tracheopulmonary aspergillosis is an unusual form of *Aspergillus* infection, that is confined entirely or predominantly to the bronchitis. Tracheopulmonary aspergillosis is characterized by thick mucus in bronchitis. Patients who are susceptible to tracheopulmonary aspergillosis are lung transplant recipients, patients with AIDS and generally patients with cancer.[7] According to bronchoscopic appearance, tracheopulmonary aspergillosis is classified as pseudomembranous, ulcerative and obstructive. Pseudomembranous tracheopulmonary aspergillosis is the excessive invasion of tracheobronchial tree along with mucosa compromising *Aspergillus* species. It is the most dreadful as compared to other types. In case of obstructive tracheopulmonary aspergillosis, thick mucus and inflammation of tracheobronchial tree is observed. Ulcerative type is generally observed in the recipients of lung transplantation.[7] Presence of *Aspergillus* colonization in the cavities or the airways in the bronchitis leads tracheopulmonary aspergilloma. It is widely spread in the immunocompromised patients, such as, in case of AIDS and those who have had lung transplantation. The most common symptoms of this aspergillosis are fever, cough, chest pain and hemoptysis.[29]

2.4 INVASIVE ASPERGILLOSIS OR INVASIVE PULMONARY ASPERGILLOSIS

Invasive aspergillosis or invasive pulmonary aspergillosis (IA or IPA) is an infection of the pulmonary, parenchyma, which is affected by the growing hyphae of *Aspergillus*. Furthermore, the term invasive aspergillosis can be modified as angioinvasive aspergillosis, if there is vascular invasion by the hyphae of aspergillosis. *A. fumigatus* is the most common aetiological agent of IPA. *A. fumigatus* is an omnipresent fungus and its conidia are airborne, so, its exposure is constant and universal to almost all human beings.[30] IPA is one of the most widespread forms of invasive disease.[16] A defect in the host defense mechanism is the main reason of invasive aspergillosis. The phagocytic function is greatly affected; hence, it is more in case of patients with a haematological patients like acute leukemia and in anaemic patients.[7] It is a mycotic infection of immunosuppressed patients who have undertaken chemotherapy and in those hosts who have taken the doses of steroids for malignant or haemopoietic diseases. IPA mainly causes infection in the lungs, which is important reason for the mortality of patients associated hematological malignancy and in solid organ and stem cell transplant recipients,[30] allogeneic bone marrow transplantation, and in HIV infected patients with its last stage, chronic granulomatosis.[20,31] Patients, who have undergone bone marrow transplantation are susceptible to IPA and in those cases 90% of the patients die.[32] Depending on the situation of the patients, it was found that 24% of patients with acute leukemia and 7% of patients with lymphoid malignancies are susceptible to IPA.[33] Traditionally, it is reported that the patients who had undergone hematopoietic stem cell transplantation IPA was observed after 10–14 days.[34] In the last few years, it has been reported that, IPA is increasing in the patients with chronic obstructive pulmonary disease (COPD).[35,36] IPA has been observed in a patient with neutropenia, having neutrophil less than 500/μL or less.[36]

Various immunological problems are reported for immunocompromised patients who are associated with diabetes mellitus and with alcohol abuse or who may be patients with renal failure, such as neutrophil proliferation, maturation and its life span is affected very badly and these all are the characteristic features of patients with invasive aspergillosis.[37] Various clinical and pathological forms of IPA occur, which include acute bronchopneumonia, angioinvasive aspergillosis, acute tracheobronchitis, and pleural aspergillosis.[7] Clinical staging of IPA comprises of pleuritic chest pain, dry cough, fever, and dyspnea. Severe and pleuritic chest pain are the symbols of angioinvasive aspergillosis.[29] The most frequently observed radiological imaging symbols of IPA include: diffuse pulmonary nodular infiltrates, cavitary invasions, and pleural effusions along with pleural wedge-shaped densities in the pulmonary nodules. One of the main signs of IPA is the low attenuation around the pulmonary nodule which gradually gives rise to cavities. Hitopathologically, a large number of fungal hyphae invasion and discrete nodular destruction are observed along the blood vessels.[36]

Diagnostically, as per Japanese guidelines, IPA is categorized into three types: "proven infection", "clinically documented infection or probable infection" and "possible infection". Proven infection is positive, when presence of *Aspergillus* hyphae infection is found mycologically and histopathologically. The detection of *Aspergillus* infection can be done using the sputum and bronchoalveolar lavage specimen. Probable or clinically documented infection is diagnosed when galactomannon antigen and β-D glucan is present in serum or gene is diagnosed by using polymerase chain reaction and such diagnosis is positive. Clinically documented infection is known as probable infection by the European Organization for Research and Treatment of Cancer (EORTC). Possible infection is positive when serum, genetic, and imaging finding is positive.[7]

3 RELATIONSHIP OF PULMONARY ASPERGILLOSIS WITH IMMUNOCOMPROMISED PATIENTS

Immunocompromised patients are prone to pulmonary aspergillosis. IPA is the most dreadful type of pulmonary aspergillosis, which results in high mortality and morbidity in immunocompromised patients.[38] IPA is observed most frequently in immunocompromised patients with neutropenia, organ transplantation, hematopoieitc stem cell transplant, hematological malignancy, chemotherapy and corticosteroid therapy, chronic granulomatous disease (CGD), advanced stage of AIDS, acute leukemias and in anaemic patient, diabetes mellitus and with alcohol abuse with renal failure.[37] From the previous report, it was found that the mortality risk was 50% in patients with neutropenia and 90% mortality risk in patients with hematopoietic stem cell transplants.[37] When the neutrophil count is less than 500 cells/mm^3, it results in neutopenia as the number of neutrophil is decreased the immunity is decreased, and it results in the susceptibility to IPA. As far as the risk of IPA in neutropenia is concerned, it is estimated to be 1% until the first 3 weeks and then it is 4% per day. Organ transplantation, especially the lungs, hematopoietic stem transplant is also risk issue for IPA.[36] When allogeneic and autologous hematopoietic transplants are compared, it was found that allogeneic transplants have a higher risk for IPA, as compared to autologous transplants. For allogeneic transplants, the risk factor is 2.5–15%, whereas, for autologous transplants it is 0.5–4%.[18] From the previous study, it was found that the risk factor is higher in the case of allogenic transplants and the percent of the risk factor in such patients increase with time. It is 5% at 2 months, near about 9% at 6 months, whereas 10% in a year and after 3 years of transplantation risk percentage rises to 11%.[18] Reports have revealed that the patients with severe COPD are prone to IPA. Long-lasting use of sterocorticoids, changes in the structural architecture of the lungs, antibiotic doses and repeated hospitalization along with other associated diseases with COPD, such as diabetes mellitus, alcoholism. All these reasons are responsible for the increase of the susceptibility of COPD patients.[37]

Patients suffering from immunocompromised disease such as HIV, lung transplant recipient and those with nutropenia are prone to tracheobronchial pulmonary aspergillosis.[3] ABPA is the most common type of pulmonary aspergillosis and is observed in patients suffering from cystic fibrosis and in asthma. About 1–2 % of asthma patients, 7–14% in steroid dependent patients, and 2–15 % cystic fibrosis patients are supposed to be susceptible to ABPA.[17]

4 MORPHOLOGICAL AND MOLECULAR IDENTIFICATION OF ASPERGILLI

Even though many novel methods are available for identification of different species of *Aspergillus*, morphological features are essential tools in the identification of Aspergilli.[39] Morphological characteristics are subdivided into two sections: macro-morphological and micro-morphological characteristics. The micro-morphological features, which are used to distinguish *Aspergillus* species involve, the shape of the conidia head, vesicle shape and diameter, stipe length, width, texture and color, conidial size, shape, texture and color, seriation, size. Macromorphological characteristics depend on various features such as colony color, texture, production of soluble pigments by the fungi in the media, formation of sclerotia, cleisthethocia, and reverse color of the plate. Depending on the seriation genus *Aspergillus* is either uniseriate or biseriate. The morphological characteristics of *Aspergillus* vary with the use of growth media.[40] Culture and microscopy are preliminary factors which are useful in the morphological studies. For the identification of the fungus color of the colonies is very significant. A section of the colony color can be used in the identification of *Aspergillus* species. Yellowish-green, green, and deep-green color colony sections were observed for sections of *Flavi, Fumigati,* and *Nidulantes,* respectively. *A. niger* showed a black or brownish shade.[39,41]

Morphological characteristics of *Aspergillus* in culture are very important for the identification and classification up to the genus level (Fig. 12.1). Microscopic characteristics for the identification is based on conidial heads, stipes, color and length vesicles shape and seriation, metula covering, conidia size, shape, and roughness, also colony features including diameter after 7 days, color of conidia, mycelia, exudates and reverse, colony texture, and shape. Morphological characters of the isolate are compared with morphological characters of the pure *Aspergillus* culture. After morphological classification of the isolates, pure cultures of *Aspergillus* isolates can be maintained on potato dextrose agar. After proper growth and sporulation cultures of *Aspergillus* can be examined using lactophenol cotton-blue mount for the sporulation after 10, 20, and 30 days. All the morphological and microscopic identification of *Aspergillus* species can be confirmed by comparing with the characters given in the Atlas of Clinical Fungi.[42]

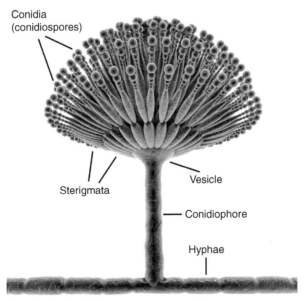

FIGURE 12.1 Structure of Aspergillus

The molecular diagnosis technique is excellent because it facilitates easy, early, and correct identification of aspergillosis which can help in the antifungal treatment. Single nucleotide polymorphism multiplex (SNaPAfu) assay is a novel technique which involves identification, detection, and genotyping of clinical isolates of *Aspergillus* species. In this assay, only Single Nucleotide Polymorphism (SNP) gene present in Multilocus sequence typing MLST is amplified and targeted in a multiplex reaction and its analysis depends on the mini-sequence reaction. Detailed information of clinical sample can be obtained from the single amplification reaction. MLST alignment created by the multiplex reaction is required for the SNaPfu assay. SNaPfu assay requires less than 8 hours to get the results of clinical samples and the cost per sample of the assay is very low. The significant advantage of SNaPAfu assay is that, it can identify the genotype of *A. fumigatus* in a single step reaction.[43]

A real time PCR-based innovative commercialized technique known as MycAssay is developed for the detection of *Aspergillus* DNA in case of lower respiratory disease. This assay is generally performed in bronchoalveolar lavage (BAL) samples from patients without hematological cancer. MycAssay™ *Aspergillus* is a potential technique that requires less time for the diagnosis as compared to the conventional fungal culture. It provides much more accurate and additional information related to the fungal species responsible for the aspergillosis.[44] A 2-PCT assay system was used to target 18 S ribosomal DNA of *Aspergillus*. In this assay, a tissue sample of the fungi was embedded in paraffin wax to detect ribosome of *A. fumigatus*. This technique involves the use of deparaffinized tissue section. Specific primers are used for the detection of aspergillosis. New molecular diagnostic techniques are potential and promising, which can improve the detection of fungal infection in critically ill patients.[45] Random amplification of polymorphic DNA (RAPD-PCR) is a simple, rapid, and useful tool used for the detection of *Aspergillus* clinical isolates and can be useful in the identification of aspergillosis. This method is crucial for the detection of hospital isolates of *Aspergillus*.[46]

5 CLINICAL DIAGNOSIS OF ASPERGILLOSIS

5.1 MICROSCOPY

For clinical diagnosis of aspergillosis, clinical sample like sputum is collected with precautionary measures. After the collection of sample, direct microscopy is performed. This is the simplest approach to diagnose the clinical samples. Gram staining of the sample should be performed. Usually, 10% KOH is used for wet mounting and the sample is examined under microscope. This method does not specifically differentiate *Aspergillus* species because the hyphae of many filamentous fungi seem to be similar and it is difficult to differentiate between them.[30,47]

5.2 HISTOPATHOLOGY

This clinical identification is noticed in the case of IA. Histopathological study is performed in case of biopsy or autopsy sample of IA. This study is used to demonstrate the tissue invasion by filamentous fungi, but it is not possible to identify the mycelium of the fungus merely by observing it until the time the positive culture of fungus used is not available.[47] A few clinical tests that can be used in the identification of *Aspergillus* species will be described in the following sections.

5.3 GALACTOMANNAM ANTIGEN TEST

The cell wall of *Aspergillus* spp. is composed of galactomannam (GM) antigens. This assay gives information about the antigen present in the clinical samples. Serum and plasma samples of patients are required for performing this assay for the clinical testing of invasive aspergillosis patients. The specificity of the results obtained by serum and plasma is up to 55–95% to 75–96%, for serum and plasma, respectively.

Another assay called platelia GM assay is similar to enzyme-linked immuno assay. The Food and Drug Administration (FDA) approved that platelia GM assay can be used in the clinical evaluation of invasive aspergillosis in the serum and BAL of patients.[48,49]

5.4 BETA-D-GLUCAN TEST

Beta-D-Glucan is one of the components of fungal cell walls. This assay identifies the presence of *Aspergillus* and *Candida* merely by the presence of beta-D-glucan in the patient's serum. Antigen present on the surface of *Aspergillus* sp. can be determined by this assay.[30,50]

The FDA has approved another diagnosis assay for determining invasive fungal infection, which is known as "Fungitell". The specificity of the results changes with the variation of clinical settings. This is one of the assays used for the clinical testing of *Aspergillus* sp.[51]

5.5 CHEST RADIOGRAPHY AND COMPUTED TOMOGRAPHY

Chest radiography is the basic and predominant method for the detection of aspergllosis in the early stages of disease. Generally, ABPA and IA affected patients are examined by using chest radiography. Rounded densities, accumulation of infiltration and cavitation can be investigated by chest radiography. By using chest radiography, tuberculosis like symptoms which are seen similar to aspergillosis can be differentiated on the basis of the chest radiograph. Although this is an accurate method of identification, at the chronic stage it cannot be differentiated perfectly. This gives better results in ABPA patient along with asthma. Computed tomography (CT) scan examination gives the basic information regarding the aspergillosis. It detects the early stage development of diseases. When a CT scan is performed on an IA patient, it shows a characteristic "halo" sign. The halo sign is a special type of chest imaging seen in IA affected patients. This is a feature characterized by invasion of hemorrhagic nodules, tumors, and other inflammatory processes. Most often the halo sign is observed in neutropenic individuals. A CT scan, when used in combination with high resolution images (HRCT), provide us with much clearer and better results. In general, plain chest radiography does not give a clear picture about the inflammation, but HRCT can. By using HRCT, a mucous plug may be noted very clearly in ABPA patients.[47,52]

5.6 LATERAL FLOW DEVICE

This is an innovative technique which uses JF5, the monoclonal antibodies from mice. A lateral flow device (LFD) detects a glycoprotein and antigen present in the blood sample and BAL of IA patients. This device is a new and potent tool, which can be used in the clinical analysis of the IA patient. The main advantages of this technique are that it requires less time to perform and no expertise is required for handling it. The specificity of the result is much higher in LFD as compared to the other results. A recent study has shown that in the case of BAL, it has a specificity up to 100% and false results can be interpreted in case of *Penicillium* spp.[48,49]

5.7 VOLATILE ORGANIC COMPOUNDS

Recently, it has been discovered that patients suffering from IA infection exhale a few organic compounds that are volatile in nature. Such volatile compounds can be detected when a patient infected with *A. fumigatus* breathes. These volatile compounds are called as volatile organic compounds (VOC).[53] Individuals infected with IA produce 2-pentylfuran at the time of breathing and such compounds can be easily and rapidly detected. This technique shows accuracy until 81% in case of critically ill IA patients.[54] This is supposed to be one of the most promising tools that can be used for early detection of *Aspergillus* spp. However, more study needs to be done on the concept of the volatile compounds released by infected patients.[49]

6 TREATMENT OF ASPERGILLOSIS

A number of antifungal drugs are available for the treatment of aspergillosis. Various antifungal compounds have been clinically active against *Aspergillus* spp. and are approved for treatment of pulmonary aspergillosis. The antifungal drugs include Amphotericin B and its lipid formulations, itraconazole, voriconazole, posaconazole, and caspofungin.[3,55,56] Voriconazole and Amphotericin B are mostly preferred for primary treatment of IPA with a recommended dose 1–1.5 mg/kg per day.[7,57] Voriconazole is a promising alternative to Amphotericin B [58] and is currently the most frequently recommended therapy for invasive aspergillosis specially.[59] Posaconazole has been used for prophylaxis of IA in neutropenic patients with leukemia and myelodysplasia. Posaconazole is also recommended for treatment of IA that is refractory to an Amphotericin B formulation or to itraconazole. Micafungin and anidulafungin, members of the class of echinocandins have significant activity against aspergillosis. Antifungal drug treatment for pulmonary aspergillosis is summarized in Table 12.1.

Table 12.1 Antifungal Agents Used for the Treatment of Pulmonary Aspergillosis

Disease	Antifungal Drug	References
Invasive pulmonary aspergillosis	Voriconazole Liposomal Amphotericin B Itraconazole Posaconazole	[3,60–63]
Aspergilloma	Itraconazole Voriconazole	[64,65]
Chronic necrotizing aspergillosis	Voriconazole Itraconazole Amphotericin B or Amphotericin lipid formulation Posaconazole	[17,56,66,67]
Allergic bronchopulmonary aspergillosis	Itraconazole Voriconazole Fluconazole	[68–70]
Tracheobronchial aspergillosis	Itraconazole Intravenous Amphotericin B Voriconazole	[71,72]

6.1 AMPHOTERICIN B

Amphotericin B binds to ergosterol, which is a primary sterol content in the fungal cell membrane. It is released from the phospholipids, when it is in close proximity to ergosterol, allowing for delivery of enough amphotericin to the site of infection.[29] Amphotericin B activity has been demonstrated against most of the *Aspergillus* species, which is a common causative agent of pulmonary aspergillosis. The side effects of use of Amphotericin B include fever, chills, rigors, myalgias, arthralgias, bronchospasm, nausea, and vomiting, have been observed in infusion-related reactions. The nephrotoxic side effects of Amphotericin B include azotemia, renal tubular acidosis, hypokalemia, and hypomagnesemia. The morbidity and mortality associated with the use of Amphotericin B have prompted the search for improved antifungal agents with activity against *Aspergillus*.[73] Previous research has shown Amphotericin B used for the treatment of invasive aspergillosis given in doses of 5 mg/kg per day. It has been used for salvage therapy of IA at doses of 3–5 mg/kg per day, or 3 mg/kg per day.[3,74] A recent study has found that the efficacy and toxicity of drug showed similar results in both, suggesting that in this specific population of patients with early pulmonary aspergillosis higher doses are not beneficial when compared with initial therapy.

6.2 VORICONAZOLE

Voriconazole is attractive as an efficient antifungal agent and is considered to be the treatment of choice for IA and ABPA.[75,76] It has high efficacy against CPA, and is available in oral and intravenous formulation.[77] *In vitro* study has revealed that voriconazole is effective at lower dosages as compared with Amphotericin B against *Aspergillus* spp. In addition, voriconazole exhibits superior fungicidal activity against *A. fumigatus*, the leading cause of aspergillosis.[77] Voriconazole penetrates into tissues and body fluids, including the lung tissue and the pulmonary epithelial lining fluid.[78] Treatment of IA with voriconazole is initiated with a loading dose of 6 mg/kg IV every 12 h for two doses, followed by 4 mg/kg every 12 h. Oral voriconazole used 200 mg twice a day for CPA for a period of 6–12 months.[79]

6.3 ITRACONAZOLE

Itraconazole is lipophilic compound available in form of capsules, oral formulations. It is metabolized in liver and its metabolite forms are excreted into urine and biles. These excreted metabolites possess antifungal activity and effective against *Aspergillus* infections.[80] Itraconazole is most preferably used for IA treatment specifically in standard antifungal therapy intolerant patients. It plays an adjective role in management of allergic bronchopulmonary aspergillosis.[81] Recommended doses of itraconazole for IA are 400 mg/day in case of capsules and 2.5 mg/kg two times a day in case of oral formulations. Intravenous formulation of itraconazole dose of 200 mg/day is preferably recommended.[82]

6.4 POSACONAZOLE

Posaconazole is given for treatment of IA to patients with aplastic anemia and CPA.[83] Posaconazole was also approved for treatment of invasive aspergillosis that is resistance to Amphotericin B formulation and to itraconazole. It is administrated orally in dose of 200 mg, four times a day.[3,84] Felton et al.[67] reported the efficacy and safety of posaconazole for the treatment of choronic pulmonary aspergillosis and described response rates of 61% at 6 months and 46% at 12 months, with a relatively low incidence of side effects.

7 MULTIDRUG RESISTANCE

Drug resistance to azoles is most common in pulmonary aspergillosis cases as compared with other antifungal drugs used for treatment. Azole resistance in patients has been developed, either during treatment or due to resistant spores present in environment. Although there are many antifungal agents available for the treatment of pulmonary aspergillosis, the patients still developed resistance. In order to develop new antifungal treatment we need to understand how antifungal drug resistance occurs. Antifungal drug resistance is detected in fungal cell, when it remains unaffected by antifungal drug even at curative concentration. The antifungal drug resistance might be due to alteration in target site of drug, biofilm formation, and increased efflux of drug and may be due to mutation that lowers the toxic effect of drug. Patients suffering from IPA, aspergilloma, chronic necrotizing aspergillosis, ABPA, tracheobronchial aspergillosis acquired drug resistance because of long-term exposure to the antifungal drugs.

Infections caused by *Aspergillus* spp. are responsible for a high rate of morbidity and mortality in patients at risk. Mostly azoles are used for the treatment and prevention of *Aspergillus* infections. Increasing study on pulmonary aspergillosis showed that *Aspergillus* has developed resistance against antifungal agents specially azoles.[85] In *A. fumigatus*, azole resistance occurs due to mutation of genes leads to alteration in target protein.[86]

Resistance to triazole azole antifungal drugs in *A. fumigatus* is now a major clinical problem in a number of European locations, in China, Canada, and the US, with particularly high frequencies from the North-West of the UK, and the Netherlands.[87] The treatment failure and continuous increase in occurrence of azole-resistant *A. fumigatus* which is a causative agent of pulmonary aspergillosis has been reported through many studies.[21,88–92]

The clinical study of van der Linden et al.[93] showed that *A. fumigatus* is highly resistant to voriconazole. They screened clinically isolated *A. fumigatus* species for azole resistance in different patient types in the Netherlands. The results reveal that voriconazole followed by itraconazole and posaconazole showed resistance in patients who receive azole as primary treatment for invasive aspergillosis. Now, Itraconazole resistance is also increasing in India, Canada, China, United States, and so on. Patients treated with itraconazole also showed cross resistance to voriconazole and posaconazole.[94] The survey on development of azole resistance in *Aspergillus* reveals severity of the resistance particularly in patients residing in Japan 11%, China 4%, United States 3.6% and India 2%.

Chowdhary et al.[95] studied the widespread of resistance in clinical *Aspergillus* isolates in a referral chest hospital in Delhi, India during 4 years. About 1.75% of clinical isolates of *Aspergillus* were found resistant to azoles. One year nationwide survey carried out by Vermeulen et al.[96] to study *Aspergillus* resistance among patients suffering from aspergillosis in Belgium. The epidemiological data was evaluated up to one year after incorporation of drug in patients. Azole resistance was observed in 7% of patients suffering from aspergilloma or chronic aspergillosis, 5.5% in patients suffering from ABPA, and 4.6% in patients with IA. A study carried out by Choukri et al.[97] in France reported emerging problem of azole resistance in *A. fumigatus*. Their study observed azole resistance among the tested isolates of *A. fumigatus* was 1.8% among total patients enrolled.

Voriconazole is still primary treatment of choice for pulmonary aspergillosis. The high level of resistance was observed in patients treated with azoles, particularly itraconazole, and in some patients cross resistance was observed between itraconazole and posaconazole.[98]

8 CONCLUSIONS

Aspegillus-related infections are increasing day by day, causing many diseases including classic aspergillosis, infections of the auditory canal, skin, nails, eyes, sinuses, meninges and bones, etc. It is the most common fungus causing allergic sinusitis. Among these, pulmonary aspergillosis is predominant. ABPA has been reported in people who are allergic to the *Aspergillus*. In IPA, pulmonary parenchyma is affected by *Aspergillus* sp. It has been reported that pulmonary aspergillosis is common in immunocompromised hosts. Considering this fact, it is imperative to identify the pathogens by morphological and molecular markers in order to diagnose the authentic pathogen responsible for the diseases. In addition, clinical identification of pulmonary aspergillosis is also important, which is commonly based on histopathology, galactomannam antigen test, Beta-D-Glucan test, chest radiography, CT, volatile compound and LFD. Many drugs are available for the treatment of aspergillosis including Amphotericin B and its lipid formulations, itraconazole, voriconazole, posaconazole, and caspofungin. Unfortunately, the multidrug resistance problem has been increasing rapidly; for example, azole resistance is common in pulmonary aspergillosis. In fact, there is a greater need to study the reasons and mechanism of drug resistance in pulmonary aspergillosis. It has been found that *A. fumigatus* is highly resistant to voriconazole. The drug resistance in fungi, particularly in Aspergilli, is a matter of great concern and warrants thorough investigation to understand the real mechanism involved in drug resistance. The emergence of nanotechnology has promise to tackle this problem by using metal nanoparticles in combination with antifungal agents like azoles.

REFERENCES

1. Visagie CM, Hirooka Y, Tanney JB, Whitfield E, Mwange K, Meijer M, Amend AS, Seifert KA, Samson RA. *Aspergillus*, *Penicillium* and *Talaromyces* isolated from house dust samples collected around the world. *Stud Mycol* 2014;**78**:63–139.
2. Nikumbh DB, Kanthikar SN, Desale SS, Rajeshwari K. Cytodiagnosis of sinonasal aspergillosis—a case report with brief review of literature. *World J Pathol* 2015;**4**:39–43.
3. Walsh TJ, Anaissie EJ, Denning DW, Herbrecht R, Kontoyiannis DP, Marr VA, et al. Treatment of Aspergillosis: clinical practice guidelines of the Infectious Diseases Society of America. *Clin Infect Dis* 2008;**46**(3):327–60.
4. Badawy M, Badawy B, Yousef LM, Sherief N. Evaluation of pulmonary fungal diseases in patients with fungal rhino-sinusitis. *Egypt J Chest Dis Tubercul* 2013;**62**(3):493–500.
5. Peral B, Redondo-González L, Verrie-Hernández A. Invasive maxillary sinus aspergillosis: a case report successfully treated with voriconazole and surgical debridement. *Clin Exp Dent* 2014;**64**:448–51.
6. Sharma D, Mahajan N, Rao S, Khurana N, Jain S. Invasive maxillary aspergillosis masquerading as malignancy in two cases: utility of cytology as a rapid diagnostic tool. *Cytol* 2012;**29**(3):194–6.
7. Kousha M, Tadi R, Soubani AO. Pulmonary aspergillosis: a clinical review. *Eur Respir Rev* 2011;**20**(121):156–74.
8. Person AK, Chudgar SM, Norton, Tong BC, Stout JE. *Aspergillus niger*: an unusual cause of invasive pulmonary aspergillosis. *J Med Microbiol* 2010;**59**(7):834–8.
9. Bjorkholm M, Kalin M, Grane P, Celsing F. Long-term treatment of invasive sinus, tracheobroncheal, pulmonary and intracerebral aspergillosis in acute lymphoblastic leukaemia. *Infection* 2012;**40**(1):81–5.
10. Hall GL, Siles EV, Borzykowski RM, Gruson KI, Dorfman HD, Geller DS. *Aspergillus* osteomyelitis of the proximal humerus: a case report. *Skeletal Radiol* 2012;**41**(8):1021–5.
11. Karaman I, Karaman A, Bodurolu EC, Erdoan D, Tanir G. Invasive *Aspergillus* infection localized to the gastric wall: report of a case. *Surg Today* 2013;**43**(6):682–4.

12. Bergeron A, Porcher R, Sulahian A, de Bazelaire C, Chagnon K, Raffoux E, et al. The strategy for the diagnosis of invasive pulmonary aspergillosis should depend on both the underlying condition and the leukocyte count of patients with hematologic malignancies. *Blood* 2012;**119**(8):1831–7.

13. Tunnicliffe G, Schomberg L, Walsh S, Tinwell B, Harrison T, Chua F. Airway and parenchymal manifestations of pulmonary aspergillosis. *Respir Med* 2013;**107**(8):1113–23.

14. Gupta R, Chandra A, Gautam P. Allergic bronchopulmonary Aspergillosis—a clinical review. *J Assoc Physicians India* 2012;**60**:46–51.

15. Patterson K, Strek ME. Allergic bronchopulmonary aspergillosis. *Proc American Thorac Soc* 2010;**73**:237–44.

16. Hernandez SF, Cortez AK, Varon J. Invasive pulmonary aspergillosis in an immunocompetent host. *Crit Care Shock* 2015;**18**:43–5.

17. Aggarwal R, Chakrabarti A, Shah A, Gupta D, Meis JF, Guleria R, Moss R, Denning DW. Allergic bronchopulmonary aspergillosis: review of literature and proposal of new diagnostic and classification criteria. *Clin Exp Allergy* 2013;**43**:850–73.

18. Zmeili OS, Soubani AO. Pulmonary Aspergillosis: a clinical update. *QJM An Int J Med* 2007;**100**:317–24.

19. Agarwal R, Khan A, Aggarwal AN, Saikia B, Gupta D, Chakrabarti A. Role of inhaled corticosteroids in the management of serological allergic bronchopulmonary aspergillosis ABPA. *Intern Med* 2011;**50**:855–60.

20. Chakrabarti A, Chatterjee SS, Das A, Shivaprakash MR. Invasive aspergillosis in developing countries. *Med Mycol* 2011;**49**(Suppl. 1):35–47.

21. Chowdhary A, Agarwal K, Kathuria S, Gaur SN, Randhawa HS, Meis JF. Allergic bronchopulmonary mycosis due to fungi other than *Aspergillus*: a global overview. *Crit Rev Microbiol* 2013;**40**(1):30–48.

22. Hinson K, Moon V, Plummer N. Broncho-pulmonary aspergillosis. *Thorax* 1952;**7**:317–33.

23. Svirshchevskaya E, Zubkov D, Mouyna I, Berkova N. Innate immunity and the role of epithelial barrier during *Aspergillus fumigatus* infection. *Current Immunol Rev* 2012;**8**:254–61.

24. Gefter WB, Weingrad TR, Epstein DM, Ochs RH, Miller WT. Semi-invasive pulmonary aspergillosis: a new look at the spectrum of *Aspergillus* infections of the lung. *Radiology* 1981;**140**(2):313–21.

25. Lovrenski A, Panjkovic V, Eri Z, Klem I, Povazan D, Ilincic D, Milic M. Chronic necrotizing pulmonary aspergillosis. *Vojnosanit Pregle* 2011;**68**(11):988–91.

26. Nam HS, Jeon K, Um SW, Suh GY, Chung MP, Kim H, Kwon OJ, Koh WJ. Clinical characteristics and treatment outcomes of chronic necrotizing pulmonary aspergillosis: a review of 43 cases. *Int J Infect Dis* 2010;**14**:479–82.

27. Izumikawa K, Takazono T, Kohno S. Chronic *Aspergillus* infections of the respiratory tract: diagnosis, management and antifungal resistance. *Curr Opin Infect Dis* 2010;**23**:584–9.

28. Smith NL, Denning DW. Underlying conditions in chronic pulmonary aspergillosis, including simple aspergilloma. *Eur Respir J* 2010;**37**:865–72.

29. Thompson GR, Patterson TF. Pulmonary Aspergillosis. *Semin Respir Crit Care Med* 2008;**282**:103–10.

30. Barton RC. Laboratory diagnosis of invasive Aspergillosis: from diagnosis to prediction of outcome. *Scientifica* 2013;**1**:1–29.

31. Marchetti O, Lamoth F, Mikulska M, Viscoli C, Verweij P, Bretagne S. European conference on infections in leukemia ecil laboratory working groups: Ecil recommendations for the use of biological markers for the diagnosis of invasive fungal diseases. *Bone Marrow Transplant* 2012;**47**:846–54.

32. Bodey G, Boueltman D, Duguid W, Gibbs D, Hanak H, Hotchi M, et al. Fungal infection in cancer patients an international autopsy survey. *Eur J Clin Microbiol* 1992;**11**:09–99.

33. Maschmeyer G, Haas A, Cornely OA. Invasive aspergillosis: epidemiology, diagnosis and management in immunocompromised patients. *Drugs* 2007;**67**(11) 1567–1501.

34. Gerson SL, Talbot GH, Hurwitz S, George D, Talbot H, Hurwitz S, et al. Prolonged granulocytopenia: the major risk factor for invasive pulmonary aspergillosis in patients with acute leukemia. *Ann Intern Med* 1984;**100**:345–51.

35. Guinea J, Torres-Narbona M, Gijon P, Munoz P, Pozo F, Pelaez T, et al. Pulmonary aspergillosis in patients with chronic obstructive pulmonary disease: incidence, risk factors, and outcome. *Clin Microbiol Infec* 2010;**16**:870–7.

36. Xu H, Li L, Huang WJ, Wang LX, Li WF, Yuan WF. Invasive pulmonary aspergillosis in patients with chronic obstructive pulmonary disease: a case control study from China. *Clin Microbiol Infec* 2012;**18**:403–8.

37. Zhang S, Wang S, Wan Z, Li R, Yu J. The diagnosis of invasive and noninvasive pulmonary aspergillosis by serum and bronchoalveolar lavage fluid galactomannan assay. *BioMed Res Int* 2015;**43691**:1–5.

38. Ved P, Mishra P, Verma S, Sinha S, Sharma M. Prevalence and fungal profile of pulmonary aspergillosis in immunocompromised and immunocompetent patients of a tertiary care hospital. *Int J Med Res Health Sci* 2013;**31**:92–7.

39. Nyongesa B, Okoth S, Ayug V. Identification key for *Aspergillus* species isolated from maize and soil of Nandi County, Kenya. *Adv Microbiol* 2015;**5**:205–29.

40. Silva DM, Batista LR, Rezende EF, Helena M, Fungaro P, Sartori D, Alve E. Identification of fungi of the genus *Aspergillus* section *Nigri* using polyphasic taxonomy. *Braz J Microbiol* 2011;**42**:761–73.

41. Bandh SA, Kamili AN, Ganai BA. Dentification of some *Aspergillus* species isolated from Dal Lake, Kashmir by traditional approach of morphological observation and culture. *Afr J Microbiol Res* 2012;**6**(29):5824–7.

42. Gajjar D, Pal A, Ghodadra B, Vasavada A. Microscopic evaluation, molecular identification, antifungal susceptibility, and clinical outcomes in *Fusarium*, *Aspergillus* and dematiaceous keratitis. *BioMed Res Int* 2013;1–10.

43. Caramalho R, Gusma L, Lackner M, Amorim A, Araujo R. SNaPAfu: a novel single nucleotide polymorphism multiplex assay for *Aspergillus fumigatus* direct detection, identification and genotyping in clinical specimens. *Plos One* 2013;**8**(10):1–9.

44. Guinea J, Padilla C, Escribano P, Munoz P, Padilla B, Gijon P, et al. Evaluation of MycAssayTM *Aspergillus* for diagnosis of invasive pulmonary aspergillosis in patients without hematological cancer. *Plos One* 2013;**8**(4):1–7.

45. Hofman V, Dhouibi A, Butori C, Padovan B, Toussaint M, Hermoso D, et al. Usefulness of molecular biology performed with formaldehyde-fixed paraffin embedded tissue for the diagnosis of combined pulmonary invasive mucormycosis and aspergillosis in an immunocompromised patient. *Diagn Pathol* 2010;**5**:1–7.

46. Diba K, Makhdoomi K, Mirhendi H. Molecular characterization of *Aspergillus* infections in an Iranian educational hospital using RAPD-PCR method. *Iran J Basic Med Sci* 2014;**17**(9):646–50.

47. Sherif R, Segal B. Pulmonary aspergillosis: clinical presentation, diagnostic tests management and complications. *Curr Opin Pulm Med* 2010;**16**(3):242–50.

48. White PL, Parr C, Thornton C, Barnes RA. Evaluation of real-time PCR, galactomannan enzyme-linked immunosorbent assay ELISA, and a novel lateral-flow device for diagnosis of invasive aspergillosis. *J Clin Microbiol* 2013;**51**:1510–6.

49. Arvanitis M, Mylonakis E. Diagnosis of invasive aspergillosis: recent developments and ongoing challenges. *Eur J Clin Invest* 2015;**45**:647–62.

50. Karageorgopoulos DE, Vouloumanou EK, Ntziora F, Michalopoulos A, Rafailidis PI, Falagas ME. beta-D-glucan assay for the diagnosis of invasive fungal infections: a meta-analysis. *Clin Infect Dis* 2011;**52**:750–70.

51. Wright WF, Overman SB, Ribes JA. 1-3-β-D-Glucan assay: a review of its laboratory and clinical application. *Lab Med* 2011;**42**(11):679–85.

52. Shah A, Panjabi C. Allergic aspergillosis of the respiratory tract. *Eur Respir Rev* 2014;**23**:8–29.

53. Heddergott C, Calvo AM, Latge JP. The volatome of *Aspergillus fumigatus*. *Eukaryot cell* 2014;**13**(8):1014–25.

54. Chambers ST, Syhre M, Murdoch DR, McCartin F, Epton MJ. Detection of 2-pentylfuran in the breath of patients with *Aspergillus fumigatus*. *Med Mycol* 2009;**47**:468–76.

55. Patterson TF. Advances and challenges in management of invasive mycoses. *Lancet* 2005;**366**:1013–25.

56. Sambatakou H, Dupont B, Lode H, Denning DW. Voriconazole treatment for subacute invasive and chronic pulmonary aspergillosis. *Am J Med* 2006;**119**(6) 527.e17–e24.

57. Chai LYA, Kullberg BJ, Earnest A, Johnson EM, Teerenstra S, Vonk AG, et al. Voriconazole or Amphotericin b as primary therapy yields distinct early serum galactomannan trends related to outcomes in invasive aspergillosis. *PLoS ONE* 2014;**92**:e90176.

58. Herbrecht R, Denning DW, Patterson TF, Bennett JE, Greene RE, Oestmann JW, et al. Voriconazole versus amphotericin B for primary therapy of invasive aspergillosis. *N Engl J Med* 2002;**347**:408–15.

59. Ashbee HR, Barnes RA, Johnson EM, Richardson MD, Gorton R, Hope WW. Therapeutic drug monitoring TDM of antifungal agents: guidelines from the British Society for Medical Mycology. *J Antimicrob Chemother* 2014;**69**:1162–76.

60. Cornely OA, Maertens J, Bresnik M, Ebrahimi R, Ullmann AJ, Bouza E, et al. Liposomal amphotericin B as initial therapy for invasive mold infection: a randomized trial comparing a high-loading dose regimen with standard dosing AmBiLoad trial. *Clin Infect Dis* 2007;**44**:1289–97.

61. Lass-Florl C. Triazole antifungal agents in invasive fungal infections: a comparative review. *Drugs* 2011;**7118**:2405–19.

62. Richardson MD, Warnock DW. *Fungal infection: diagnosis and management.* 4th Edition Chichester: Wiley-Blackwell; 2012.

63. Bassetti M, Pecori D, Della Siega P, Corcione S, De Rosa FG. Current and future therapies for invasive aspergillosis. *Pulm Pharmacol Ther* 2015;**32**:155–65.

64. De Beule K, De Doncker P, Cauwenbergh G, Koster M, Legendre R, Blatchford N, et al. The treatment of aspergillosis and aspergilloma with itraconazole, clinical results of an open international study 1982-1987. *Mycoses* 1988;**31**:476–85.

65. Campbell JH, Winter JH, Richardson MD, Shankland GS, Banham SW. Treatment of pulmonary aspergilloma with itraconazole. *Thorax* 1991;**46**:839–41.

66. Caras WE, Pluss JL. Chronic necrotizing pulmonary aspergillosis: pathologic outcome after itraconazole therapy. *Mayo Clin Proc* 1996;**71**(1):25–30.

67. Felton TW, Baxter C, Moore CB, Roberts SA, Hope WW, Denning DW. Efficacy and safety of posaconazole for chronic pulmonary aspergillosis. *Clin Infect Dis* 2010;**51**(12):1383–91.

68. Stevens DA, Schwartz HJ, Lee JY, Moskovitz BL, Jerome DC, Catanzaro A, et al. A randomized trial of itraconazole in allergic bronchopulmonary aspergillosis. *N Engl J Med* 2000;**342**:756–62.

69. Bandres Gimeno R, Munoz Martinez MJ. Prolonged therapeutic response to voriconazole in a case of allergic bronchopulmonary aspergillosis. *Arch Bronconeumol* 2007;**431**:49–51.

70. Erwin GE, Fitzgerald JE. Case report: allergic bronchopulmonary aspergillosis and allergic fungal sinusitis successfully treated with voriconazole. *J Asthma* 2007;**44**:891–5.

71. Conte I, Riva G, Obert R, Lucchini A, Bechis G, De Rosa G, et al. Tracheobronchial aspergillosis in a patient with AIDS treated with aerosolized amphotericin B combined with itraconazole. *Mycoses* 1996;**39**(9–10):371–4.

72. Zhang CR, Li M, Lin JC, Wen Ming XU, Niu YY. Voriconazole used for treatment of tracheobronchial aspergillosis: a report of two cases. *J Clin Trials* 2012;**2**:117.

73. Bates DW, Su L, Yu DT, Chertow GM, Seger DL, Gomes DR, et al. Mortality and costs of acute renal failure associated with amphotericin B therapy. *Clin Infect Dis* 2001;**32**:686–93.

74. Denning DW. Therapeutic outcome in invasive aspergillosis. *Clin Infect Dis* 1996;**23**:608615.

75. Glackin L, Leen G, Elnazir B, Greally P. Voriconazole in the treatment of allergic bronchopulmonary aspergillosis in cystic fibrosis. *Ir Med J* 2009;**102**:29.

76. Chishimba L, Niven RM, Cooley J, Denning DW. Voriconazole and posaconazole improve asthma severity in allergic bronchopulmonary aspergillosis and severe asthma with fungal sensitization. *J Asthma* 2012;**49**:423–33.

77. Perfect JR, Marr KA, Walsh TJ, Greenberg RN, DuPont B, de la Torre-Cisneros J, et al. Voriconazole treatment for less common, emerging, or refractory fungal infections. *Clin Infect Dis* 2003;**36**:1122–31.

78. Capitano B, Potoski BA, Husain S, Zhang S, Paterson DL, Studer SM, et al. Intrapulmonary penetration of voriconazole in patients receiving an oral prophylactic regimen. *Antimicrob Agents Chemother* 2006;**50**:1878–80.

79. Cadranel J, Philippe B, Hennequin C, Bergeron A, Bergot E, Bourdin A, et al. Voriconazole for chronic pulmonary aspergillosis:a prospective multicenter trial. *Eur J Clin Microbiol Infect Dis* 2012;**31**:3231–9.
80. Slain D, Rogers PD, Cleary JD, Chapman SW. Intravenous itraconazole. *Ann Pharmacother* 2001;**35**:720–9.
81. Rai SP, Panda BN, Bhargava S. Treatment of allergic bronchopulmonary aspergillosis with fluconazole and itraconazole. *Med J Armed Force India* 2004;**60**:128–30.
82. Caillot D. Intravenous itraconazole followed by oral itraconazole for the treatment of Amphotericin B refractory invasive pulmonary aspergillosis. *Acta Haematol* 2003;**109**(3):111118.
83. Kohno S, Izumikawa K. Posaconazole for chronic pulmonary aspergillosis: the next strategy against the threat of azole-resistant *Aspergillus* infection. CID 2010:51.
84. Khan SA. Antifungal therapy for invasive aspergillosis. *US Pharm* 2013;**384**:2–5.
85. Lelièvre L, Groha M, Angebault C, Maherault AC, Didier E, Bougnoux ME. Azole resistant *Aspergillus fumigatus*: an emerging problem. *Médecine et maladies infectieuses* 2013;**43**:139–45.
86. Verweij PE, Howard SJ, Melchers WJ, Denning DW. Azole resistance in *Aspergillus*: proposed nomenclature and breakpoints. *Drug Resist Updat* 2009;**12**:141–7.
87. Bowyer P, Moore CB, Rautemaa R, Denning DW, Richardson MD. Azole antifungal resistance today: focus on *Aspergillus. Curr Infect Dis Rep* 2011;**6**:485–91.
88. Burgel PR, Baixench MT, Amsellem M, Audureau E, Chapron J, Kanaan R, et al. High prevalence of azole-resistant *Aspergillus fumigatus* in adults with cystic fibrosis exposed to itraconazole. *Antimicrob Agents Chemother* 2012;**56**:869–74.
89. Chowdhary A, Kathuria S, Randhawa HS, Gaur SN, Klaassen CH, Meis JF. Isolation of multiple-triazole-resistant *Aspergillus fumigatus* strains carrying the TR/L98H mutations in the cyp51A gene in India. *J Antimicrob Chemother* 2012;**67**:362–6.
90. Escribano P, Peláez T, Muñoz P, Bouza E, Guinea J. Is azole resistance in *Aspergillus fumigatus* a problem in Spain? *Antimicrob Agents Chemother* 2013;**57**:2815–20.
91. Lavergne RA, Morio F, Favennec L, Dominique S, Meis JF, Gargala G, et al. First description of azole-resistant *Aspergillus fumigatus* due to TR46/Y121F/T289A mutation in France. *Antimicrob Agents Chemother* 59(7):4331–5.
92. Steinmann J, Hamprecht A, Vehreschild MJ, Cornely OA, Buchheidt D, Spiess B, et al. Emergence of azole-resistant invasive aspergillosis in HSCT recipients in Germany. *J Antimicrob Chemother* 2015;**70**:1522–6.
93. van der Linden JW, Camps SM, Kampinga GA, Arends JP, Debets-Ossenkopp YJ, Haas PJ, et al. Aspergillosis due to voriconazole highly resistant *Aspergillus fumigatus* and recovery of genetically related resistant isolates from domestic homes. *Clin Infect Dis* 2013;**57**:513–20.
94. Denning DW, Bowyer P. Voriconazole resistance in *Aspergillus fumigatus*: should we be concerned? *Clin Infect Dis* 2013;**574**:521–3.
95. Chowdhary A, Sharma C, Kathuria S, Hagen F, Meis JF. Prevalence and mechanism of triazole resistance in *Aspergillus fumigatus* in a referral chest hospital in Delhi, India and an update of the situation in Asia. *Front Microbiol* 2015;**6**:428.
96. Vermeulen E, Maertens J, Bel AD, Nulens E, Boelens J, Surmont I, et al. Nationwide surveillance of azole resistance in *Aspergillus* disease. *Antimicrob Agents Chemother* 2015;**59**(8):4569–76.
97. Choukri F, Botterel F, Sitterlé E, Bassinet L, Foulet F, Guillot J. Prospective evaluation of azole resistance in *Aspergillus fumigatus* clinical isolates in France. *Med Mycol* 2015;**53**(6):593–6.
98. Bader O, Weig M, Reichard U, Lugert R, Kuhns M, Christner M, et al. *cyp51A*-based mechanisms of *Aspergillus fumigatus* azole drug resistance present in clinical samples from Germany. *Antimicrob Agents Chemother* 2013;**57**(8):3513–7.

LABORATORY DIAGNOSIS OF *PNEUMOCYSTIS JIROVECII* PNEUMONIA

O. Matos*, F. Esteves**

*Medical Parasitology Unit, Group of Opportunistic Protozoa/HIV and Other Protozoa, Global Health and Tropical Medicine, Instituto de Higiene e Medicina Tropical, Universidade Nova de Lisboa, Portugal;
**Department of Genetics, Toxicogenomics & Human Health (ToxOmics), NOVA Medical School/Faculdade de Ciências Médicas, Universidade Nova de Lisboa, Portugal

1 INTRODUCTION

Pneumocystis jirovecii (formerly *Pneumocystis carinii* f. sp. *hominis*) is an atypical fungus exhibiting pulmonary tropism and highly defined host specificity. The story of *Pneumocystis* began in Brazil in 1909 where Carlos Chagas, a young physician working at Oswaldo Cruz Institute, in Rio de Janeiro, mistakenly identified *Pneumocystis* as a stage of the life cycle of *Trypanosoma cruzi*, in the lungs of guinea pigs.[1] In 1910, Antonio Carini, an Italian scientist living in the city of São Paulo also in Brazil, found similar cysts in the lungs of *Rattus norvegicus*, but doubted that the cysts were a part of the life cycle of *T. cruzi*.[2] Because of this doubt, he sent his data and specimens to the Pasteur Institute in Paris, where in 1912, the researcher duo of Pierre and Marie Delanoë described the organism identified as a new biological entity, and suggested naming it *Pneumocystis carinii* because cysts were found only in the lungs of hosts and Carini had provided the specimens for study.[3]

In the early 20th century, this pathogen was considered an enigmatic lung pathogen, but the year 1981 became a turning point where *P. carinii*, was no longer seen as a relatively obscure human pathogen but became one of the leading causes of death in the global epidemic of acquired immunodeficiency syndrome (AIDS).[4]

Classified at first as a protozoan, it was latter reclassified as a fungus based on greater DNA sequence homology with fungal organisms.[5,6] *Pneumocystis* infects a variety of mammalian hosts. The human form, because of its host specificity, was renamed *Pneumocystis jirovecii* in honor of Otto Jirovec, which linked it to epidemics of interstitial plasma cell pneumonia in neonates in Europe,[7] while *P. carinii* is now reserved for the rat form of *Pneumocystis*.

Pneumocystis pneumonia (PcP or pneumocystosis) is an opportunistic disease with airborne transmission predominantly reported in patients with impaired immunity, mainly among human immunodeficiency virus (HIV)-infected persons.[8,9] In developed countries, widespread use of PcP chemoprophylaxis and potent combination antiretroviral therapy (cART) have reduced the incidence of this pathology.[10] Despite these advances, in the 21st century, PcP continues to be a serious problem for HIV-infected patients, especially for those who are undiagnosed or who are noncompliant with preventive medication.[11] Recently, PcP has been reported as a serious emerging problem in non-HIV-infected

The Microbiology of Respiratory System Infections. http://dx.doi.org/10.1016/B978-0-12-804543-5.00013-0

persons who are undergoing immunosuppressive treatments related to malignancies, connective tissue diseases, or organ transplantation. Reports on pulmonary colonisation with *P. jirovecii* in patients presenting diverse levels of immunodeficiency, primary respiratory disorders, or even in the immunocompetent general population, is also an important epidemiological issue, especially in terms of transmission.[9,12–14] In addition, in developing countries, where most HIV-infected persons reside, PcP is an emerging disease with high prevalence and is poorly controlled since the access to cART and PcP prophylaxis is still limited; and possibly also due to lack of PcP diagnostic resources and expertise.[15–18]

Therefore, prevention of PcP and control of existing cases through early diagnosis remains a very important objective from a public health point of view in all countries, and especially in the developing countries.

2 LABORATORY DIAGNOSIS OF PCP

Pneumocystis jirovecii is an opportunistic pathogen that is usually found in the lungs of humans, but which has also been found in extrapulmonary sites.[19] In general, laboratory findings are less severe in HIV-infected patients than in non-HIV immunosuppressed patients.[20]

The presumptive diagnosis of PcP is based on: clinical manifestations, pulmonary function testing, arterial blood gas testing (ABG) at rest and after exercise, and nonspecific radiological and laboratory tests. The history and physical examination are part of the evaluation of the patient with pulmonary symptoms. Over the years it was noticed that the presentation of PcP depends on the underlying disease. Thus, in HIV-infected patients the duration of the clinical manifestations is longer and the diagnosis is more difficult than in patients with other immune deficiencies.[21]. Pulmonary manifestations are the most frequent in the natural history of PcP. Symptoms and signs typical of a patient with PcP are fever between 38 and 40°C in more than 80% of patients, nonproductive cough in more than 50% of patients which may be accompanied by sputum in 20–30% of the cases, progressively worsening dyspnea in more than 60% of patients, and sometimes discrete crackling rales. This condition occurs over a period of one to 3 weeks. In addition, tachypnea, tachycardia, and occasionally cyanosis can also be observed, but the auscultation may also be normal.[22,23] The chest X-ray characteristics of a patient with PcP can, in 40–80% of cases, show a classic image of diffuse interstitial bilateral infiltrates. However, in 6–20% of the cases of PcP, the radiological patterns may be atypical or even normal.[22,23] Taking into account that PcP is an interstitial disease, and as such, causes serious difficulties in gas exchange occurring in the lungs, another parameter used in the presumptive diagnosis of PcP is ABG at rest and after exercise. A partial pressure of oxygen (PaO_2) in peripheral blood $\leq$ 9.3 kPa (70 mmHg) is indicative of PcP inspite of the fact that 10–20% of cases of PcP have normal levels of PaO_2.[23] The measurement of lactate dehydrogenase (LDH), which increases in the beginning phase of the infection, is a nonspecific laboratory test for PcP diagnosis that can be used as a prognostic tool, and to assess response to PcP therapy.[24] In addition, the evaluation of the immune status of the patient is important to determine the risk of developing PcP. This disease occurs most frequently in patients with $CD4^+$ T cells count $\leq$ 200/μL blood. This information applies to HIV-infected patients and to non-HIV-infected patients with other immunodeficiency, such as cancer patients receiving chemotherapy.

All these elements are useful but nonspecific, neither confirming or nor denying the diagnosis of PcP. For many years *P. jirovecii* could not be cultured—a culture system to propagate *P. jirovecii* in vitro was only developed in 2014. Since this culture system still needs to be validated, disseminated, and shown to be cost-effective for diagnostic purposes,[25] microscopic visualization of cysts or trophic forms in respiratory specimens with cytochemical staining or immunofluorescent staining with monoclonal antibodies (IF/Mab) are the standard procedures to identify this microorganism and diagnose the disease.

2.1 CURRENT METHODS FOR DIAGNOSIS OF PCP

A comparative analysis of the most applied/studied laboratory diagnostic method for PcP diagnosis is summarized in Table 13.1. This table briefly shows the most important variables, such as sensitivity and specificity, operational and time costs, useful specimens, benefits, and disadvantages implicated in the laboratory diagnostic procedures available for *P. jirovecii* identification/detection and PcP diagnosis in a clinical laboratory.

2.1.1 Biological specimens

The diagnosis of PcP is determined by cyto-histopathological examination of respiratory specimens obtained by invasive techniques, such as open lung biopsy (LB), transbronchial biopsy (TBB), bronchoalveolar lavage fluid (BALF), and bronchial secretions (BS), and specimens obtained by less invasive techniques such as spontaneous sputum (SS), nasopharyngeal aspirate (NA), and oropharyngeal washing (OW). Currently, BALF and IS are the most widely used clinical specimens for the diagnosis of PcP. In general, noninvasive tests should be tried in order to make the initial diagnosis, and invasive techniques should be used only when necessary and clinically feasible.[34] Invasive diagnosis should be considered when dealing with non-HIV immunosuppressed patients with suspicion of PcP.

2.1.1.1 Specimens obtained by invasive methods

LB represents the gold standard laboratory diagnostic method in the assessment of lung inflammatory processes in immunosuppressed hosts.[34] The diagnosis of lung tissue fragments allows the observation of the microorganism in more than 95% of cases of infection.[35] Transbronchial biopsy (TBB) also allows the identification of *P. jirovecii* in 95% of cases, and should be considered mainly in non-HIV immunosuppressed patients with strong suspicion of PcP and negative BALF.[36,37] Bronchoalveolar lavage (BAL) is performed by fiberoptic bronchoscopy and instillation of 150 mL of physiologic saline solution preheated at 37°C, divided into three syringes of 50 mL each, into the middle lobe and then slowly aspirating.[38] According to some studies, the BALF analysis allows diagnosis in more than 80% of all patients with PcP, and in more than 95% of patients with HIV co-infection.[37,39] It is, however, an expensive and invasive procedure, which involves secondary hazards, such as pneumothorax.[40] Given its high sensitivity, this is the most frequently used technique in the diagnosis of fungal infections, particularly of PcP.[41–44] The BS are obtained by the aspiration of secretions from the main bronchi of the bronchial tree during bronchoscopy. However, in some cases, there may be a need to instill sterile saline solution in order to assist in the removal of secretions and to stimulate coughing.[45] This is especially performed in very young children.[46]

Sputum is a biological sample which may be obtained spontaneously by a noninvasive and inexpensive method that carries a low risk of complications, which, if any, are mostly transient and devoid of severity.[47] The SS collected routinely for culturing bacterial and fungal agents, is rarely diagnostic in cases of PcP.[48] In turn, induced sputum (IS) is based on the concept that changes in the microenvironment of the airways such as pH and osmolarity, as well as the activation of inflammatory mediators,

Table 13.1 Laboratory Methods for PcP Clinical Diagnosis. Includes Worldwide Microscopic, Molecular and Serologic Standard Technical Procedures for *P. jirovecii* Identification/Detection

Method	Technique	Sensitivity	Specificity	Estimated Cost/Sample USD (€)	Approximate Time Load (h)	Most Suitable Specimens (Also Available Specimens)	Observations	References
Microscopy	GMS	79% (BALF)	99% (BALF)	112.1 (102.9) (BALF)	3	BALF or biopsy	Needs experienced/qualified microscopist; needs invasive and expensive samples; cumbersome protocol; identification of cysts; recommended combination with Giemsa or Giemsa-like stains; allows semi-quantification methods; optical microscope	[26]
	TBO	68% (BALF)	100% (BALF)	103.2 (94.7) (BALF)	2	BALF or biopsy	Needs experienced/qualified microscopist; needs invasive and expensive samples; identification of cysts; recommended combination with Giemsa or Giemsa-like stains; allows semi-quantification methods; optical microscope	[27]
	Giemsa (or DQ)	68% (BALF)	88% (BALF)	104.2 (95.7) (BALF)	1	BALF or biopsy	Needs experienced/qualified microscopist (very difficult to read); needs invasive and expensive samples; rapid/easy protocol; identification of trophic forms and spores; recommended combination with GMS or TBO; allows semi-quantification methods; optical microscope	[28]
	IF	97% (BALF)	100% (BALF)	110.7 (101.6) (BALF)	2	BALF, IS or biopsy	Excellent sensitivity/specificity (robustness); most accurate/robust microscopic method; easy to read; needs invasive and expensive samples; identification of cysts and/or trophic forms; allows semi-quantification methods; needs expansive/specific equipment (fluorescence microscope)	[29]

Molecular / Serologic	Method					Sample	Comments	Ref
Molecular	nPCR	76–100% (BALF)	53–86% (BALF)	114.1 (104.7) (BALF)	8	BALF, IS or biopsy (OW, SS, NA)	Needs experienced/qualified staff; needs invasive and expensive samples; alternative non-invasive samples may be used; detection of low fungal burdens (eg, colonised patients); possible false positives; allows further genotyping; needs expensive/specific equipment (thermocycler)	[30]
	RT-qPCR	94–99% (BALF)	89–96% (BALF)	116.6 (107.0) (BALF)	4	BAL, IS or biopsy (OW, SS, NA)	Highthroughput format; needs experienced/qualified staff; needs invasive and expensive samples; alternative non-invasive samples may be used; possible false positives; detection of low fungal burdens (eg, colonised patients); allows quantification; needs expensive/specific equipment (real-time apparatus)	[31]
Serologic	BG	91% (serum)	77% (serum)	26.1 (24.0) (serum)	3	Serum	Highthroughput format; minimally invasive or inexpensive samples; suitable for screening; positive results in other fungal infections (false positives); recommended confirmation of results with GMS/Giemsa, TBO/Giemsa or IF; not quantitative; allows indirect quantification; needs expensive/specific equipment (microplate reader)	[32,33]
	KL-6	72% (serum)	79% (serum)	26.6 (24.4) (serum)	2.5	Serum	Highthroughput format; minimally invasive or inexpensive samples; positive results in other interstitial lung diseases (false positives); needs combination with GMS/Giemsa, TBO/Giemsa or IF; not quantitative; needs expensive/specific equipment (microplate reader)	

(*Continued*)

Table 13.1 Laboratory Methods for PcP Clinical Diagnosis. Includes Worldwide Microscopic, Molecular and Serologic Standard Technical Procedures for *P. jirovecii* Identification/Detection (*cont.*)

Method	Technique	Sensitivity	Specificity	Estimated Cost/Sample USD (€)	Approximate Time Load (h)	Most Suitable Specimens (Also Available Specimens)	Observations	References
	LDH	80% (serum)	52% (serum)	4.8 (4.4) (serum)	2	Serum	Highthroughput format; minimally invasive or inexpensive samples; positive results in organ damage cases (false positives); very low specificity; needs combination with GMS/Giemsa, TBO/Giemsa or IF; not quantitative; low cost but needs expensive/specific equipment (microplate reader)	
	SAM	68% (serum)	52% (serum)	14.3 (13.1) (serum)	2	Serum (plasma)	Highthroughput format; minimally invasive or inexpensive samples; very low robustness/accuracy; needs combination with GMS/Giemsa, TBO/Giemsa or IF; not quantitative; needs expensive/specific equipment (microplate reader)	
	BG/KL-6	94% (serum)	90% (serum)	49.4 (45.4) (serum)	5.5	Serum	Highthroughtput format; minimally invasive or expensive samples; most accurate serologic method; suitable for screening; not quantitative; needs combination with GMS/Giemsa, TBO/Giemsa or IF; not quantitative; needs expensive/specific equipment (microplate reader)	

BG/LDH	97% (serum)	72% (serum)	30.9 (28.4) (serum)	5	Serum	Highthroughput format; minimally invasive or expensive samples; needs combination with GMS/Giemsa, TBO/Giemsa or IF; not quantitative; needs expensive/specific equipment (microplate reader)
LDH/KL-6	89% (serum)	74% (serum)	31.4 (28.8) (serum)	4.5	Serum	Highthroughput format; minimally invasive or expensive samples; low specificity; needs combination with GMS/Giemsa, TBO/Giemsa or IF; not quantitative; needs expensive/specific equipment (microplate reader)

GMS, Grocott's Methenamine Silver stain; TBO, Toluidine Blue O; DC, Diff-Quick; IF, immunofluorescence staining; RT-qPCR, real-time quantitative PCR; BG, (1-3)-β-D-Glucan quantification assay; KL-6, Krebs von den Lungen-6 antigen quantification assay; LDH, lactate dehydrogenase quantification assay; SAM, S-adenosylmethionine quantification assay; BG/KL-6, combination test using BG and KL-6 quantification assays; BG/LDH, combination test using BG and LDH quantification assays; LDH/KL-6 combination test using LDH and KL-6 quantification assays; BALF, bronchoalveolar lavage fluid; IS, induced sputum; OW, oropharyngeal washing; SS, spontaneous sputum; NA, nasopharyngeal aspirate.

Serologic combination tests (BG/KL-6, BG/LDH, LDH/KL-6) are considered positive when both biomarkers levels are indicative of PcP, negative when both biomarkers levels are below the cutoff level for PcP, and undetermined when either one of the two biomarker assays yields contradictory results. Estimated costs per sample include sample collection (BALF or serum) and laboratorial technical procedure. Approximate time load was estimated based on previous data.[33]

can acutely increase secretions and make it possible to obtain specimens in patients who originally had unproductive cough.[49] The IS is obtained by inhaling 1.8% saline with the aid of an ultrasonic nebulizer for 10–15 min, which promotes transudation and tracheobronchial exfoliation. Patients should have taken nothing by mouth for 6–8 h prior to induction.[50] This is a less invasive method with less discomfort and risk for the patient than BALF collection, but it should be obtained only in infectious diseases clinics that carefully control sputum induction (having a single room with a ventilation system that allows for the total exhausting of air from the room to the external environment to avoid infectious droplets, if present, produced by coughing to be expelled into the room air).[51] There are also contraindications for sputum induction, such as, recent history of pneumothorax, extreme asthenia, or active tuberculosis. Sputum induction for the diagnosis of PcP is widely used for patients with AIDS, but its utility for patients with other forms of immunosuppression is less defined. Non-HIV immunosuppressed patients with PcP have lower burden of organisms, and sputum induction may consequently have lower diagnostic yield in these patients.[52] This technique allows *Pneumocystis* detection in 30–55% of cases of infection, after staining with nonspecific dyes, sometimes with difficult interpretation.[34,50,53] The problem of low sensitivity can be overcome with IS liquefaction with dithiothreitol, followed by cell sedimentation and staining. Mucus liquefaction allows the concentration and better visualization of clusters of cysts and trophic forms of *P. jirovecii* existent in the specimen observed under the microscope.[54] The application of immunofluorescence with monoclonal antibodies (MAb-IF) anti-*P. jirovecii* or PCR methodologies to liquefied IS, increases the sensitivity of detection from 60 to 97%.[50,52,55] In addition to the risk of associated complications, bronchoscopy as well as sputum induction are expensive and they require specialized personnel, rooms, and equipment.

2.1.1.2 Specimens obtained by noninvasive methods

As an alternative to BALF and IS, upper respiratory samples (NA and OW) have been studied to obtain biological specimens in a minimally invasive manner and with good sensitivity.[56–59] For the NA, a disposable catheter connected to a mucus extractor is inserted into the nasopharynx to a depth of 5–7 cm and drawn back while applying gentle suction with an electric suction device.[60] OWs are collected after patients rinse their oral cavities with 10 mL sterile saline and gargling for about 1 min. Before the collection of specimens, patients should have a good oral hygiene.[58] These specimens when studied by PCR technologies, have a sensitivity above 75%.[56,61]

The diagnosis of extrapulmonary pneumocystosis is possible by the demonstration of *P. jirovecii* cystic or trophic forms in affected tissues using a nonspecific or specific staining method, or a PCR method to detect DNA of the pathogen in those tissues.

2.1.2 Staining methodologies

Several staining methods have been described for the diagnosis of PcP. Gomori-methenamine-silver (GMS), Giemsa, or rapid Giemsa-like stains such as Diff-Quik, toluidine blue O (TBO), cresyl echt violet, calcofluor white and immunofluorescent stains with monoclonal anti-*P. jirovecii* antibodies (IF-MAb) reveal the characteristic morphology of the cystic and/or trophic forms. Mature cysts of *Pneumocystis* spp. are about 5 to 7 μm in diameter, round, and thick-walled and have eight spores inside, and the trophic forms can be round to ellipsoid or irregular in shape and size (1 to 10 μm long) (Fig. 13.1).

2.1.2.1 Cytochemical staining methods
2.1.2.1.1 Methods that stain the cysts wall
GMS described by Gomori and modified by Grocott[62] and Musto and collaborators[63] has been considered the gold standard technique for the diagnosis of PcP for many years. This reagent, initially

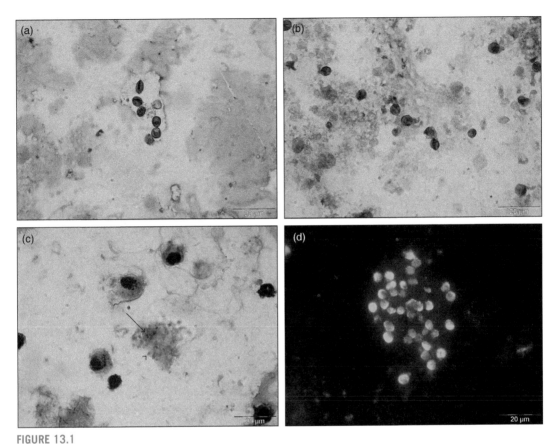

FIGURE 13.1

Respiratory Specimens with *Pneumocystis* after Appropriate Staining (Magnification ×1000)
(a) Cluster of *P. jirovecii* cysts in bronchoalveolar lavage fluid (BALF) stained with gomori-methenamine-silver
(GMS). (b) Rat-derived *Pneumocystis* cysts stained with toluidine blue O (TBO). (c) Clustered trophic and cystic
forms of *P. jirovecii* in BALF stained with Giemsa. A mature cyst (*arrow*), and several trophic forms (*arrowhead*)
are quite visible. (d) Cluster of *P. jirovecii* cysts in BALF stained with immunofluorescent-MAbs anti-*P. jirovecii*.

developed to stain fungi, selectively stains the *Pneumocystis* cyst wall, which appears dark brown[62]
(Fig. 13.1a).

TBO stain also has a good affinity for components of the cyst wall, which stains metachromatically
in reddish violet (Fig. 13.1b). GMS, TBO and cresyl echt violet, similar to TBO, all identify *Pneumocystis* cysts but not its developmental forms. They can be used in any kind of clinical specimens, but
they also stain the cell wall of yeasts and other fungi, which requires an experienced technician to read
the microscopic slides.

Calcofluor white is a chemifluorescent agent that nonspecifically binds to β-linked polysaccharide
polymers of the cell wall of *Pneumocystis*.[64] It also stains other microorganisms such as *Cryptospo-
ridium*, Microsporidia, and other fungi.

2.1.2.1.2 Methods that stain the nuclei of the developing forms

Giemsa and Diff-Quik stain the nuclei of all *Pneumocystis* life cycle stages. They do not stain cystic or sporocytic walls, which appear like a clear peripheral halo around these forms[38,65] (Fig. 13.1c). These panoptic stains, in addition to identifying trophic and sporocytic forms that are not identified by GMS and TBO, are also used for the diagnosis of other pathogens in the lungs. However, the reading of the microscopic slides is difficult, especially when the fungal burden is scarce, which requires observation by an experienced technician.

Overall, in respiratory specimens with *Pneumocystis,* the number of nuclei are about tenfold higher than the number of cysts.[66] Taking this into consideration, and since these staining methods have low sensitivity and are not specific, the best strategy for a more accurate PcP diagnosis in laboratories where they are the only methods implemented is the parallel use of a method that stains the cyst wall and a panoptical method that stains the nuclei of the developing forms, in different smears from the same specimen.

Histologic examination of biopsy specimens for recognition of foamy eosinophilic material in which *P. jirovecii* cysts or trophic forms are embedded can be achieved by performing GMS, Giemsa or Diff-Quik, TBO, or IF-Mab anti-*P. jirovecii*. This reveals the characteristic morphology of the cysts, which are easier to identify than the trophic forms in paraffin-embedded tissues.

2.1.2.2 Immunohistochemical staining methods

With the development of monoclonal antibodies (MAbs) specific for *P. jirovecii* in 1986, immunofluorescent techniques began to be implemented in the diagnosis of PcP.[67,68] They are considered more sensitive and specific than other cytochemical staining methods, which makes them the preferred method to apply in less-than-optimal specimens (IS, BS, NA, and OW).[50,55,68–70] Kits of direct and indirect immunofluorescent assays (DFA and IFA, respectively) with MAbs anti-*P. jirovecii* are commercially available for clinical use. The DFA identifies cysts and trophic forms while IFA identifies only cysts. The MAb anti-*P. jirovecii* conjugated with fluorescein isothiocyanate (FITC) bind the antigen specific of *P. jirovecii* which, when exposed to a given wavelength, emit characteristic apple-green fluorescence (Fig. 13.1d). Actually, because of its reliability, IF-MAb is the most commonly used technique in the diagnosis of PcP.[71–73]

2.1.3 Molecular diagnosis of PCP

2.1.3.1 Usefulness of molecular detection of *Pneumocystis jirovecii* in PcP diagnosis

In the last two decades, the detection of *P. jirovecii* DNA in clinical specimens by using molecular tools has brought important advances in PcP diagnosis, epidemiology, and management. PCR assays show high efficiency to amplify *P. jirovecii* DNA in clinical specimens.[10,14,74–79]

The absence, until 2014, of well established culture systems allowing routine isolation of *P. jirovecii* in clinical specimens for diagnostic purposes led to the development of highly efficient protocols that relied on PCR-based approaches to detect and characterize *P. jirovecii* worldwide.[42,74,75,79–87]

DNA-based methods for molecular diagnosis of PcP in blood specimens, such as serum or plasma, have been shown to be inefficient, leading to contradictory results. Despite efforts of several investigators, *P. jirovecii* DNA detection in blood did not demonstrate value for PcP diagnosis.[8,88–90] Invasive methods such as bronchoscopy, the first-line procedure to obtain reliable respiratory specimens for PcP diagnosis (BALF), cannot be used for monitoring the response to the anti-PcP treatment and disease progression or follow-up.[18,33,91] In general, invasive respiratory specimens are not easy to obtain in patients with respiratory failure or in children and are impossible to collect in asymptomatic carriers. Additionally, the apparatus needed for BAL or sputum induction are expensive and especially difficult to implement in developing countries.[18,92] For these reasons the highly specific molecular methods were mainly developed in order to increase the sensitivity of *P. jirovecii* detection with direct impact in the diagnosis of PcP patients presenting with low pathogen burdens (early PcP cases, non-HIV-infected patients with PcP, and PcP patients under treatment) and detection of *P. jirovecii* DNA in noninvasive clinical specimens.[58,77,93] PCR is also successfully applied to detect *P. jirovecii* colonisation, either in susceptible individuals or in apparently healthy people, including the staff of healthcare facilities.[13,52,94–96] Still, a highly sensitive technique is required for the detection of *P. jirovecii* because false negative tests by staining methods may occur in specific clinical conditions owing to the low sensitivity of these assays, especially in patients receiving chemoprophylaxis and aggressive cART who become predisposed to a decreased number of organisms in respiratory specimens and a clinical condition with parameters that are difficult to interpret.[8,30,76–78]

2.1.3.2 Molecular detection of *Pneumocystis jirovecii* DNA in respiratory specimens

The extent of efficiency of the amplification methods is dependent on the type of specimen to analyze, such as, a highly invasive biopsy specimen (rarely used), an invasive BALF (standard specimen), a less invasive alternative such as IS (first alternative to BAL), or noninvasive alternative specimens like NA and OW, which will determine the final overall sensitivity of the full molecular PcP diagnostic protocol.[30,42,58,76,77,91,94,97] Molecular techniques play an important role when used in association with less invasive or even noninvasive respiratory specimens (IS, NA, OW).[56,93,98–100] Moreover, the analytical specificity of highly accurate PCR protocols for *P. jirovecii* DNA detection, such as the mitochondrial large subunit rRNA (*mtLSUrRNA*) nested PCR (nPCR), when applied to noninvasive respiratory specimens can be equivalent to the same protocols in BAL specimens (100%).[89,98] However, the sensitivity of these PCR assays is directly affected by the implicit lower sensitivity of the noninvasive respiratory specimens, due to their lower fungal burden. In BALF, the sensitivity of PCR assays for detection of *P. jirovecii* DNA, such as the *mtLSUrRNA* nPCR, is usually very satisfactory (higher than 95%), with a detection threshold that can reach values of 0.5–1 organism/μL of sample.[77,89] Yet, in noninvasive specimens, those PCR assays result in lower sensitivity values (less than 80%).[56,89,101,102] Nevertheless, noninvasive respiratory specimens, such as OW, can be easily obtained and repeated in order to monitor the therapeutic response and evolution of disease.[101]

The PCR tools have greatly improved the diagnosis of PcP, allowing the early detection of *P. jirovecii* infection. In fact, PCR assays can detect *P. jirovecii* DNA in respiratory specimens from PcP patients presenting negative staining results.[33,42,58,77,81] The use of these techniques may lead to a decreased time period from onset of symptoms to treatment which provides a recognized positive effect on prognosis by avoiding the evolution of early PcP cases to a severe disease with associated respiratory failure requiring medical ventilation, which is directly associated with significant mortality.[24] However, the role of PCR in respiratory specimens remains controversial, mainly due to the ambiguity created by the fact that a positive PCR test associated with a negative microscopy may be the result of either PcP or *P. jirovecii* colonisation.[10,30,77,94] Studies suggest that *P. jirovecii* colonisation may induce local or systemic inflammation, which could aggravate chronic pulmonary diseases and other concomitant pulmonary infections.[56,94,103] If a colonized patient or asymptomatic carrier of *P. jirovecii* is diagnosed as having a PcP episode, or a PcP patient is profiled as colonized with *P. jirovecii*, then the patient may be started on an inadequate treatment with compromised outcome. Taking this into account, a trustworthy distinction between PcP case and *P. jirovecii* colonisation can be achieved by combining the test results with the etiologic features of *P. jirovecii* as opportunistic pathogen and the patient's pathophysiologic manifestations.[23,33] Only the careful assessment of the patient's clinical signs and symptoms, radiological and laboratorial complementary tests, in combination with the PCR data analysis, will shed light on the effective state of a specific *P. jirovecii* infection.

In clinical practice, colonisation or subclinical carriage of *P. jirovecii* is usually identified when the DNA of the pathogen is detected by a PCR assay in respiratory specimens from immunosuppressed or immunocompetent persons demonstrating neither symptoms nor signs of PcP, who do not develop a PcP episode.[10,23,104] On these occasions, the microscopic examinations are rarely positive due to the very low fungal burdens that are normally detected in these subjects.[105]

2.1.3.3 Molecular techniques for *Pneumocystis jirovecii* DNA detection

In 1990, Wakefield and coworkers developed a PCR assay specific for the amplification of mitochondrial ribosomal RNA sequences of *Pneumocystis* in both biological samples from animals (rat) and humans, and environmental samples (air), demonstrating for the first time that molecular detection of DNA of *Pneumocystis* organisms was possible.[74,75] These findings had a great impact, advancing the PcP diagnosis in respiratory specimens and opening new possibilities for *P. jirovecii* epidemiology. Since then, PCR-based approaches have been developed and used to detect *P. jirovecii* and describe differences between organisms.[79,80,82,84,86,87,106–108]

The adequate selection of target genes and the correct design of primers or probes will contribute to the specificity of *P. jirovecii* DNA detection techniques, applied to respiratory specimens, which can reach 100%.

Several PCR protocols for the detection of *P. jirovecii* DNA in clinical samples have been reported, using different PCR techniques and targeting a variety of *P. jirovecii* genes, such as, the mitochondrial large subunit rRNA (*mtLSUrRNA*), mitochondrial small subunit rRNA (*mtSSUrRNA*), the internal transcribed spacers (*ITS*) regions of the rRNA operon, 5S ribosomal RNA (*5S rRNA*), 18S ribosomal RNA (*18S rRNA*), dihydropteroate synthase (*DHPS*), dihydrofolate reductase (*DHFR*), thymidylate synthase (*TS*), *cdc2* gene, major surface glycoprotein (*MSG*), upstream conserved sequence of the MSG (*UCS*), kexin-like serine protease (*KEX*1), cytochrome *b* (*CYB*), superoxide dismutase (*SOD*), α and β tubulin (α- β-*TUB*), thioredoxin reductase (*TRR*1) and *arom*.[74,75,109–125]

There are few comparative evaluation studies, essentially because they are very difficult to perform, mainly due to the fact that the final efficiency score of each PcP molecular diagnosis is directly dependent on the main steps of laboratorial strategies chosen. Sampling methods, specific protocols, and laboratory reagents used for specimen processing and DNA extraction, amplification techniques, specific DNA fragments analyzed, level of expertise of technicians/personnel performing the protocols, and different clinical contexts are major factors with a direct impact on the final PcP molecular diagnosis.[10,23,32,33,126]

A comparative study showed that the PCR assay directed to the amplification of the mitochondrial multicopy gene of *P. jirovecii mtLSU rRNA* demonstrated higher sensitivity than the nuclear *5S rDNA* and *DHFR* genes.[89] Another study demonstrated that a PCR targeting the nuclear *ITS* regions of the rRNA operon was the most effective method for molecular detection of *P. jirovecii* in BALF from HIV-infected patients, when compared with PCR assays amplifying the *18S rRNA, mt rRNA, 5S rRNA, TS,* and *DHFR* loci.[126] However, the detection of *P. jirovecii* DNA through *mtLSU rRNA* nPCR was demonstrated to be more sensitive than the *ITS* nPCR, probably due to the fact that the *ITS* regions of the rRNA operon are present in a single copy in the *P. jirovecii* genome, while the *mtLSU rRNA* is a multicopy gene.[8,101,127] Another study concluded that the *mtLSU rRNA* nPCR was the most sensitive method among nine PCR assays evaluated for detection of *P. jiroveii* DNA. The *mtLSU rRNA* nPCR produced less false negative results, and displayed a higher degree of concordance with microscopic data than *mtLSU rRNA* single PCR, *ITS* regions nPCR, *DHPS* single- and nPCR, *DHFR* nPCR, *MSG* heminested-PCR, *18S rRNA* 1-tube nPCR and *5S rRNA* real-time quantitative PCR (RT-qPCR).[127] The low diagnostic specificity of the *DHPS* and *DHFR* PCR assays demonstrates its inability to amplify all *P. jirovecii* positive respiratory specimens, as performed in several studies.[84,128,129] Several genes used for molecular detection of *P. jirovecii* (eg, *DHPS*, *DHFR*, *ITS*) are present as single nuclear encoded genes, whereas the mitochondrial *mtLSU rRNA* is a multicopy gene in the *P. jirovecii* genome, which contributes largely to the higher successful amplification rates demonstrated by the *mtLSU rRNA* PCR assays, especially the nPCR technique.[75,87,130]

The clinical significance of nPCR and IF-MAb assays for detection/identification of *P. jirovecii* and PcP diagnosis in BALF or IS was assessed in a study, which concluded that *mtLSU rRNA* nPCR has a very high sensitivity value (sensitivity 96%, specificity 59%). Nevertheless, IF-MAb was the most specific (better correlated with the clinical diagnosis) method for diagnosis of PcP (sensitivity 60%, specificity 97%). Furthermore, in the same study, the authors stated that additional clinical cases can be found by *mtLSU rRNA* nPCR, thereby showing a high risk of detection of *P. jirovecii* colonisation cases.[130] Moreover, in another study to evaluate the diagnostic usefulness of PCR assays in PcP, different immunosuppressed and immunocompetent patient groups were examined by microscopy, and single and nPCR. In BALF from HIV-infected patients, there was 100% sensitivity and specificity in microscopy and single PCR, whereas nPCR, although 100% sensitive, demonstrated 97.5% specificity. Among the non-HIV immunosuppressed patients, sensitivity and specificity using nPCR, single PCR and microscopy, were 100% and 84.6–94.6%, 77.8–100%, and 94.5–100%, 66.7–100%, and 100%, respectively. In this group of patients, both single and nPCR produced lower positive predictive values (PPV) than microscopy techniques, while in the immunocompetent patients, the PPV of both PCR assays were 0%. The PCR assays tested do not seem to offer any additional advantage over that of conventional microscopic examination. However, nPCR identified significant frequencies of *P. jirovecii* colonisation, in about 17–20% of immunocompetent and immunosuppressed non-HIV-infected patients.[131]

A bivariate metaanalysis and systematic review assessed the usefulness of PCR-based assays in BALF for diagnosis of PcP, included 16 studies with a total 1857 BALF samples from 1793 patients published between 1994 and 2012, pointing out for overall pooled sensitivity of 98.3% (95% CI, 91.3–99.7%) and specificity of 91.0% (95% CI, 82.7–95.5%), which predicts PCR in BALF as a very accurate method for diagnosis of PcP.[30] In the same study, the evaluation of the subgroup within nPCR based *P. jirovecii* detection demonstrated high sensitivity of 98% (95% CI, 76–100%) and a relatively medium specificity of 73% (95% CI, 53–86%).[30] These findings are consistent with another study in which simple and nPCR both had 100% sensitivity and 98 and 84% specificity respectively, compared to IF-MAb staining performed in BALF. Although nPCR seems to improve the sensitivity of the detection of *P. jirvocii* in respiratory specimens, the clinical relevance of a positive result remains to be investigated.[132] The lower specificity rates found can be largely explained by the asymptomatic carriers or colonized patients with positive nPCR tests incorrectly diagnosed as having PcP (false positive PcP cases). Again, this leads to the intriguing issue that positive PCR tests associated with negative microscopic examinations may correspond either to PcP or *P. jirovecii* colonisation. The use of less invasive samples from the upper respiratory tract in association with highly sensitive and specific molecular tools, such as nPCR or RT-qPCR, provide a more cost-effective option than the standard procedures using BALF.[33,133–135]

Molecular strategies for detecting and quantifying *P. jirovecii* burdens were explored by several authors. The sensitivity of conventional PCR ranged from 88.2 to 100%, with specificity from 68 to 87%,[136–138] whereas the reported sensitivity of RT-qPCR ranged from 82% to 100% and specificity from 64 to 100%.[81,139,140] RT-qPCR assay targeting the MSG multigene family of *P. jirovecii* was assessed revealing significant differences in the number of MSG copies between PcP and colonisation.[141] A comparative study consisting on the detection of *P. jirovecii* in BALF by two staining methods (TBO and calcofluor white) and two RT-qPCR assays (targeting the *β-TUB* and *KEX*1) demonstrated that taking the results of the microscopic examinations as a reference, the sensitivity of RT-qPCR targeting the *KEX*1 gene was 100%, and the specificity was 92.4%.[81] Other authors have adapted the *KEX*1 RT-qPCR for *P. jirovecii* DNA quantification in order to construct and genotype *P. jirovecii* pooled DNA samples showing that genotyping platforms for pooled DNA samples are promising methodologies for genetic characterization of infectious organisms in clinical and epidemiological studies.[87] Another study assessed the clinical significance of quantifying *P. jirovecii* DNA by using RT-qPCR targeting the *mtLSUrRNA* in BALF from immunosuppressed patients. RT-qPCR results were compared with routine IF-MAb, demonstrating 100% sensitivity and specificity. Nevertheless, the authors considered that the relevant threshold remains to be determined for this RT-qPCR and it may vary according to the underlying disease.[43] The use of a RT-qPCR commercial kit for *P. jirovecii* *mtLSUrRNA* detection in BALF was evaluated by comparison with DFA (considered by the authors as the laboratory standard), RT-qPCR (targeting the *cdc*2 gene), and single PCR (targeting the *mtLSUrRNA*). The commercial assay demonstrated 100% sensitivity, specificity, NPV, and PPV for detecting the presence of *P. jirovecii* DNA compared to the laboratory standard.[44] The comparison of two RT-qPCR assays (one *in house* and another commercial kit) for diagnosis of PcP in HIV-infected and non-HIV-infected immunosuppressed patients demonstrated 65% sensitivity and 85% specificity in the *in house* assay, while the commercial assay reached 72% and 82% of sensitivity and specificity, respectively. Concerning to PcP clinical diagnosis, RT-qPCR is probably a useful tool. However, in order to achieve more accurate tests, especially in the management of PcP in non-HIV immunosuppressed patients, thresholds should be assessed according to underlying diseases, and other clinical and radiological parameters.[142]

A metaanalysis study assessed the use of RT-qPCR for the diagnosis of PcP in immunosuppressed patients, including ten individual studies from 1990 to 2010. Overall, the sensitivity of RT-qPCR was 97% (95%CI: 93–99%) and specificity was 94% (95%CI: 90–96%). In the subgroup of HIV-infected patients, the sensitivity and specificity were 97% (95%CI: 93–99%) and 93% (95%CI: 89–96%). Regarding *P. jirovecii* DNA detection in BALF, the sensitivity was 98% (95%CI: 94–99%) and specificity 93% (95%CI: 89–96%).[31] Although RT-qPCR demonstrated good diagnostic accuracy, further studies are needed in order to identify any differences in the diagnostic performance of RT-qPCR in HIV-infected and non-HIV-infected immunosuppressed patients and in differentiating colonisation from disease.

In conclusion, several studies have consistently pointed to the nPCR and the RT-qPCR assays, especially the ones targeting the *mtLSUrRNA* gene, as the most sensitive and specific molecular tools for detection of *P. jirovecii*.[43,44,74,77,81,129,143,144] The general characteristics of the nPCR and RT-qPCR assays for *P. jirovecii* detection and genetic characterization, targeting the most studied *P. jirovecii* loci are depicted in Table 13.2.

Table 13.2 List of the General Characteristics of the Nested PCR and RT-qPCR Assays (Primers, Probes, Amplification Fragments and Clinical or Epidemiological Applications) Targeting the Most Studied *P. jirovecii* loci

Locus (PCR method)	Primers and probes	Sequence (5′ → 3′)	Amplicon (pb)	References	Specific Applications
mtLSU rRNA (nPCR)	pAZ102-E pAZ102-H (1st round)	GATGGCTGTTTC-CAAGCCCA GTGTACGTTG-CAAAGTACTC	346	[74]	Diagnostic (standard method) Subtyping Colonisation detection Transmission
	pAZ102-X pAZ102-Y (2nd round)	GTGAAATACAAATC-GGACTAGG TCACTTAATATTA-ATTGGGGAGC	263	[83]	
ITS regions (nPCR)	1724F2 ITS2R (1st round)	AGTTGAT-CAAATTTGGTCATT-TAGAG CTCGGACGAG-GATCCTCGCC	157–161	[120]	Subtyping Transmission Clinical severity
	ITS-FX ITS-RT2 (2nd round)	TTCCGTAGGT-GAACCTGCG CTGATTTGAGATTA-AAATTCTTG	177–192		
DHPS (nPCR)	Dp15 Dp800 (1st round)	TCTGAATTTTATA-AAGCGCCTACAC ATTTCATAAACAT-CATGAACCCG	785	[80]	Subtyping Antimicrobial resistance Clinical severity Transmission
	DHPR-NF DHPS-NR (2nd round)	AAATGCAGGGGC-GACGATAAT GCCTTAATTGCTT-GTTCTGCAA	186	[145]	
DHFR (nPCR)	FR208 FR1038 (1st round)	GCAGAAAGTAGGTA-CATTATTACGAGA AACCAGTTACCTA-ATCAAACTATATTGC	858	[124]	Subtyping Antimicrobial resistance Clinical severity
	FR242 FR1018 (2nd round)	GTTTGGAATAGAT-TATGTTCATGGTG-TACG GCTTCAAACCTTGT-GTAACGCG	798		
mtLSU rRNA (RT-qPCR)	PNC-LSU3 PNC-LSU4 (primers)	TGGTAAGTAGT-GAAATACAAATCGG ACTCCCTCGAGA-TATTCAGTGC	152	[43]	Diagnostic DNA quantification Colonisation detection

(Continued)

Table 13.2 List of the General Characteristics of the Nested PCR and RT-qPCR Assays (Primers, Probes, Amplification Fragments and Clinical or Epidemiological Applications) Targeting the Most Studied _P. jirovecii_ loci (_cont._)

Locus (PCR method)	Primers and probes	Sequence (5′→3′)	Amplicon (pb)	References	Specific Applications
	PNC-LSU5 (probe)	TTCGCAGAAAAC-CAGCTATATCCTAGT [5′ LCRed640 labeled, 3′ Ph labeled]			
	PNC-LSU6 (probe)	AGAGGAATAAA-CAATTTGCCAAAA-CAA [3′FITC labeled]			
_KEX_1 (RT-qPCR)	Forward Reverse (primers)	CAACCCTGTTC-CAATGCCTAA CAACACCGATTCCA-CAAACAGT	101	[81,87]	Diagnostic DNA quantification DNA pooling Colonisation detection
	MGB Probe (probe)	TGCTGGTGAAG-TAGCTGCCGTTCGA [dye-labeled (FAM) minor groove binder]			

2.1.3.4 Molecular typing for disease and transmission tracking

The advances in the understanding of the genetic diversity of _P. jirovecii_ have raised important questions, such as whether some organisms are more pathogenic than others, and whether resistant or more virulent genotypes spread between hosts.[80,82,86,107,146] The sequencing of selected DNA PCR amplified fragments has been widely used to identify _P. jirovecii_ genotypes. In addition to their usefulness in diagnosis, several genetic regions of _P. jirovecii_ have been recognized as markers of choice to study the population genetics, geographical distribution, modes of transmission, and drug susceptibility or resistance of specific _P. jirovecii_ genotypes.[14,79,80,83,84,86,87,108] Presently, the genetic characterization of clinical specimens is consensually considered to be the best procedure to identify _P. jirovecii_ organisms (Table 13.2).

The majority of polymorphisms studied in _P. jirovecii_ targeted mutations of the _DHPS_ gene, potentially associated with sulfa resistance, which may interfere with trimethoprim-sulfamethoxazole (TMP-SMZ) prophylaxis or treatment in PcP. Significant associations between the single-nucleotide polymorphisms (SNP) at bases 165 and 171 of _DHPS_ and the use of sulfa drugs for PcP prophylaxis[16,80,82,84,147] or failure of both TMP-SMZ treatment[148] and TMP-SMZ or dapsone prophylaxis,[149] were reported. Although it is still uncertain if specific _DHPS_ mutations are an effective cause of drug resistance in PcP patients, _DHPS_ point mutations were shown to be associated with TMP-SMZ or dapsone (DHPS inhibitors) usage.[84,124,147–149]

Other studies have associated specific genotypes with more pathogenic _P. jirovecii_ organisms. _ITS_ regions genotypes of the rRNA operon were associated with the severity of disease.[119,120,122,133] Multilocus genotyping approaches related the SNP at base 85 of the _mtLSUrRNA_ gene with _P. jirovecii_ burden levels and follow-up of infection, the SNP at position 312 of the *DHFR* gene with PcP infection burden, the SNP at positions 110 and 215 of the *SOD* locus with linkage disequilibrium and severity of PcP episodes.[86,87,146,150]

Recently, the *de novo* assembly of the *P. jirovecii* genome was published, opening the way to solve some critical issues, such as the identification of nutritional supplements for development of reliable and cost-effective culture in vitro systems, and detection of new targets for development of anti-PcP drugs and vaccines.[85,87]

2.2 NEW ALTERNATIVES FOR THE DIAGNOSIS OF PCP

2.2.1 Usefulness of blood in the diagnosis of PcP

The effectiveness of microscopy in the diagnosis of PcP depends primarily on the available resources and technology of the laboratory, as well as the type of biological specimen analyzed, the staining methods used, and the microscopist's experience. Also, the detection of *P. jirovecii* by PCR techniques in persons with negative microscopy has been observed in several studies, complicating the interpretation of results obtained for the clinicians.[68,151]

As an alternative to respiratory specimens, blood and serum specimens have been tested since the 1990s, using PCR methodologies for the diagnosis of PcP, without showing promising results. Most studies (five out of seven) showed low to very low sensitivity (0–30%) with the exceptions of the other two studies showing sensitivity of 86 and 100%, depending on the *locus* in analysis.[88,90,152–156] *P. jirovecii* DNA was found in the blood or serum of AIDS patients with *P. jirovecii* infection localized in the lungs, and without any evidence of dissemination,[152] and in cases of disseminated infection,[153] which made some authors suggest that the appearance of DNA of the pathogen in blood could represent a blood-borne phase of the infection, which was never demonstrated.[152] New alternative strategies for the diagnosis of PcP (Table 13.1) through the measurement of blood biomarkers, *P. jirovecii* molecules and host molecules reflecting the host-pathogen interaction, have been proposed as noninvasive interventions.[150,157,158]

2.2.2 Blood biomarkers for diagnosis of PcP

Alternative serum testing for the diagnosis of PcP has not yet been established.[150,158] Previous reports suggested that in patients with PcP, the serologic levels of $(1-3)$-β-D-Glucan (BG), Krebs von den Lungen-6 antigen (KL-6) and lactate dehydrogenase (LDH) are increased, while the levels of S-adenosylmethionine (SAM) are low[33,150,159] (Table 13.1). KL-6 antigen, a mucin-like glycoprotein prominent on type II alveolar pneumocytes, was reported to be a potentially sensitive indicator for interstitial pneumonitis.[150,160] The high serum levels of LDH in PcP are likely to be a reflection of the underlying lung injury.[32,150] A few studies suggest that SAM, a metabolic intermediate in methylation reactions and polyamine synthesis, is a promising serologic biomarker for PcP diagnosis. In contrast to almost all other pathogenic microorganisms, *Pneumocystis* species seem to have a need for exogenous SAM. However, conflicting data have been published reporting that *Pneumocystis* encodes a functional SAM synthetase gene.[161,162]

The measurement of serum BG, the main structural component of the cell wall of *Pneumocystis* cysts,[163] is the most promising procedure for establishing the serologic diagnosis of PcP.[32,150,158–161] Two metaanalysis studies estimated its sensitivity at 96% (95% CI, 92–98%) and 95% (95% CI, 91–97%) and specificity at 84% (95% CI, 83–86%) and 86% (95% CI, 82–90%), respectively.[164,165] In both metaanalysis, the majority of cases were from HIV-infected patients but the diagnostic accuracy for these patients and those non-HIV-infected was not significantly different. This data was also corroborated in another study conducted recently.[33]

Two studies in which BG was assessed in combination with other serum markers, LDH and KL-6, proposed serologic assays as a complementary/indicative test for PcP in patients with advanced immunosuppression (HIV-infected and non-HIV-infected) and clinical signs and symptoms indicative of PcP.[32,33] These combination tests should not be used as a single determinant for PcP diagnosis and they must be interpreted within the clinical context. The combination tests BG/KL-6 were demonstrated to be the most accurate serologic approach for PcP diagnosis. In a cost-effectiveness analysis the authors also came to the conclusion that although less sensitive/specific rather than the classic methods (BAL followed by microscopy or DNA detection), the BG/KL-6 combination tests may provide a less onerous procedure for PcP diagnosis, using a minimally invasive and inexpensive sample (blood), which may also be a major benefit to the patient's care, especially when dealing with patients with respiratory failure, or in children in whom invasive procedures for specimen collection are not easy to perform, with its associated risk of complications[32,33] (Table 13.1).

3 CONCLUSIONS

Pneumocystis jirovecii is an opportunistic agent that causes significant morbidity and mortality in immunosuppressed patients, usually in the form of a severe pneumonia. In the immunosuppressed subject suspected of PcP, early diagnosis is essential for the early implementation of treatment and for a good clinical outcome. Since the 1990s, the methodologies for the diagnosis of PcP have improved dramatically. Although several findings can suggest the diagnosis of PcP by routine history, physical examination, and laboratory tests, none of these are specific. Diagnosis of *P. jirovecii* infection still depends on the confirmation of its presence in the affected tissues. A wide variety of methods, based on staining or molecular biology assays applied to BALF or IS are available, with varying sensitivity and specificity, as well as varying cost and complexity. The choice of the method to apply depends, among other factors, on the local incidence of PcP, on the biological specimens to be analyzed, as well as on the level of local expertise. New diagnostic tools based on the measurement of blood biomarkers of infection (*P. jirovecii* molecules and host molecules reflecting the host-pathogen interaction) have been evaluated as non-invasive methodologies. BG has shown the best performance until now, being an attractive laboratory tool for the diagnosis of PcP, especially in patients where bronchoscopy is not feasible or difficult to perform.

Future studies are still needed to increase the accuracy of *P. jirovecii* detection and PcP diagnosis in non-invasive biological specimens, and distinguishing PcP cases from colonisation cases. New technological platforms also need to be developed based upon nanotechnology (microfluidics and lab on paper) to allow early detection of PcP biomarkers, which is detrimental for the early implementation of therapeutic curative/control measures, especially in resource-limited settings.

REFERENCES

1. Chagas C. Nova tripanosomiae humana. *Mem Inst Oswaldo Cruz* 1909;**1**:159–218.
2. Carini A. Formas de eschizogonia do *Trypanosoma lewisi*. Soc De Med et Chir de São Paulo 1910. *Bull Inst Pasteur* 1910;**IX**:973–8.

3. Delanoë P, Delanoë M. Sur les rapports des kystes de Carini du poumon des rats avec le *Trypanosoma lewisi*. *CR Hebd Seances Acad Sci* 1912;**155**:658–9.

4. Hughes WT. Historical overview. In: Walzer PD, Cushion MT, editors. *Pneumocystis pneumonia*. 3rd ed. NY, USA: Marcel Dekker, Inc.; 2005 [Chapter 1].

5. Edman JC, Kovacs JA, Masur H, Santi DV, Elwood HJ, Sogin ML. Ribosomal RNA sequence shows Pneumocystis carinii to be a member of the fungi. *Nature* 1988;**334**:519–22.

6. Stringer SL, Hudson K, Blase MA, Walzer PD, Cushion MT, Stringer JR. Sequence from ribosomal RNA of *Pneumocystis carinii* compared to those of four fungi suggests an ascomycetous affinity. *J. Protozool* 1989;**36**:6S–14S.

7. Stringer JR, Beard CB, Miller RF, Wakefield AE. A new name for *Pneumocystis* from humans and new perspectives on the host-pathogen relationship. *Emerg Infect Dis* 2002;**8**:891–6.

8. Matos O, Lundgren B, Caldeira L, Mansinho K, Aguiar P, Forte M, Antunes F. Evaluation of two nested polymerase chain reactions for diagnosis of *Pneumocystis carinii* pneumonia in immunocompromised patients. *Clin Microbiol Infect* 2000;**6**:149–51.

9. Roux A, Canet E, Valade S, Gangneux-Robert F, Hamane S, Lafabrie A, et al. *Pneumocystis jirovecii* pneumonia in patients with or without AIDS. *France. Emerg Infect Dis* 2014;**20**:1490–7.

10. Morris A, Lundgren JD, Masur H, Walzer PD, Hanson DL, Frederick T, et al. Current epidemiology of *Pneumocystis* pneumonia. *Emerg Infect Dis* 2004;**10**:1713–20.

11. Lundberg BE, Davidson AJ, Burman WJ. Epidemiology of Pneumocystis carinii pneumonia in an era of effective prophylaxis: the relative contribution of non-adherence and drug failure. *AIDS* 2000;**14**:2559–66.

12. Totet A, Latouche S, Lacube P, Pautard JC, Jounieaux V, Raccurt C, et al. *Pneumocystis jirovecii* dihydropteroate synthase genotypes in immunocompetent infants and immunosuppressed adults, Amiens, France. *Emerg Infect Dis* 2004;**10**:667–73.

13. Medrano FJ, Montes-Cano M, Conde M, de la Horra C, Respaldiza N, Gasch A, et al. *Pneumocystis jirovecii* in general population. *Emerg Infect Dis* 2005;**11**:245–50.

14. Esteves F, Montes-Cano MA, de la Horra C, Costa MC, Calderón EJ, Antunes F, et al. *Pneumocystis jirovecii* multilocus genotyping profiles in patients from Portugal and Spain. *Clin Microbiol Infect* 2008;**14**:356–62.

15. French N, Kaleebu P, Pisani E, Whitworth JAG. Human immunodeficiency virus (HIV) in developing countries. *Ann Trop Med Parasit* 2006;**100**:433–54.

16. Matos O, Esteves F. Epidemiology and clinical relevance of *Pneumocystis jirovecii* Frenkel, 1976 dihydropteroate synthase gene mutations. *Parasite* 2010;**17**:219–32.

17. de Armas Rodríguez Y, Wissmann G, Müller AL, Pederiva MA, Brum MC, Brackmann RL, et al. Pneumocystis jirovecii pneumonia in developing countries. *Parasite* 2011;**18**:219–28.

18. Matos O. *Pneumocystis jirovecii* pneumonia in Africa: impact and implications of highly sensitive diagnostic technologies. *N Am J Med Sci* 2012;**4**:486–7.

19. Raviglione MC. Extrapulmonary pneumocystosis: the first 50 cases. *Rev Infect Dis* 1990;**12**:1127–38.

20. Hughes WT. *Pneumocystis* pneumonitis in non-HIV-infected patients: update. In: Walzer PD, Cushion MT, editors. *Pneumocystis pneumonia*. 3rd ed. NY, USA: Marcel Dekker, Inc.; 2005 [Chapter 16].

21. Baughman RP. Current methods of diagnosis. In: Walzer PD, editor. Pneumocystis carinii *Pneumonia*. 2nd ed. NY, USA: Marcel Dekker, Inc.; 1994 [Chapter 19].

22. Moe AA, Hardy WD. *Pneumocystis carinii* infection in the HIV-seropositive patient. *Infect. Dis. Clin. North Am* 1994;**8**:331–64.

23. Barry SM, Johnson MA. *Pneumocystis carinii* pneumonia: a review of current issues in diagnosis and management. *HIV Med* 2001;**2**:123–32.

24. Huang L. Clinical presentation and diagnosis of *Pneumocystis* pneumonia in HIV-infected patients. In: Walzer PD, Cushion MT, editors. *Pneumocystis Pneumonia*. 3rd ed. NY, USA: Marcel Dekker, Inc.; 2005 [Chapter 15].

25. Schildgen V, Mai S, Khalfaoui S, Lüsebrink J, Pieper M, Tillmann RL, et al. *Pneumocystis jirovecii* can be productively cultured in differentiated CuFi-8 airway cells. *MBio* 2014;**5** e01186–e01114.

26. Procop GW, Haddad S, Quinn J, Wilson ML, Henshaw NG, Reller LB, et al. Detection of *Pneumocystis jiroveci* in respiratory specimens by four staining methods. *J Clin Microbiol* 2004;**42**:3333–5.

27. Aderaye G, Woldeamanuel Y, Asrat D, Lebbad M, Beser J, Worku A, et al. Evaluation of Toluidine Blue O staining for the diagnosis of *Pneumocystis jiroveci* in expectorated sputum sample and bronchoalveolar lavage from HIV-infected patients in a tertiary care referral center in Ethiopia. *Infection* 2008;**36**:237–43.

28. Raab SS, Cheville JC, Bottles K, Cohen MB. Utility of Gomori methenamine silver stains in bronchoalveolar lavage specimens. *Mod Pathol* 1994;**7**:599–604.

29. Galan F, Oliver JL, Roux P, Poirot JL, Bereziat G. Detection of *Pneumocystis carinii* DNA by polymerase chain reaction compared to direct microscopy and immunofluorescence. *J Protozool* 1991;**38**:199S–200S.

30. Fan LC, Lu HW, Cheng KB, Li HP, Xu JF. Evaluation of PCR in bronchoalveolar lavage fluid for diagnosis of *Pneumocystis jirovecii* pneumonia: a bivariate meta-analysis and systematic review. *PLoS one* 2013;**8**:e73099.

31. Summah H1, Zhu YG, Falagas ME, Vouloumanou EK, Qu JM. Use of real-time polymerase chain reaction for the diagnosis of *Pneumocystis pneumonia* in immunocompromised patients: a meta-analysis. *Chin Med J* 2013;**126**:1965–73.

32. Esteves F, Lee CH, de Sousa B, Badura R, Seringa M, Fernandes C, et al. (1–3)-Beta-D-Glucan in association with lactate dehydrogenase as biomarkers of *Pneumocystis pneumonia* (PcP) in HIV-infected patients. *Eur J Clin Microbiol Infect Dis* 2014;**33**:1173–80.

33. Esteves F, Calé SS, Badura R, de Boer MG, Maltez F, Calderón EJ, et al. Diagnosis of Pneumocystis pneumonia: evaluation of four serologic biomarkers. *Clin Microbiol Infect* 2015;**21**:379.e1–379.e10.

34. Rodríguez M, Fishman JA. Prevention of infection due to Pneumocystis spp. in human immunodeficiency virus-negative immunocompromised patients. *Clin Microbiol Rev* 2004;**17**:770–82.

35. Rossiter SJ, Miller DC, Churg AM, Carrington CB, Mark JB. Open lung biopsy in the immunosuppressed patient. Is it really beneficial? *J Thorac Cardiovasc Surg* 1979;**77**:338–45.

36. Broaddus C, Dake MD, Stulbarg MS, Blumenfeld W, Hadley WK, Golden JA, et al. Bronchoalveolar lavage and transbronchial biopsy for the diagnosis of pulmonary infections in the acquired immunodeficiency syndrome. *Ann Int Med* 1985;**102**:747–52.

37. Fishman JA. *Pneumocystis carinii* and parasitic infection in the immunocompromised host. In: Rubin RH, Lowell SY, editors. *Clinical approach to infection in the compromised host*. New York: Kluwer Academic/ Plenum Publishers; 2002. p. 265–334.

38. Gosey LL, Howard RM, Witebsky FG, Ognibene FP, Wu TC, Gill VJ, et al. Advantages of a modified toluidine blue O stain and bronchoalveolar lavage for the diagnosis of *Pneumocystis carinii* pneumonia. *J Clin Microbiol* 1985;**22**:803–7.

39. Stover DE, Zaman MB, Hajdu SI, Lange M, Gold J, Armstrong D. Bronchoalveolar lavage in the diagnosis of diffuse pulmonary infiltrates in the immunocompromised host. *Ann Intern Med* 1984;**101**:1–7.

40. Baughman RP, Dohn M, Frame P. Generalised immune response to *Pneumocystis carinii* infection in the lung. *J Protozool* 1991;**38**:187–8.

41. Miller RF, Kocjan G, Buckland J, Holton J, Malin A, Semple S. Sputum induction for the diagnosis of pulmonary disease in HIV patients. *J Infect* 1991;**23**:5–15.

42. Arcenas RC, Uhl JR, Buckwalter SP, Limper AH, Crino D, Roberts GD, et al. A real-time polymerase chain reaction assay for detection of *Pneumocystis* from bronchoalveolar lavage fluid. *Diagn Microbiol Infect Dis* 2006;**54**:169–75.

43. Botterel F, Cabaret O, Foulet F, Cordonnier C, Costa JM, Bretagne S. Clinical significance of quantifying *Pneumocystis jirovecii* DNA by using real-time PCR in bronchoalveolar lavage fluid from immunocompromised patients. *J Clin Microbiol* 2012;**50**:227–31.

44. McTaggart LR, Wengenack NL, Richardson SE. Validation of the mycassay Pneumocystis kit for detection of Pneumocystis jirovecii in bronchoalveolar lavage specimens by comparison to a laboratory standard of direct immunofluorescence microscopy, real-time PCR, or conventional PCR. *J Clin Microbiol* 2012;**50**:1856–9.

45. Martín WR, Albertson TE, Siegel B. Tracheal catheters in patients with acquired immunodeficiency syndrome for the diagnosis of *Pneumocystis carinii* pneumonia. *Chest* 1990;**98**:29–32.
46. Zajac-Spychała O, Gowin E, Fichna P, Wysocki J, Fichna M, Kowala-Piaskowska A, et al. *Pneumocystis* pneumonia in children—the relevance of chemoprophylaxis in different groups of immunocompromised and immunocompetent paediatric patients. *Cent Eur J Immunol* 2015;**40**:91–5.
47. Speich R. Diagnosis of pulmonary problems in HIV-infected patients. *Monaldi Arch Chest Dis* 1993;**48**:221–32.
48. Lau WK, Young LS, Remington JS. *Pneumocystis carinii* pneumonia. Diagnosis by examination of pulmonary secretions. *JAMA* 1976;**236**:2399–402.
49. Leigh TR, Kirby K, Gazzard B-GLUCAN, Collins JV. Effect of sputum induction on arterial oxygen saturation and spirometry in HIV-infected patients. *Eur Respir J* 1994;7:453–458.
50. Kovacs JA, Ng VL, Masur H, Leoung G, Hadley WK, Evans G, et al. Diagnosis of *Pneumocystis carinii* pneumonia: improved detection in sputum with use of monoclonal antibodies. *New Engl J Med* 1988;**318**:589–93.
51. HB 260-2003 Hospital acquired infections—engineering down the risk. Published by Standards Australia International Ltd GPO Box 5420, Sydney, NSW, 2001.
52. LaRocque RC, Katz JT, Perruzzi P, Baden LR. The utility of sputum induction for diagnosis of *Pneumocystis* pneumonia in immunocompromised patients without human immunodeficiency virus. *Clin Infect Dis* 2003;**37**:1380–3.
53. Ng VL, Gartner I, Weymouth LA, Goodman CD, Hopewell PC, Hadley WK. The use of mucolysed induced sputum for the identification of pulmonary pathogens associated with human immunodeficiency virus infection. *Arch Pathol Lab Med* 1989;**113**:488–93.
54. Zaman MK, Wooten OJ, Suprahmanya B, Ankobiah W, Finch PJ, Kamholz SL. Rapid noninvasive diagnosis of *Pneumocystis carinii* from induced liquefied sputum. *Ann Intern Med* 1988;**1**(109):7–10.
55. Ng VL, Virani NA, Chaisson RE, Yajko DM, Sphar HT, Cabrian K, et al. Rapid detection of *Pneumocystis carinii* using a direct fluorescent monoclonal antibody stain. *J Clin Microbiol* 1990;**28**:2228–33.
56. Wakefield AE, Miller RF, Guiver LA, Hopkin JM. Oropharyngeal samples for detection of *Pneumocystis carinii* by DNA amplification. *Int J Med* 1993;**86**:401–6.
57. Helweg-Larsen J, Jensen JS, Lundgren B. Non-invasive diagnosis of *Pneumocystis carinii* pneumonia by PCR on oral washes. *Lancet* 1997;**350**:1363.
58. Matos O, Costa MC, Lundgren B, Caldeira L, Aguiar P, Antunes F. Effect of oral washes on the diagnosis of Pneumocystis carinii pneumonia with a low parasite burden and on detection of organisms in subclinical infections. *Eur J Clin Microbiol Inf Dis* 2001;**20**:573–5.
59. Samuel CM, Whitelaw A, Corcoran C, Morrow B, Hsiao N-Y, Zampoli M, et al. Improved detection of *Pneumocystis jirovecii* in upper and lower respiratory tract specimens from children with suspected pneumocystis pneumonia using real-time PCR: a prospective study. *BMC Infect Dis* 2011;**11**:329.
60. Heikkinen T, Marttila J, Salmi AA, Ruuskanen O. Nasal swab versus nasopharyngeal aspirate for isolation of respiratory viruses. *J Clin Microbiol* 2002;**40**:4337–9.
61. To KK, Wong SC, Xu T, Poon RW, Mok KY, Chan JF, et al. Use of nasopharyngeal aspirate for diagnosis of *Pneumocystis* pneumonia. *J Clin Microbiol* 2013;**51**:1570–4.
62. Grocott RG. A stain for fungi in tissue sections and smears using Gomori's methenamine-silver nitrate technic. *Am J Clin Pathol* 1955;**25**:975–9.
63. Musto L, Flanigan M, Elbadawi A. Ten-minute silver stain for *Pneumocystis carinii* and fungi in tissue sections. *Arch Pathol Lab Med* 1982;**106**:292–4.
64. Baselski VS, Robison MK, Pifer LW, Woods DR. Rapid detection of *Pneumocystis carinii* in bronchoalveolar lavage samples by using Cellufluor staining. *J Clin Microbiol* 1990;**28**:393–4.
65. Holten-Andersen W, Kolmos HJ. Comparison of methenamine silver nitrate and Giemsa stain for detection of *Pneumocystis carinii* in bronchoalveolar lavage specimens from HIV infected patients. *APMIS* 1989;**97**:745–7.

66. Cushion MT, Walzer PD. Development of candidate anti-*Pneumocystis* drugs: in vitro and in vivo approaches. In: Walzer PD, Cushion MT, editors. *Pneumocystis Pneumonia*. 3rd ed. NY, USA: Marcel Dekker, Inc.; 2005 [Chapter 25].

67. Kovacs JA, Gill V, Swan JC, Ognibene F, Shelhamer J, Parrillo JE, et al. Prospective evaluation of a monoclonal antibody in diagnosis of *Pneumocystis carinii* pneumonia. *Lancet* 1986;**2**:1–3.

68. Kovacs JA, Gill VJ, Meshnick S, Masur H. New insights into transmission, diagnosis, and drug treatment of *Pneumocystis carinii* pneumonia. *J Am Med Assoc* 2001;**286**:2450–60.

69. Ng VL, Yajko DM, Mcphaul LW, Gartner I, Byford B, Goodman CD, et al. Evaluation of an indirect fluorescent-antibody stain for detection of *Pneumocystis carinii* in respiratory specimens. *J Clin Microbiol* 1990;**28**:975–9.

70. Cregan P, Yamamoto A, Lum A, VanDerHeide T, MacDonald M, Pulliam L. Comparison of four methods for rapid detection of *Pneumocystis carinii* in respiratory specimens. *J Clin Microbiol* 1990;**28**:2432–6.

71. Baughman RP, Strohoper SS, Clinton BA, Nickol AD, Frame PI. The use of an indirect fluorescent antibody test for detecting *Pneumocystis carinii*. *Arch Pathol Lab Med* 1989;**113**:1062–5.

72. Lautenschlager I, Lyytikainen O, Jokipii L, Jokipii A, Maiche A, Ruutu T, Tukiainen P, Ruutu P. Immunodetection of *Pneumocystis carinii* in bronchoalveolar lavage specimens compared with methenamine silver stain. *J Clin Microbiol* 1996;**34**:728–30.

73. Bava AJ, Cattaneo S, Bellegarde E. Diagnosis of pulmonary pneumocystosis by microscopy on wet mount preparations. *Rev Inst Med Trop S Paulo* 2002;**44**:279–82.

74. Wakefield AE, Pixley FJ, Banerji S, Sinclair K, Miller RF, Moxon ER, et al. Detection of *Pneumocystis carinii* with DNA amplification. *Lancet* 1990;**336**:451–3.

75. Wakefield AE, Pixley FJ, Banerji S, Sinclair K, Miller RF, Moxon ER, Hopkin JM. Amplification of mitochondrial ribosomal RNA sequences from *Pneumocystis carinii* DNA of rat and human origin. *Mol Biochem Parasitol* 1990;**43**:69–76.

76. Olsson M, Elvin K, Löfdahl S, Linder E. Detection of *Pneumocystis carinii* DNA in sputum and bronchoalveolar lavage samples by polymerase chain reaction. *J Clin Microbiol* 1993;**31**:221–6.

77. Durand-Joly I, Chabé M, Soula F, Delhaes L, Camus D, Dei-Cas E. Molecular diagnosis of *Pneumocystis* pneumonia (PcP). *FEMS Immunol. Med. Microbiol* 2005;**45**:405–10.

78. Azoulay E, Bergeron A, Chevret S, Bele N, Schlemmer B, Menotti J. Polymerase chain reaction for diagnosing *Pneumocystis* pneumonia in non-HIV immunocompromised patients with pulmonary infiltrates. *Chest* 2009;**135**:655–61.

79. Beard CB, Carter JL, Keely SP, Huang L, Pieniazek NJ, Moura IN, et al. Genetic variation in *Pneumocystis carinii* isolates from different geographic regions: implications for transmission. *Emerg Infect Dis* 2000;**6**:265–72.

80. Helweg-Larsen J, Benfield TL, Eugen-Olsen J, Lundgren JD, Lundgren B. Effects of mutations in *Pneumocystis carinii* dihydropteroate synthase gene on outcome of AIDS-associated *P. carinii* pneumonia. *Lancet* 1999;**354**:1347–51.

81. Rohner P, Jacomo V, Studer R, Schrenzel J, Graf JD. Detection of *Pneumocystis jirovecii* by two staining methods and two quantitative PCR assays. *Infection* 2009;**37**:261–5.

82. Crothers K, Beard CB, Turner J, Groner G, Fox M, Morris A, et al. Severity and outcome of HIV-associated Pneumocystis pneumonia containing Pneumocystis jirovecii dihydropteroate synthase gene mutations. *AIDS* 2005;**19**:801–5.

83. Tsolaki AG, Beckers P, Wakefield AE. Pre-AIDS era isolates of *Pneumocystis carinii* f. sp. *hominis*: high genotype similarity with contemporary isolates. *J Clin Microbiol* 1998;**36**:90–3.

84. Huang L, Beard CB, Creasman J, Levy D, Duchin JS, Lee S, et al. Sulfa or sulfone prophylaxis and geographic region predict mutations in the *Pneumocystis carinii* dihydropteroate synthase gene. *J Infect Dis* 2000;**182**:1192–8.

85. Cissé OH, Pagni M, Hauser PM. De novo assembly of the *Pneumocystis jirovecii* genome from a single bronchoalveolar lavage fluid specimen from a patient. *MBio* 2012;**4** e00428-12.

86. Esteves F, Gaspar J, De Sousa B, Antunes F, Mansinho K, Matos O. Clinical relevance of multiple single-nucleotide polymorphisms in *Pneumocystis jirovecii* Pneumonia: development of a multiplex PCR-single-base-extension methodology. *J Clin Microbiol* 2011;**49**:1810–5.

87. Esteves F, Gaspar J, de Sousa B, Antunes F, Mansinho K, Matos O. *Pneumocystis jirovecii* multilocus genotyping in pooled DNA samples: a new approach for clinical and epidemiological studies. *Clin Microbiol Infect* 2012;**18**:E177–84.

88. Tamburrini E, Mencarini P, Visconti E, Zolfo M, De Luca A, Siracusano A, et al. Detection of Pneumocystis carinii DNA in blood by PCR is not of value for diagnosis of P. carinii pneumonia. *J Clin Microbiol* 1996;**34**:1586–8.

89. Tamburrini E, Mencarini P, Visconti E, Zolfo M, Marinaci S, Zinzi D, et al. Potential impact of *Pneumocystis* genetic diversity on the molecular detection of the parasite in human host. *FEMS Immunol Med Microbiol* 1998;**22**:37–49.

90. Rabodonirina M, Cotte L, Boibieux A, Kaiser K, Mayencon M, Raffenot D, et al. Detection of Pneumocystis carinii DNA in blood specimens from human immunodeficiency virus-infected patients by nested PCR. *J Clin Microbiol* 1999;**37**:127–31.

91. Esteves F, Medrano FJ, de Armas Y, Wissmann G, Calderón EJ, Matos O. *Pneumocystis* and Pneumocystosis: first meeting of experts from Latin-American and Portuguese-speaking countries - a mini-review. *Expert Rev Anti Infect Ther* 2014;**12**:545–8.

92. Miller RF, Huang L, Walzer PD. *Pneumocystis* pneumonia associated with human immunodeficiency virus. *Clin Chest Med* 2013;**34**:229–41.

93. Respaldiza N, Montes-Cano MA, Friaza V, Muñoz-Lobato F, Medrano FJ, Varela JM, et al. Usefulness of oropharyngeal washings for identifying *Pneumocystis jirovecii* carriers. *J Eukaryot Microbiol* 2006;**53**:100–1.

94. Durand-Joly I, Soula F, Chabé M, Dalle JH, Lafitte JJ, Senechal M, et al. Longterm colonisation with *Pneumocystis jirovecii* in hospital staffs: a challenge to prevent nosocomial pneumocystosis. *J Eukaryot Microbiol* 2003;**50**:614–5.

95. Nevez G, Chabé M, Rabodonirina M, Virmaux M, Dei-Cas E, Hauser PM, et al. Nosocomial *Pneumocystis jirovecii* infections. *Parasite* 2008;**15**:359–65.

96. Matos O, Costa MC, Correia I, Monteiro P, Monteiro M, Soares J, et al. *Pneumocystis jirovecii* carriage in Portuguese immunocompetent patients: preliminary results. *J Eukaryot Microbiol* 2003;**50**:647–8.

97. Turner D, Schwarz Y, Yust I. Induced sputum for diagnosing *Pneumocystis carinii* pneumonia in HIV patients: new data, new issues. *Eur Respir J* 2003;**21**:204–8.

98. Helweg-Larsen J, Jensen JS, Lundgren B. Noninvasive diagnosis of *Pneumocystis carinii* pneumonia in haematological patients using PCR on oral washes. *J Eukaryot Microbiol* 1997;**44**:59S.

99. Khan MA, Farrag N, Butcher P. Diagnosis of *Pneumocystis carinii* pneumonia: immunofluorescence staining, simple PCR or nPCR. *J Infect* 1999;**39**:77–80.

100. Richards CG, Wakefield AE, Mitchell CD. Detection of *Pneumocystis* DNA in nasopharyngeal aspirates of leukaemic infants with pneumonia. *Arch Dis Child* 1994;**71**:254–5.

101. Tsolaki AG, Miller RF, Wakefield AE. Oropharyngeal samples for genotyping and monitoring response to treatment in AIDS patients with *Pneumocystis carinii* pneumonia. *J Med Microbiol* 1999;**48**:897–905.

102. Larsen HH, Huang L, Kovacs JA, Crothers K, Silcott VA, Morris A, et al. A prospective, blinded study of quantitative touch-down polymerase chain reaction using oral-wash samples for diagnosis of *Pneumocystis* pneumonia in HIV-infected patients. *J Infect Dis* 2004;**189**:1679–83.

103. Calderón EJ, Rivero L, Respaldiza N, Morilla R, Montes-Cano MA, Friaza V, et al. Systemic inflammation in patients with chronic obstructive pulmonary disease who are colonised with *Pneumocystis jirovecii*. *Clin Infect Dis* 2007;**45**:17–9.

104. Morris A, Wei K, Afshar K, Huang L. Epidemiology and clinical significance of *Pneumocystis* colonisation. *J Infect Dis* 2008;**197**:10–7.

105. Vidal S, de la Horra C, Martín J, Montes-Cano MA, Rodríguez E, Respaldiza N, et al. *Pneumocystis jirovecii* colonisation in patients with interstitial lung disease. *Clin Microbiol Infect* 2006;**12**:231–5.

106. Matos O, Esteves F. *Pneumocystis jirovecii* multilocus gene sequencing: findings and implications. *Future Microbiol* 2010;**5**:1257–67.

107. Hauser PM, Sudre P, Nahimana A, Francioli P. Prophylaxis failure is associated with a specific *Pneumocystis carinii* genotype. *Clin Infect Dis* 2001;**33**:1080–2.

108. Miller RF, Evans HE, Copas AJ, Cassell JA. Climate and genotypes of *Pneumocystis jirovecii*. *Clin Microbiol Infect* 2007;**13**:445–8.

109. Kaiser K, Rabodonirina M, Mayencon M, Picot S. Evidence for cdc2 gene in *Pneumocystis carinii* hominis and its implication for culture. *AIDS* 1999;**13**:419–20.

110. Wada M, Nakamura Y. Unique telomeric expression site of major-surface-glycoprotein genes of *Pneumocystis carinii*. *DNA Res* 1996;**3**:55–64.

111. Kutty G, Kovacs JA. A single copy gene encodes Kex1, a serine endoprotease of *Pneumocystis jiroveci*. *Infect Immun* 2003;**71**:571–4.

112. Hunter JA, Wakefield AE. Genetic divergence at the mitochondrial small subunit ribosomal RNA gene among isolates of *Pneumocystis carinii* from five mammalian host species. *J Eukaryot Microbiol* 1996;**43**:24S–5S.

113. Walker DJ, Wakefield AE, Dohn MN, Miller RF, Baughman RP, Hossler PA, et al. Sequence polymorphisms in the *Pneumocystis carinii* cytochrome b gene and their association with atovaquone prophylaxis failure. *J Infect Dis* 1998;**178**:1767–75.

114. Denis CM, Mazars E, Guyot K, Odberg-Ferragut C, Viscogliosi E, Dei-Cas E, et al. Genetic divergence at the SODA locus of six different formae speciales of *Pneumocystis carinii*. *Med Mycol* 2000;**38**:289–300.

115. Stringer JR, Stringer SL, Zhang J, Baughman R, Smulian AG, Cushion MT. Molecular genetic distinction of *Pneumocystis carinii* from rats and humans. *J Eukaryot Microbiol* 1993;**40**:733–41.

116. Edlind TD, Bartlett MS, Weinberg GA, Prah GN, Smith JW. The beta-tubulin gene from rat and human isolates of *Pneumocystis carinii*. *Mol Microbiol* 1992;**6**:3365–73.

117. Kutty G, Huang SN, Kovacs JA. Characterization of thioredoxin reductase genes (trr1) from *Pneumocystis carinii* and *Pneumocystis jiroveci*. *Gene* 2003;**310**:175–83.

118. Banerji S, Lugli EB, Miller RF, Wakefield AE. Analysis of genetic diversity at the arom locus in isolates of *Pneumocystis carinii*. *J Eukaryot Microbiol* 1995;**42**:675–9.

119. Lee CH, Lu JJ, Bartlett MS, Durkin MM, Liu TH, Wang J, et al. Nucleotide sequence variation in *Pneumocystis carinii* strains that infect humans. *J Clin Microbiol* 1993;**31**:754–7.

120. Lee CH, Helweg-Larsen J, Tang X, Jin S, Li B, Bartlett MS, et al. Update on *Pneumocystis carinii* f. sp. *hominis* typing based on nucleotide sequence variations in internal transcribed spacer regions of rRNA genes. *J Clin Microbiol* 1998;**36**:734–41.

121. Liu Y, Rocourt M, Pan S, Liu C, Leibowitz MJ. Sequence and variability of the 5.8S and 26S rRNA genes of Pneumocystis carinii. *Nucleic Acids Res* 1992;**20**:3763–72.

122. Lu JJ, Bartlett MS, Shaw MM, Queener SF, Smith JW, Ortiz-Rivera M, et al. Typing of *Pneumocystis carinii* strains that infect humans based on nucleotide sequence variations of internal transcribed spacers of rRNA genes. *J Clin Microbiol* 1994;**32**:2904–12.

123. Mazars E, Odberg-Ferragut C, Dei-Cas E, Fourmaux MN, Aliouat EM, Brun-Pascaud M, et al. Polymorphism of the thymidylate synthase gene of *Pneumocystis carinii* from different host species. *J Eukaryot Microbiol* 1995;**42**:26–32.

124. Ma L, Borio L, Masur H, Kovacs JA. *Pneumocystis carinii* dihydropteroate synthase but not dihydrofolate reductase gene mutations correlate with prior trimethoprim-sulfamethoxazole or dapsone use. *J Infect Dis* 1999;**180**:1969–78.

125. Garbe TR, Stringer JR. Molecular characterization of clustered variants of genes encoding major surface antigens of human *Pneumocystis carinii*. *Infect Immun* 1994;**62**:3092–101.

126. Walzer PD, Evans HE, Copas AJ, Edwards SG, Grant AD, Miller RF. Early predictors of mortality from Pneumocystis jirovecii pneumonia in HIV-infected patients: 1985–2006. *Clin Infect Dis* 2008;**46**:625–33.

127. Totet A, Pautard JC, Raccurt C, Roux P, Nevez G. Genotypes at the internal transcribed spacers of the nuclear rRNA operon of *Pneumocystis jiroveci* in nonimmunosuppressed infants without severe pneumonia. *J Clin Microbiol* 2003;**41**:1173–80.
128. Lu JJ, Chen CH, Bartlett MS, Smith JW, Lee CH. Comparison of six different PCR methods for detection of *Pneumocystis carinii*. *J Clin Microbiol* 1995;**33**:2785–8.
129. Robberts FJ, Liebowitz LD, Chalkley LJ. Polymerase chain reaction detection of *Pneumocystis jiroveci*: evaluation of nine assays. *Diagn Microbiol Infect Dis* 2007;**58**:385–92.
130. Olsson M, Strålin K, Holmberg H. Clinical significance of nested polymerase chain reaction and immunofluorescence for detection of *Pneumocystis carinii* pneumonia. *Clin Microbiol Infect* 2001;**7**:492–7.
131. Sing A, Trebesius K, Roggenkamp A, Rüssmann H, Tybus K, Pfaff F, et al. Evaluation of diagnostic value and epidemiological implications of PCR for *Pneumocystis carinii* in different immunosuppressed and immunocompetent patient groups. *J Clin Microbiol* 2000;**38**:1461–7.
132. Moonens F, Liesnard C, Brancart F, Van Vooren JP, Serruys E. Rapid simple and nested polymerase chain reaction for the diagnosis of Pneumocystis carinii pneumonia. *Scand J Infect Dis* 1995;**27**:358–62.
133. Thomas CF, Limper AH. *Pneumocystis* pneumonia. *N Engl J Med* 2004;**350**:2487–98.
134. Fillaux J, Malvy S, Alvarez M, Fabre R, Cassaing S, Marchou B, et al. Accuracy of a routine real-time PCR assay for the diagnosis of *Pneumocystis jirovecii* pneumonia. *J Microbiol Methods* 2008;**75**:258–61.
135. Harris JR, Marston BJ, Sangrujee N, DuPlessis D, Park B. Cost-effectiveness analysis of diagnostic options for *Pneumocystis* pneumonia (PCP). *PLoS One* 2011;**6**:e23158.
136. Flori P, Bellete B, Durand F, Rahebin H, Cazorla C, Hafid J, et al. Comparison between real-time PCR, conventional PCR and different staining techniques for diagnosing *Pneumocystis jiroveci* pneumonia from bronchoalveolar lavage specimens. *J Med Microbiol* 2004;**53**:603–7.
137. Huggett JF, Taylor MS, Kocjan G, Evans HE, Morris-Jones S, Gant V, et al. Development and evaluation of a real-time PCR assay for detection of *Pneumocystis jirovecii* DNA in bronchoalveolar lavage fluid of HIV-infected patients. *Thorax* 2008;**63**:154–9.
138. Fujisawa T, Suda T, Matsuda H, Inui N, Nakamura Y, Sato J, et al. Real-time PCR is more specific than conventional PCR for induced sputum diagnosis of *Pneumocystis* pneumonia in immunocompromised patients without HIV infection. *Respirology* 2009;**14**:203–9.
139. Brancart F, Rodríguez-Villalobos H, Fonteyne PA, Peres-Bota D, Liesnard C. Quantitative TaqMan PCR for detection of *Pneumocystis jiroveci*. *J Microbiol Methods* 2005;**61**:381–7.
140. Bandt D, Monecke S. Development and evaluation of a real-time PCR assay for detection of *Pneumocystis jiroveci*. *Transpl Infect Dis* 2007;**9**:196–202.
141. Larsen HH, Masur H, Kovacs JA, Gill VJ, Silcott VA, Kogulan P, et al. Development and evaluation of a quantitative, touchdown, real-time PCR assay for diagnosing *Pneumocystis carinii* pneumonia. *J Clin Microbiol* 2002;**40**:490–4.
142. Montesinos I, Brancart F, Schepers K, Jacobs F, Denis O, Delforge ML. Comparison of two real-time PCR assays for diagnosis of *Pneumocystis jirovecii* pneumonia in human immunodeficiency virus (HIV) and non-HIV immunocompromised patients. *Diagn Microbiol Infect Dis* 2015;**82**:143–7.
143. Wakefield AE. DNA sequences identical to *Pneumocystis carinii* f. sp. *carinii* and *Pneumocystis carinii* f. sp. *hominis* in samples of air spora. *J Clin Microbiol* 1996;**34**:1754–9.
144. Lu JJ, Lee CH. *Pneumocystis* pneumonia. *J Formos Med Assoc* 2008;**107**:830–42.
145. Costa MC, Gaspar J, Mansinho K, Esteves F, Antunes F, Matos O. Detection of *Pneumocystis jirovecii* dihydropteroate synthase polymorphisms in patients with *Pneumocystis* pneumonia. *Scand J Infect Dis* 2005;**37**:766–71.
146. Esteves F, Gaspar J, Marques T, Leite R, Antunes F, Mansinho K, et al. Identification of relevant single-nucleotide polymorphisms in *Pneumocystis jirovecii*: relationship with clinical data. *Clin Microbiol Infect* 2010;**16**:878–84.
147. Costa MC, Esteves F, Antunes F, Matos O. Genetic characterization of the dihydrofolate reductase gene of *Pneumocystis jirovecii* isolates from Portugal. *J Antimicrob Chemother* 2006;**58**:1246–9.

148. Mei Q, Gurunathan S, Masur H, Kovacs JA. Failure of co-trimoxazole in *Pneumocystis carinii* infection and mutations in dihydropteroate synthase gene. *Lancet* 1998;**351**:1631–2.

149. Kazanjian P, Locke AB, Hossler PA, Lane BR, Bartlett MS, Smith JW, et al. Pneumocystis carinii mutations associated with sulfa and sulfone prophylaxis failures in AIDS patients. *AIDS* 1998;**12**:873–8.

150. Tasaka S, Hasegawa N, Kobayashi S, Yamada W, Nishimura T, Takeuchi T, et al. Serum indicators for the diagnosis of *Pneumocystis* pneumonia. *Chest* 2007;**131**:1173–80.

151. Huang L, Cattamanchi A, Davis JL, den Boon S, Kovacs J, Meshnick S, et al. International HIV-associated Opportunistic Pneumonias (IHOP) Study; Lung HIV Study. HIV-associated *Pneumocystis* pneumonia. *Proc Am Thorac Soc* 2011;**8**:294–300.

152. Schluger N, Sepkowitz K, Armstrong D, Bernard E, Rifkin M, Cerami A, et al. Detection of *Pneumocystis carinii* in serum of AIDS patients with *Pneumocystis* pneumonia by the polymerase chain reaction. *J Protozool* 1991;**38**:240S–2S.

153. Lipschik GY, Gill VJ, Lundgren JD, Andrawis VA, Nelson NA, Nielsen JO. Improved diagnosis of *Pneumocystis carinii* infection by polymerase chain reaction on induced sputum and blood. *Lancet* 1992;**340**:203–6.

154. Roux P, Lavrard I, Poirot JL, Chouaid C, Denis M, Olivier JL, et al. Usefulness of PCR for detection of *Pneumocystis carinii* DNA. *J Clin Microbiol* 1994;**32**:2324–6.

155. Atzori C, Lu JJ, Jiang B, Bartlett MS, Orlando G, Queener SF, et al. Diagnosis of *Pneumocystis carinii* pneumonia in AIDS patients by using polymerase chain reactions on serum specimens. *J Infect Dis* 1995;**172**:1623–6.

156. Matos O, Lundgren B, Caldeira L, Mansinho K, Aguiar P, Forte M, et al. Evaluation of a nested PCR for detection of *Pneumocystis carinii* in serum from imunocompromised patients. *J Euk Microbiol* 1999;**46**:104S–5S.

157. Finkelman MA. *Pneumocystis jirovecii* infection: Cell wall (1-3)-D-glucan biology and diagnostic utility. *Crit Rev Microbiol* 2010;**36**:271–81.

158. Morris AM, Masur H. A serologic test to diagnose Pneumocystis pneumonia: are we there yet? *Clin Inf Dis* 2011;**53**:203–4.

159. Desmet S, Van Wijngaerden E, Maertens J, Verhaegen J, Verbeken E, De Munter P, et al. Serum (1-3)-beta-D-glucan as a tool for diagnosis of *Pneumocystis jirovecii* pneumonia in patients with human immunodeficiency virus infection or hematological malignancy. *J Clin Microbiol* 2009;**47**:3871–4.

160. Nakamura H, Tateyama M, Tasato D, Haranaga S, Yara S, Higa F, et al. Clinical utility of serum β-D-glucan and KL-6 levels in *Pneumocystis jirovecii* pneumonia. *Intern Med* 2009;**48**:195–202.

161. de Boer MG, Gelinck LB, van Zelst BD, van de Sande WW, Willems LN, van Dissel JT, et al. β-D-glucan and S-adenosylmethionine serum levels for the diagnosis of *Pneumocystis* pneumonia in HIV-negative patients: a prospective study. *J Infect* 2011;**62**:93–100.

162. Skelly MJ, Holzman RS, Merali S. S-Adenosylmethionine levels in the diagnosis of *Pneumocystis carinii* pneumonia in patients with HIV infection. *Clin Infect Dis* 2008;**46**:467–71.

163. Matsumoto Y, Matsuda S, Tegoshi T. Yeast glucan in the cyst wall of *Pneumocystis carinii*. *J Protozool* 1989;**36**:21S–2S.

164. Onishi A, Sugiyama D, Kogata Y, Saegusa J, Sugimoto T, Kawano S, et al. Diagnostic accuracy of serum 1,3-beta-D-glucan for *Pneumocystis jiroveci* pneumonia, invasive candidiasis, and invasive aspergillosis: systematic review and meta-analysis. *J Clin Microbiol* 2012;**50**:7–15.

165. Karageorgopoulos DE, Qu JM, Korbila IP, Zhu YG, Vasileiou VA, Falagas ME. Accuracy of beta-D-glucan for the diagnosis of *Pneumocystis jirovecii* pneumonia: a meta-analysis. *Clin Microbiol Infect* 2013;**19**:39–49.

ANTIMICROBIAL APPROACHES AGAINST BACTERIAL PATHOGENS WHICH CAUSE LOWER RESPIRATORY SYSTEM INFECTIONS

14

A.F. Jozala*, D. Grotto*, L.C.L. Novaes**, V. de Carvalho Santos-Ebinuma[†], M. Gerenutti*, F.S. Del Fiol*

*University of Sorocaba, Sorocaba, SP, Brazil; **RWTH Aachen University, Aachen, Germany; [†]São Paulo State University, Araraquara, SP, Brazil

1 INTRODUCTION

Lower respiratory infections are a leading cause of morbidity and mortality in both children and adults worldwide, especially when caused by Gram-negative pathogens in hospitalized patients. The most common infections usually include bronchitis, bronchiolitis, and especially pneumonia, which is the cause of the most deaths, with four million killed each year worldwide.[1-3] The inappropriate doses (for the broad-spectrum coverage) and the extended time of therapy have generated acute problems, such as the emergence of multidrug-resistant microorganisms.[4,5]

The main problems are the inadequate treatment of multiresistant Gram-negative agents and *Staphylococcus aureus* oxacillin-resistant. Considering these facts, the clinical characterization of the host, the prevalence of bacterial agents, and their sensitivity profile through quantitative cultures are basic elements applied in order to promote a rational use of antimicrobials.

According to WHO (2015), Antibiotic resistance is occurring everywhere in the world, compromising the treatment of infectious diseases and undermining many other advances in health and medicine. It represents one of the biggest threats to global health today, and can affect any one, of any age, in any country. It leads to longer hospital stays, higher medical costs and increased mortality. Antibiotic resistance occurs naturally, but misuse of antibiotics in humans and animals is accelerating the process. Tackling antibiotic resistance is a high priority for the WHO. As part of implementation of objective of the global action plan on antimicrobial resistance, WHO is coordinating a global campaign to raise awareness and encourage best practices among the public, policymakers, health and agriculture professionals. This survey provides a snapshot of current public awareness and common behaviors related to antibiotics in a range of countries.

On the other hand, efforts should be made on alternative therapy. In fact, there is a need to develop new strategies to subtly manipulate bacterial antimicrobial behavior. There is evidence that

The Microbiology of Respiratory System Infections. http://dx.doi.org/10.1016/B978-0-12-804543-5.00014-2

antimicrobials do not help acute bronchitis because they often do not require it. Therefore, antimicrobials can be administered to patients with acute exacerbations of chronic bronchitis.

Studies have shown increasing evidence of the important role played by the resident microbiota in offering protection against infectious diseases. Rather than continuing the traditional approach of killing bacteria wherever they occur, there is a need to develop new antimicrobial strategies aimed at subtle manipulation of bacterial behavior. Such therapies would favor natural host defenses and the maintenance of the normal microbiota to keep growth of pathogenic species in check. At the outset the design of strategies for novel antibiotics should include exploration of strategies for exploiting beneficial and commensal bacteria in fighting infections in sites where normal microbiota reside.[6,7]

In this chapter, we address different aspects of antimicrobial application against pathogens which cause lower respiratory infections. Moreover, natural therapy by plants to treat these infections will also be discussed.

2 BACTERIAL PATHOGENS WHICH CAUSE LOWER RESPIRATORY SYSTEM INFECTIONS

Pneumonia is caused by a variety of bacteria, fungi, viruses, and parasites. Numerous factors, including environmental contaminants and autoimmune diseases may cause pneumonia. The pathogens that cause pneumonia are categorized in many ways for the purpose of laboratory testing, epidemiologic study, and choice of therapy. Infections arise while a patient is hospitalized or living in an institution such as a nursing home, which are called hospital-acquired or nosocomial pneumonia. Etiologic pathogens associated with community-acquired and hospital-acquired pneumonia are somewhat different. However, many organisms can cause both types of infections.[8]

The community-acquired bacterial pneumonia may occur spontaneously or as a complication of viral infections. Its most common agents are *Streptococcus pneumoniae, Mycoplasma pneumoniae, Haemophilus influenzae,* and *Legionella* spp. Bacterial pneumonia acquired in hospitals are the most frequent agents caused by *Staphylococcus aureus, Klebsiella pneumoniae, Pseudomonas aeruginosa* and other enterobacteria. In chronic pneumonia, the most frequent cause is *Mycobacterium tuberculosis* and then other mycobacteria.[9]

These syndromes, especially pneumonia, can be severe or fatal. For this reason, the initiation of broad-spectrum antimicrobial therapy is important after clinical diagnosis because it is associated with lower mortality rates.[10] On the other hand, it is estimated that up to 50% of antimicrobials prescribed may be unnecessary, being important to conduct a clinical trial to evaluate the efficacy, especially for assessment of their broad-spectrum sensibility.[11]

Bronchitis appears to be caused by a combination of environmental factors, such as smoking, and bacterial infection by *S. pneumonia*.[2] *S. pneumoniae* is found in 80% of cases and is one of the most dominant species. Recurrence of infections associated with bacterial persistence results in frequent antibiotic courses. This favors the emergence of multidrug resistant species.[12,13]

Bronchiolitis is a viral respiratory disease of infants and is caused primarily by respiratory syncytial virus. Other viruses, including parainfluenza viruses, influenza viruses and adenoviruses are also known to cause bronchiolitis. *Bordetella pertussis, Mycoplasma pneumoniae,* and *Chlamydia pneumoniae* may cause the severe form of the disease, while chronic forms are generally found in association with smoking or pollution. The exacerbation of these processes is often associated with *Haemophilus influenzae* and *S. pneumoniae*.[14]

2.1 WHO'S DATA ABOUT ANTIMICROBIAL RESISTANCE

WHO's 2014 report on global surveillance of antimicrobial resistance revealed that antibiotic resistance is no longer a prediction for the future; it is happening right now, across the world, and is putting at risk the ability to treat common infections in the community and hospitals. Without urgent, coordinated action, the world is heading toward a postantibiotic era, in which common infections and minor injuries, which have been treatable for decades, can once again kill. Resistance to first-line drugs to treat infections caused by *Staphlylococcus aureus*—a common cause of severe infections acquired both in health care facilities and in the community—is also widespread.

3 ANTIMICROBIAL THERAPIES

Inappropriate antibiotics therapy and overuse of antibiotics may predispose patients to increased resistance and hospital mortality.[15] In 2001, WHO published a global strategies for containment of antimicrobial resistance as follows:

> The emergence of antimicrobial resistance is a complex problem driven by many interconnected factors, in particular the use and misuse of antimicrobials. Antimicrobial use, in turn, is influenced by an interplay of the knowledge, expectations and interactions of prescribers and patients, economic incentives, characteristics of the health system(s) and the regulatory environment. In the light of this complexity, coordinated interventions are needed that simultaneously target the behaviour of providers and patients and change important features of the environments in which they interact.

More expensive therapies must be used when infections become resistant to first-line drugs. A longer duration of illness and treatment, often in hospitals, increases health-care costs as well as the economic burden on families and societies. Antimicrobial resistance jeopardizes health-care gains to society. The achievements of modern medicine are put at risk by antimicrobial resistance. Without effective antimicrobials for prevention and treatment of infections, the success of organ transplantation, cancer chemotherapy, and major surgery would be compromised (WHO, 2015).

Antimicrobial prescription is affected by acutely ill patient populations, diagnostic uncertainty, a lack of appreciation for the potential harms of antimicrobials, and prescribing by clinically inexperienced physicians in training. The empirical prescription behavior of trainees can be shaped by the practices of senior physicians, who rely on experience and often consider themselves exempt from evidence-based guidelines. To limit inappropriate empirical prescriptions of broad-spectrum agents, interventions can include: the direct education provider; selective formulary restriction; development, dissemination, and enforcement of evidence-based therapeutic guidelines; preprinted order sheets; and use of an audit and feedback program. These strategies are important tools that are available to institutional antimicrobial stewardship programs to improve rational empirical antibiotic prescribing.[5]

3.1 ANTIMICROBIAL APPROACHES FOLLOWING CLINICAL GUIDELINES

Pneumonia: This treatment is community acquired pneumonia (CAP) specific.[16] Bacteria and viruses are responsible for the majority of cases. Preponderating: *Streptococcus pneumonia* and *Mycoplasma.*[17] Empiric therapy. Use CURB65 score.[18]

Score 1 = Outpatient
Score 2 and over = Impatient treatment

For adults: First choice: Azithromycin 500 mg PO (first dose) then 250 mg/day for 5 days. If the patient has had antibiotic therapy within past three months:

(Azithromycin) + Amoxicillin 1g PO 3 times/day; or,
(Azithromycin) + Levofloxacin 750 mg PO once/day.
Alternatives Doxycycline 100 mg PO twice/daily (5–7 days).

For children. (>3 months) antibiotic therapy is not routinely required for preschool-aged children with CAP because of viral etiology. If suspected bacterial infection:

Amoxicillin 90 mg/kg per day (divided in two daily doses)[19]

Alternatives: Azithromycin 10 mg/kg PO (first dose "maximum 500 mg") then 5 mg/kg (maximum 250 mg) PO for 4 days.[20]

Bronchitis: Self-limited inflammation of the upper airways due viral infection (majority), bacterial (5–10%) or irritants. Pharmacological treatment: The etiology of acute bronchitis is almost completely viral, so the treatment should be symptomatic, aiming to guarantee respiratory comfort and removal of secretions. Thus, there is use of cough suppressants or bronchodilators, mucolytics, or corticosteroids. The use of antibiotics is very controversial[17,21,22] and there is no indication as recent revisions.

The Cochrane Review from 2014 stated:

> There is limited evidence to support the use of antibiotics in acute bronchitis. Antibiotics may have a modest beneficial effect in some patients such as frail, elderly people with multimorbidity who may not have been included in trials to date. However, the magnitude of this benefit needs to be considered in the broader context of potential side effects, medicalization for a self-limiting condition, increased resistance to respiratory pathogens and cost of antibiotic treatment.[21]

The US Center for Disease Control and Prevention, the American Academy of Family Physicians, the American College of Physicians/American Society of Internal Medicine, and the Infectious Diseases Society of America all recommend not prescribing antibiotics for acute bronchitis.[22]

Bronchiolitis: Etiology—respiratory syncytial virus (RSV) (50%), parainfluenza (25%); human metapneumovirus.[23] Treatment is supportive aiming moisture and oxygen, so the majority of patients with bronchiolitis can be treated as outpatients. In some situations, when there is an association of malnutrition in children, hypoxia, tachypnea, or other immunosuppressive condition, should proceed to hospitalization. The administration of bronchodilators and corticosteroids are not indicated as well as the use of antibacterial agents, unless there are clear signs of bacterial infections.[24–27] Children with severe respiratory impairment should fed intravenously or nasogastric tube because of the high risk of food aspiration.[28] Only in severe cases, there is an indication of the use of antiviral drugs (ribavirin), aiming to RSV. Ribavirin shows broad antiviral activity that interferes with the RNA metabolism. Typically, ribavirin is administered aerosolized 6 g vial (20 mg/mL) in sterile water by SPAG-2 generator over 18–20 h/day, for 3–5 days.[23]

Prophylaxis: Immunization with the anti-RSV monoclonal antibody palivizumab. Palivizumab is a humanized monoclonal antibody that binds to the F protein of RSV and inhibits viral infection and replication.[29] The prophylaxis indications are as follows:

Preterm infants with chronic lung disease: Prophylaxis may be considered during the RSV season during the first year of life for preterm infants who develop chronic lung disease of prematurity defined as gestational age <32 weeks, 0 days and a requirement for >21% oxygen for at least the first 28 days after birth.

Infants with hemodynamically significant CHD: Infants with haemodynamically significant congenital heart disease can be considered for prophylaxis during the first year of life.

Children with anatomic pulmonary abnormalities or neuromuscular disorder: Infants with neuromuscular disease or congenital anomaly that impairs the ability to clear secretions from the upper airway because of ineffective cough are known to be at risk for a prolonged hospitalization related to lower respiratory tract infection, and therefore, may be considered for prophylaxis during the first year of life.

Immunocompromised children: Prophylaxis may be considered for children younger than 24 months of age, who are profoundly immunocompromised during the RSV season.

4 PROBIOTIC TREATMENT

Instead of continuing the traditional approach to kill bacteria wherever they occur, the new criteria for successful antiinfective chemotherapeutics are to preserve the efficacy of each agent as long as possible by delaying the emergence of drug resistance and to spare the normal microbiota as much as possible. The normal microbiota is viewed as containing invaluable allies in combating microbial pathogenesis by protecting niches against new microbial competitors and sustaining the species diversity that impedes virulence.[4,5]

Antibiotic development has focused on the identification of "essential" targets whose inhibition is lethal under conditions of maximal microbial proliferation. A fresh approach would be to revise the operational definition of essentiality so that it more accurately reflects the biological reality: Which genes are essential to the pathogen in vitro under conditions that are relevant in the host? Which genes are essential to the pathogen in specific host environments, including polymicrobial communities on epithelial surfaces, where the microorganism of interest may represent a relatively minor planktonic population, in monomicrobial populations deep in tissues?[4]

Araújo and colaborators[5] reviewed several studies and found that they were heterogeneous regarding strains of probiotics, the mode of administration, the time of use, and outcomes. In the same work, they identified 11 peer-reviewed, randomized clinical trials, which include a total of 2417 children up to 10 incomplete years of age. In their analysis of these studies, the reduction in new episodes of disease was a favorable outcome for the use of probiotics in the treatment of respiratory infections in children. It is noteworthy that most of these studies were conducted in developed countries with basic sanitation, health care and strict, well-established, and well-organized guidelines on the use of probiotics. Adverse effects were rarely reported, demonstrating probiotics to be safe. They concluded that the encouraging results—that is, reducing new episodes of respiratory infections—emphasize the need for further research, especially in developing countries, where rates of respiratory infections in children are higher when compared to the high per capita-income countries identified.

5 NATURAL MEDICINES

Complementary and alternative therapies have grown rapidly over the last two decades to increment or replace the traditional medical practice in both America and Europe,[30,31] especially to multidrug resistant strains of bacteria such as *Escherichia coli* and *K. pneumoniae*, which are for example widely distributed in hospitals. Thus, herbal medicines have often been recommended in the treatment of many diseases, including in lower respiratory system infections, as follows.

Regarding bronchitis, a large number of plants were found in the literature indicating its treatment, which are presented in Table 14.1. Bronchitis treatment includes, in the most studies, antitussive, expectorant, and antiinflammatory outcomes.

Table 14.1 Plants Indicated to the Bronchitis Treatment

Scientific Name	Doses/Extracts	Constituents	Observations/Results	References
Tussilago farfara L.	• Aqueous extract from flower buds (2.8 g/kg) and rachis (3.5 g/kg)	Caffeic acid, chlorogenic acid, sinapic acid, rutin and kampferol, Maleic acids, formic acid, tussilagone, and others.	• Preclinical study; ICR mice of either sex (19–24 g); • Antitussive and expectorant activities.	a
Citri grandis (L.) Osbeck	• Aqueous extract: 1005 mg/kg • 50% ethanolic extract: 568 mg/kg • 70% ethanolic extract: 247, 493, and 986 mg/kg • 90% ethanolic extract: 501 mg/kg	Not identified	• Preclinical study; NIH mice of either sex (18–22 g); • 70% ethanolic extract of *C. grandis* demonstrated the best antitussive, expectorant and antiinflammatory effects in vivo.	b
Pyrrosia petiolosa (Christ et Bar.) Ching	• Ethanol extract and fractions (petroleum ether, ethyl acetate, N-butanol and aqueous); • Test with microorganisms: ethanol extract or fractions at 0.25, 0.50, 0.625, 1.25, 2.50, 5.0, 10.0, and 20.0 mg/ mL • Antiinflammatory test: ethanol extract at 2.5, 5.0, and 10.0mg/kg	Anthraquinones, flavonoids, terpenoids, steroids, reducing sugars, and saponins.	• In vitro study with bacterial ad fungi strains, and preclinical study (Kunming mice); • The minimum inhibitory concentration (MIC) of the ethanol extract and fractions ranged from 1.25 to 10.00 mg/mL • Antibacterial activity ranging from 1.25 to 10.0 mg/mL for ethanol extract and fractions) and antiinflammatory property (ethanol extract at 5.0 and 10.0 mg/kg)	c

Table 14.1 Plants Indicated to the Bronchitis Treatment (*cont.*)

Scientific Name	Doses/Extracts	Constituents	Observations/Results	References
Hedera helix L.	• Ivy leaves extract with 50% ethanol (Hedelix—reference natural medication), 260 patients; • Ivy leaves extract with 30% (m/m) ethanol (Prospan), 258 patients; • Doses: adults and children (from 10-years old) 24 drops; children (4–10-years old) 16 drops; children (2–4-years old) 12 drops.	A minimum of 6.75% of hederacoside C	• Male or female Caucasian patients at least 2 years of age with a confirmed clinical diagnosis of acute bronchitis; • Children under 2 years of age were excluded as well as medication possibly influencing symptoms of acute bronchitis; • Patients took one of the medications 3x/ daily during 7 days; • Bronchitis severity score subscale cough, sputum, rhales/ rhonchi, chest pain during coughing, and dyspnea improved to a similar extent in both treatment groups.	d

[a]*Li ZY, Zhi HJ, Xue SY, et al., Metabolomic profiling of the flower bud and rachis of Tussilago farfara with antitussive and expectorant effects on mice. Journal of Ethnopharmacology 140 (2012) 83–90.*
[b]*Jiang K, Song Q, Wang L, Xie T, et al. Antitussive, expectorant and antiinflammatory activities of different extracts from Exocarpium Citri grandis. Journal of Ethnopharmacology 156 (2014) 97–101.*
[c]*Cheng D, Zhang Y, Gao D, Zhang H. Antibacterial and antiinflammatory activities of extract and fractions from Pyrrosia petiolosa (Christ et Bar.) Ching. Journal of Ethnopharmacology 155 (2014) 1300–1305.*
[d]*Cwientzek, U., Ottillinger, B., Arenberger, P. Acute bronchitis therapy with ivy leaves extracts in a two-arm study. A double-blind, randomised study vs. another ivy leaves extract. Phytomedicine 18 (2011) 1105–1109.*

Several bacteria are involved in pneumonia, as reported previously, and many plants have demonstrated antimicrobial activity (Table 14.2).

In addition, pneumonia treatments are sometimes associated with bronchiolitis treatments because there are common bacteria and viruses between these two diseases, such as human respiratory syncytial virus (HRSV). Chang and colleagues[32] evaluated ginger (*Zingiber officinale* Roscoe) extracts against HRSV propagated in human upper and low respiratory tract cell lines. Hot water extracts of fresh and dried ginger roots were prepared and lyophilized to dry. After that, the powders were dissolved in minimum essential medium until the final concentrations of 10, 30, 100, and 300 µg/mL. The authors found that only fresh ginger was effective against HRSV in both cell lines, even in the lowest dose, and therefore, this medicinal plant could prevent HRSV infection by inhibiting viral attachment and internalization.

Given sequence about HRSV causing bronchiolitis and pneumonia, *Pueraria lobata* was studied in both human upper and lower respiratory tract cell lines. A hot water extract of *Pueraria lobata* was

Table 14.2 Plants and Their Extracts With Antibacterial Activity, Which Could be Indicated to the Pneumonia Treatment

Caesalpinia ferrea Mart ex. Tul.	• Bark crude extracts were obtained using turbo-extraction (10%;w/v) with water or acetone:water (7:3, v/v) • Concentrations ranging from 10 to 0.078 mg/mL	Gallic acid and catechin were present in all chromatograms	• MIC (μg/mL)—Aqueous crude extract: *Staphylococcus aureus* ATCC25923 (5000 μg/mL); *S. epidermidis* INCQS 00016 (625 μg/mL); *Enterococcus faecalis* ATCC29212 (625 μg/mL); *E. coli* ATCC25922 (>10,000 μg/mL); *S. enteritidis* INCQS00258 (5000 μg/mL); *Shigella flexneri* (5000 μg/mL); *Klebsiella pneumonia* (5000 μg/mL). • MIC (μg/mL)—Acetone–water crude extract: *S. aureus* ATCC25923 (2500 μg/mL); *S. epidermidis* INCQS 00016 (150 μg/mL); *E. faecalis* ATCC29212 (1250 μg/mL); *E. coli* ATCC25922 (>10,000 μg/mL); *Salmonella enteritidis* INCQS00258 (2500 μg/mL); *Shigella flexneri* (3750 μg/mL); *Klebsiella pneumonia* (5000 μg/mL).	a
Syzygium cumini (L.) Skeels.	• Fruits crude methanolic extracts (70% methanol) • Methanol extract was enriched for its total anthocyanin. Concentrations: 0.1, 0.5 and 1 mg/ml.	N-Hexanoyl-dl-homoserine lactone, Malvidin, Petunidin, Cyanidin	• Methanol extract was enriched for its total anthocyanin. Total anthocyanin inhibited the violacein production in *C. violaceum*; biofilm formation and EPS production in *Klebsiella pneumoniae* up to 82, 79.94, and 64.29% respectively. Synergistic activity of conventional antibiotics with STA enhanced the susceptibility of *K. pneumoniae* up to 58.45%.	b
Syzygium cumini(L.) Skeels.	• Leaves extracts isolated by Soxhlet's method in: petroleum ether, chloroform, acetone, methanol and aqueous	Malic acid, jamboline, ellagic, gallic and caffeic acids, guaicol, resorcinol, corilaginin, β-sitosterol, betulinic acid, friedelin, epi-friedelanol, astragalin β-sitosterol-D glucoside, kaempferol, 17myricetin	• Water extract was active against all six bacterial strains (*Escherichia coli* (MTCC433), *Vibrio cholerae* (MTCC 3904), *Klebsiella pneumonia*, (MTCC3384), *Proteus vulgaris* (MTCC426), *Bacillus subtilis* (MTCC 441), *Salmonella typhi* (MTCC-531)) in the range of 0.750–6.00 mg/mL	c
Equisetum arvense L.	• The volatile constituents of the sterile stems	Hexahydrofarnesyl acetone, cis-geranyl acetone, thymol, and trans-phytol	• The 1:10 dilution of the essential oil of *Equisetum arvense* L. showed a strong antimicrobial activity against: bacteria *Staphylococcus aureus, Escherichia coli, Klebsiella pneumoniae, Pseudomonas aeruginosa,* and *Salmonella enteritidis*.	d

Table 14.2 Plants and Their Extracts With Antibacterial Activity, Which Could be Indicated to the Pneumonia Treatment (*cont.*)

Schinus terebinthifolius Raddi	• *S. terebinthifolius* leaf lectin (SteLL)	SteLL, a chitin-binding lectin	• Lectin was active against *Escherichia coli*, *Klebsiella pneumoniae*, *Proteus mirabilis*, *Pseudomonas aeruginosa*, *Salmonella enteritidis,* and *Staphylococcus aureus.* Highest bacteriostatic and bactericide effects were detected for *S. enteritidis* (MIC: 0.45 µg/mL) and *S. aureus* (MBC: 7.18 µg/mL), respectively.	e
Olea europaea L. *Cv.* Cobrançosa	• Olive leaf aqueous extracts • Concentrations: 0.05, 0.10, 0.50, 1.00, and 5.00 mg/ mL	Caffeic acid, verbascoside, oleuropein, luteolin 7-O-glucoside, rutin, apigenin 7-O-glucoside and luteolin 4'-O-glucoside	• Olive leaf aqueous extracts were screened for their antimicrobial activity against *B. cereus*, *B. subtilis*, *S. aureus* (Gram +), *E. coli*, *P. aeruginosa*, *K. pneumoniae* (Gram −). • The extract inhibited all the tested bacteria.	f
Eucalyptus globulus Labill.	• *Eucalyptus globulus* leaf was extracted with methanol	Not identified	• Clinical specimens from 200 patients with respiratory diseases were collected. The pathogenic bacteria isolated comprised 56 *Staphylococcus aureus*, 25 *Streptococcus pyogenes*, 12 *Streptococcus pneumoniae* and 7 *Haemophilus influenza*. • MIC_{50}, MIC_{90}, and MBC values ranged from 16 to 64, 32 to 128, and 64 to 512 mg/ L, respectively, depending on the species.	g
Copaifera spp	• Oils used here were exuded directly from the trunks of the tree	Sesquiterpenes: a-copaene, b-cariofilene, b-bisabolene, a and b-selinene, a-humulene, and d and g-cadidene	• Active against Gram-positive species (*Staphylococcus aureus*, methicillin-resistant *S. aureus*, *S. epidermidis*, *Bacillus subtilis*, and *Enterococcus faecalis*) with minimum inhibitory concentrations ranging from 31.3–62.5 µg/ml	h
Copaifera spp	• Oils used here were exuded directly from the trunks of the tree	Diterpenes 1–11	• Antimicrobial activity (MIC < 10 µg/ml) against Gram-positive bacteria (*Staphylococcus aureus* ATCC 6538; *S. epidermidis* CECT 232; *S. saprophyticus* CECT 235, *Bacillus subtilis* CECT 39), comparable with cephotaxime (used as control).	i

(Continued)

Table 14.2 Plants and Their Extracts With Antibacterial Activity, Which Could be Indicated to the Pneumonia Treatment (*cont.*)

Malva sylvestris L.	• Air-dried plant flowers and leaves were extracted using a soxhlet type apparatus with n-hexane, dichloromethane, and methanol. The extracts were dried in vacuum.	Naphthoquinone, monoterpenes, aromatic compounds, and a tetrahydroxylated acyclic diterpense	• High antibacterial effects against some human pathogen bacteria strains: *Staphylococcus aureus, Streptococcus agalactiae, Entrococcus faecalis,* with MIC value of 192, 200, and 256 µg/ml, respectively.	j

[a]Araújo AA, Soares LA, Assunção Ferreira MR, de Souza Neto MA, da Silva GR, de Araújo RF Jr, Guerra GC, de Melo MC. Quantification of polyphenols and evaluation of antimicrobial, analgesic and antiinflammatory activities of aqueous and acetone-water extracts of Libidibia ferrea, Parapiptadenia rigida, and Psidium guajava. J Ethnopharmacol. 2014 156:88–96.

[b]Gopu V, Kothandapani S, Shetty PH. Quorum quenching activity of Syzygium cumini (L.) Skeels and its anthocyanin malvidin against Klebsiella pneumoniae. Microb Pathog. 2015 79:61–69.

[c]Dhankhar S, Dhankhar S, Kumar M, Ruhil S, Balhara M, Chhillar AK. Analysis toward innovative herbal antibacterial and antifungal drugs. Recent Pat Antiinfect Drug Discov. 2012 7(3):242–248.

[d]Radulović N, Stojanović G, Palić R. Composition and antimicrobial activity of Equisetum arvense L. essential oil. Phytother Res. 2006 20(1):85–88.

[e]Gomes FS, Procópio TF, Napoleão TH, Coelho LC, Paiva PM. Antimicrobial lectin from Schinus terebinthifolius leaf. J Appl Microbiol. 2013 114(3):672–679.

[f]Pereira AP, Ferreira IC, Marcelino F, Valentão P, Andrade PB, Seabra R, Estevinho L, Bento A, Pereira JA. Phenolic compounds and antimicrobial activity of olive (Olea europaea L. Cv. Cobrançosa) leaves. Molecules. 2007 12(5):1153–1162.

[g]Salari, M.H., et al. Antibacterial effects of Eucalyptus globulus leaf extract on pathogenic bacteria isolated from specimens of patients with respiratory tract disorders. Clinical Microbiology and Infection, 2006 12(2):194–196.

[h]Santos AO, Ueda-Nakamura T, Dias Filho BP, Veiga Jr VF, Pinto AC, Nakamura CV. Antimicrobial activity of Brazilian copaiba oils obtained from different species of the Copaifera genus. Mem. Inst. Oswaldo Cruz, 2008 103, 277–281.

[i]Tincusi BM, Jiménez IA, Bazzocchi IL, Moujir LM, Mamani ZA, Barroso JP, Ravelo AG, Hernández BV. Antimicrobial terpenoids from the oleoresin of the Peruvian medicinal plant Copaifera paupera. Planta Med. 2002 68(9):808–812.

[j]Razavi SM, Zarrini G, Molavi G, Ghasemi G. Bioactivity of Malva sylvestris L., a medicinal plant from iran. Iran J Basic Med Sci. 2011 14(6):574–579.

prepared, lyophilized and then dissolved in minimum essential medium to the final concentrations of 10, 30, 100, and 300 µg/mL. The virus HRSV was propagated in both cell lines. *P. lobata* was more effective on lower respiratory tract cells than in upper respiratory tract cells, and could be used to prevent HRSV infection.[33]

6 CONCLUSIONS AND FUTURE PERSPECTIVES

The hospital community and the public should be aware of the problem of antibiotic resistance and its consequences.

Many efforts are being made by researchers to find ways to rapidly identify pathogens, to bypass the pathogen's resistance, and to find new antimicrobials or alternative ones. Complementary and alternative therapies have grown especially rapidly over the last two decades to fight the multidrug resistance problem.

"WHO (2015) has coordinated a global campaign to raise awareness of antibiotic resistance and encourage best practices among the public, policymakers, health, and agriculture professionals to avoid further emergence and spread of antibiotic resistance."

REFERENCES

1. Cappelletty D. Microbiology of bacterial respiratory infections. *Pediatr Infect Dis J*. 1998;(8 Suppl.):55–61.
2. Cappelletty D, Dasaraju PV, Liu C. Infections of the Respiratory System. In : Baron, S, editor. *Medical Microbiology*. 4th edition. Galveston (TX); 1996.[Chapter 93].
3. Simpson CR, Steiner MFC, Cezard G, Bansal N, Fischbacher C, Douglas A, Bhopal R, Sheikh A. Ethnic variations in morbidity and mortality from lower respiratory tract infections: a retrospective cohort study. *J R Soc Med* 2015;**108**(10):406–17.
4. Bartlett JG. A Call to Arms: The Imperative for Antimicrobial. *Clin Infect Dis* 2011;**53**(1):4–7.
5. Havey TC, Hull MW, Romney MG, Leung V. Retrospective cohort study of inappropriate piperacillin-tazobactam use for lower respiratory tract and skin and soft tissue infections: opportunities for antimicrobial stewardship. *AJIC* 2015;**43**:946–50.
6. Jassim SAAJ, Limoges RG. Natural solution to antibiotic resistance: bacteriophages 'The Living Drugs. *World J Microbiol Biotechnol* 2014;**30**(8):2153–70.
7. Araujo GV, Oliveira Junior MH, Peixoto DM, Sarinho ES. Probiotics for the treatment of upper and lower respiratory tract infections in children: systematic review based on randomized clinical trials. *J Pediatr (Rio J)* 2015;**91**(5):413–27.
8. Rambaud-Althaus C, Althaus F, Genton B, D'Acremont V. Clinical features for diagnosis of pneumonia in children younger than 5 years: a systematic review and meta-analysis. *Lancet* 2015;**15**(4):439–50.
9. Mandell LA. Introduction: Evolving needs in respiratory tract infections. *Clin Microbiol Infec* 1998;**4**(1):1–8.
10. Rotstein C, Evans G, Born A, Grossman R, Light RB, Magder S, et al. Clinical practice guidelines for hospital-acquired pneumonia and ventilator-associated pneumonia in adults. *Can J Infect Dis Med Microbiol* 2008;**19**(1):19–53.
11. Hayashi Y, Paterson DL. Strategies for reduction in duration of antibiotic use in hospitalized patients. *Clin Infect Dis* 2011;**52**(10):1232–40.
12. Domenech C, Ardanuy R, Pallares I, Grau S, Santos AG, et al. Some pneumococcal serotypes are more frequently associated with relapses of acute exacerbations in COPD patients. *Plos One* 2013;**8**:e59027.
13. Vandevelde NM, Tulkens PM, Diaz Iglesias Y, Verhaegen J, Rodriguez-Villalobos H, Philippart I, et al. Characterisation of a collection of *Streptococcus pneumoniae* isolates from patients suffering from acute exacerbations of chronic bronchitis: In vitro susceptibility to antibiotics and biofilm formation in relation to antibiotic efflux and serotypes/serogroups. *Int J Antimicrob Agents* 2014;**44**(3):209–17.
14. Waites KB, Talkington DF. *Mycoplasma pneumoniae* and its Role as a human pathogen. *Clin Microbiol Rev* 2004;**17**(4):697–7281.
15. Ruoxi He, Bailing Luo, Chengping Hu, Ying Li, Ruichao Niu. Differences in distribution and drug sensitivity of pathogens in lower respiratory tract infections between general wards and RICU. *J Thorac Dis* 2014;**6**(10):1403–10.
16. Mandell LA. Infectious Diseases Society of America/American Thoracic Society consensus guidelines on the management of community-acquired pneumonia in adults. *Clin Infect Dis* 2007;**44**(2):S27–72.
17. Eliopoulos, DN. The Sanford Guide to Antimicrobial Therapy 2015. 45 ed. 2015, Sperryville: Antimicrob Ther; 45th ed.; 236.
18. Lim WSDefining community-acquired pneumonia severity on presentation to hospital: an international derivation and validation study. *Thorax* 2003; 58:377–382.

19. Hazir T. Ambulatory short-course high-dose oral amoxicillin for treatment of severe pneumonia in children: a randomised equivalency trial. *Lancet* 2008;**371**:49–56.
20. Bradley JS. Executive summary: the management of community-acquired pneumonia in infants and children older than 3 months of age: clinical practice guidelines by the Pediatric Infectious Diseases Society and the Infectious Diseases. *Soc Am Clin Infect Dis* 2011;**53**:617–30.
21. Smith SM, Fahey T, Smucny J, Becker LA. Antibiotics for acute bronchitis. *Cochrane Database Syst Rev* 2014;**3**:pCd000245.
22. Smith SM, Smucny J, Fahey T. Antibiotics for acute bronchitis. *Jama* 2014;**312**(24):2678–9.
23. Eliopoulos DN. The Sanford Guide to Antimicrobial Therapy. 43rd ed. Sperryville: *Antimicrobial Therapy.*
24. Ralston SL, Lieberthal AS, Meissner HC, et al. Clinical practice guideline: the diagnosis, management, and prevention of bronchiolitis. *Pediatrics* 2014;**134**:e1474–502.
25. Gadomski AM, Scribani MB. Bronchodilators for bronchiolitis. *Cochrane Database Syst Rev* 2014;**6**:Cd001266.
26. Fernandes RM, Bialy LM, Vandermeer B, et al. Glucocorticoids for acute viral bronchiolitis in infants and young children. *Cochrane Database Syst Rev* 2013;**6**:Cd004878.
27. Pinto LA, Pitrez PM, Luisi F, et al. Azithromycin therapy in hospitalized infants with acute bronchiolitis is not associated with better clinical outcomes: a randomized, double-blinded, and placebo-controlled clinical trial. *J Pediatr* 2012;**161**:1104–8.
28. Hernandez E, Khoshoo V, Thoppil D, et al. Aspiration: a factor in rapidly deteriorating bronchiolitis in previously healthy infants? *Pediatr Pulmonol* 2002;**33**:30–1.
29. Updated guidance for palivizumab prophylaxis among infants and young children at increased risk of hospitalization for respiratory syncytial virus infection. *Pediatrics* 2014;134:415–420.
30. Eisenberg DM, Davis RB, Ettner SL, Appel S, Wilkey S, Van Rompay M, Kessler RC. Trends in Alternative Medicine Use in the United States, 1990–1997: results of a follow-up national survey. *JAMA* 1998;**280**(18):1569–75.
31. Posadzki P, Alotaibi A, Ernst E. Prevalence of use of complementary and alternative medicine (CAM) by physicians in the UK: a systematic review of surveys. *Clin Med* 2012;**12**(6):505–12.
32. Chang JS, Wang KC, Yeh CF, et al. Fresh ginger (*Zingiber officinale*) has anti-viral activity against human respiratory syncytial virus in human respiratory tract cell lines. *J Ethnopharmacol.* 2013;**145**:146–51.
33. Lin TJ, Yeh CF, Wang KC, et al. Water extract of *Pueraria lobata* Ohwi has anti-viral activity against human respiratory syncytial virus in human respiratory tract cell lines. *Kaohsiung Journal of Medical Sciences* 2013;**29**:651–7.

NANOTECHNOLOGICAL APPLICATIONS FOR THE CONTROL OF PULMONARY INFECTIONS

15

A.P. Ingle*, S. Shende*, R. Pandit*, P. Paralikar*, S. Tikar*, K. Kon**, M. Rai*

*Nanobiotechnology Laboratory, Department of Biotechnology, SGB Amravati University, Amravati, Maharashtra, India;
**Department of Microbiology, Virology and Immunology, Kharkiv National Medical University, Kharkiv, Ukraine

1 INTRODUCTION

Microbes, including viruses, bacteria and fungi, are the causal agents of pulmonary infections.[1–4] Infections of nose, sinuses, and throat are included in upper tract infections. Whereas infections of trachea, lungs and bronchial tubes are considered as lower tract infections.[5] Generally, upper tract infections are mild in nature; however, severe illness and high mortality rate are reported in lower tract infections.[6] It is estimated that about 4.2 million deaths occurs globally due to lower tract infections among all the age groups and about 1.8 million deaths have been reported in children. Children are more prone to lower tract viral infections.[3,7]

Usually the most common mode of transmission of the pulmonary infections is by contact of one person to another. Such infections are transmitted by direct or indirect contact. Individuals affected by cough and sneezing, transfer droplets consisting of microorganisms in a fraction of seconds through air or directly deposited into the mouth or nasal mucosa of another susceptible individual, for example, mouth droplet from people suffering from cough can be spread into another individual.[5] Indirect transmission involves the contact of susceptible individual with the contaminated instruments or the surface where organisms occur. Microorganisms can reside on the hand of the infected person and can be transferred, for example, by the shaking of hands. The needles and gloves of an infected person can also transmit infection to another person[5].

There are various types of pulmonary infections, such as: pneumonia, tuberculosis, pulmonary aspergillosis and nontuberculous mycobacterial infections. Pulmonary infections are commonly reported from developing and developed countries.[4,8–10] Although there are many microbiological testing methods available for the diagnosis of such infections and there are various antimicrobial treatment therapies for the management of pulmonary infections, sometimes the curing rate of these infections is reported to be low due to certain limitations. Microbiological testing methods are believed to be slow and insensitive, whereas, the uncontrolled and over use of antibiotics create the major problem of antibiotic resistance.[11,12] Recently, the ability of common pathogens to acquire resistance to widely used antimicrobial therapies has been increased.[13] The authors further suggested that the number of pathogens that exhibit resistance to one or multiple drugs (MDR) have been increasing gradually.

Currently, the problem of antimicrobial resistance in both Gram-negative and Gram-positive bacteria and other pathogens causing pulmonary infections has become a major threat. Karchmer[14] reported that antimicrobial resistance is particularly prevalent among the most common pathogens associated with community-acquired respiratory tract infections, which includes *Streptococcus pneumoniae, Haemophilus influenzae, Moraxella catarrhalis,* and *Streptococcus pyogenes.*[15] Hence, the morbidity and mortality associated with the pulmonary infection pose a growing challenge to clinical practitioners.[16] Considering the huge scope and applications of nanotechnology in biomedicine, it has been proven that nanotechnology will play a crucial role for the management of problems associated with antibiotic resistance including multiple drug resistance. Therefore, these life-threatening challenges including pulmonary infections can now be overcome with the help of nanotechnology.[12,17]

The aim of the present chapter is to discuss the global status of various pulmonary infections and the strategies for their management. Here, we have described treatment therapies available for pulmonary infections and their limitations. Given that multidrug resistance is a major challenge, novel and effective alternative treatment therapies need to be discovered. In addition, we have also discussed the role of nanotechnology for the management of pulmonary infections with a special focus on different types of nanomaterials.

2 SOME IMPORTANT PULMONARY INFECTIONS
2.1 PNEUMONIA

Pneumonia is an inflammatory disease, which affects lung air sacs (known as alveoli). This disease is mostly caused by viruses, bacteria, or fungi. Worldwide more than 450 million people suffer from pneumonia and four million people die from pneumonia annually. Common symptoms of pneumonia includes- persistent cold, fever with shaking hands, chest pain, and difficulty in breathing. Other severe sign and symbols of pneumonia are vomiting, unconsciousness, convulsions.[1] Bacteria are supposed to be the most common source of pneumonia. *Streptococcus pneumoniae* is responsible for more than 50% of the infections, *Haemophilus influenza* causes 20% of the infections. *Mycoplasma pneumoniae* also causes infection in 3% of the population and some other bacterium like *Stenotrophomonas maltophilia* also reported to cause pneumonia.[18] Certain viruses like coronaviruses, influenza virus, adenovirus, and respiratory syncytial virus causes pneumonia.[19] Fungal pneumonia is very rare. It is mostly observed in immunocompromised patients. Common fungi which cause pneumonia are *Cryptococcus neoformans, Pneumocystis jirovecii,* and *Coccidioides immitis.*[2,20]

2.2 TUBERCULOSIS

Tuberculosis (TB) is caused by the *Mycobacterium tuberculosis*. These bacteria usually infect the lungs but may also affect other parts, such as liver and kidney. When TB is observed outside of the lungs it is known as extra-pulmonary TB. Immunocompromised patients are more susceptible to this infection. The mode of transmission of disease is air. When a TB patient speaks, coughs, or sneezes, TB bacteria may be spread and infect nearby individuals. Even if a person is infected with *M. tuberculosis*, it may be in inactive form and is only activated in patients with lower immunity or who is in a later stage of life.[21] Common signs of TB are persistent cough with sputum for more than 3 weeks, chest pain, loss of appetite, chill, fever, sweating at night, fatigueness and weight loss. Most patients do not show the symptoms of TB but develop the disease at the later stage.[4,22] Not every individual become sick after this infection, they can have TB bacteria in the body but the body is able to fight against the bacteria. People with such an infection do not spread TB infection to others. This form of infection is known as Latent TB. TB bacteria are active when the immunity is lower, and in this case TB bacteria can be spread.[23]

2.3 PULMONARY ASPERGILLOSIS

Pulmonary aspergillosis is a fungal infection, which is mainly caused by *Aspergillus* species. This disease is characterized by the colonization of *Aspergillus* spores in the lungs. Aspergillosis is more commonly observed in people who are affected with TB or people with COPD. Mostly, aspergillosis is caused by *A. flavus, A. niger* and *A. terreus. A. fumigatus* is commonly found in patients suffering from invasive aspergillosis. There are three types of pulmonary infections: allergic bronchopulmonary aspergillosis (ABPA), chronic pulmonary aspergillosis (CPA), invasive pulmonary aspergillosis (IPA).[24,25] The common symptoms of aspergillosis are fever, cough, breathlessness and chest pain. ABPA infection is mainly characterized with asthmatic patients and with eosinophilia deficient patient. Generally, invasive aspergillosis is reported in immunocompromised patients with neutropenia and after organ transplantation.[26,27]

2.4 NONTUBERCULOUS MYCOBACTERIAL PULMONARY INFECTION

Nontuberculous mycobacterial (NTM) pulmonary infections are generally caused by mycobacteria other than *M. tuberculosis*. Most common causative agents of this disease are *M. kansasii, M. abscessus, M. fortuitum M. avium* and *M. intracellulare*. In the United States, the most predominant mycobacteria that are associated with pulmonary infections include *M. avium* complex, followed *by M. kansasii*.[28] More than 140 NTM species are reported. NTM pulmonary infections are noncontagious. They do not spread from individual to individual. However, these infections are transmitted by inhalation of mycobacteria. Symptoms of NTM pulmonary infections are chronic cough, sputum production, high fever and weight loss.[29] NTM pulmonary infections have been reported in immunocompromised patients, patients with lung transplantation, and those affected with cystic fibrosis (CF).[30]

3 EXISTING TREATMENTS FOR PULMONARY INFECTIONS

Different treatments are available for the management of pulmonary infections, but in case of some infections the duration of treatment may be very long, for example,TB. Macrolides drugs are commonly used in the management of various kinds of pulmonary infections and have received considerable attention because of their antiinflammatory and immunomodulatory actions, apart from the antibacterial efficacy. Min and Jang[31] studied the in vitro and in vivo efficacy of macrolides drugs (viz. azithromycin, clarithromycin, dirithromycin, erythromycin, roxithromycin, and telithromycin) against respiratory viral infections caused by rhinovirus, respiratory syncytial virus and influenza virus. The results showed significant antiviral activity against all the tested pulmonary viruses. Henry et al.[32] proposed the bacteriophage therapy for the treatment of pulmonary infections caused by antibiotic-resistant bacteria. They found that pulmonary infections caused by antibiotic-resistant *Pseudomonas aeruginosa* can be safely managed by using various bacteriophages from the *Myoviridae* and Podoviridae family. In another similar study, Semler et al.[33] demonstrated the use of phage therapy for respiratory infections caused by antibiotic-resistant *Burkholderia cepacia*. The effectiveness of the therapy was studied by establishing the *B. cenocepacia* respiratory infections in mice using a nebulizer and a nose-only inhalation device, the mice were treated with different *B. cenocepacia*-specific phages delivered as either an aerosol or intraperitoneal injection. The results obtained after 2 days of treatment suggested that aerosol phage therapy appears to be an effective method. Whereas, the mice that received phage treatment by intraperitoneal injection did not show significant activity.

As mentioned previously, pneumonia is a pulmonary infection which affects the lungs. It is caused by various bacteria (eg, *Streptococcus pneumoniae*, *Klebsiella pneumoniae*, *Haemophilus influenzae* and *Mycoplasma pneumoniae*), viruses (eg, adenoviruses, rhinovirus, influenza virus, respiratory syncytial virus and parainfluenza virus) and fungi (eg, *Histoplasma capsulatum*, *Coccidioides immitis*, *Blastomyces dermatitidis*, *Paracoccidioides brasiliensis*, *Sporothrix schenckii* and *Cryptococcus neoformans*). It is characterized primarily by inflammation of the alveoli in the lungs or by fluid filled alveoli (alveoli are microscopic sacs in the lungs that absorb oxygen). Generally, treatment of pneumonia depends on the type and severity of infections. Bacterial pneumonia is usually treated with different antibiotics (viz. azithromycin, erythromycin, doxycycline, gemifloxacin, levofloxacin, cephalexin, amoxicillin, vancomycin, etc.).[34] Efficient antifungals used for the management of fungal pneumonia include first, second and third-generation triazoles and echinocandins. In some cases Amphotericin B is less frequently used. However, viral pneumonia is treated with rest and plenty of fluids.[35]

TB is also one of the most life-threatening pulmonary infections and it is a major public health problem all over the world. According to a survey, India contributes about 26% of the global TB burden. Since ancient times, TB has been a leading cause of morbidity and mortality.[36] Over the last decade, scientists have made significant progress in treatment for TB. Regimens have been optimized and directly observed therapy short-course (DOTS) initiatives have been implemented.[37] Currently, chemotherapy used for TB commonly includes combination of first-line drugs, isoniazid, rifampin, pyrazinamide and ethambutol for about 6 months. If treatment fails because of drug resistance, then second-line drugs, such as, paraaminosalicylate, kanamycin, fluoroquinolones, capreomycin, ethionamide and cycloserine are recommended, although these drugs may have serious side effects.[38–41]

Aspergillosis in one of the most important diseases among the pulmonary infections caused by *Aspergillus* species. It is reported to be an emerging cause of life-threatening infections in immunocompromised patients having infections, such as, prolonged neutropenia, advanced HIV infection, inherited immunodeficiency and in patients who have undergone allogeneic hematopoietic stem cell transplantation (HSCT) and/or lung transplantation. Three different forms of aspergillosis have been reported so far. These include invasive aspergillosis, chronic (and saprophytic) aspergillosis and allergic forms of aspergillosis. Generally, voriconazole is superior to deoxycholate amphotericin B (D-AMB) and, hence, recommended in primary treatment for invasive aspergillosis in most patients. Apart from these, some of the FDA-approved compounds, such as, lipid formulations of amphotericin B (AMB lipid complex [ABLC], L-AMB, and AMB colloidal dispersion [ABCD]), caspofungin and antifungal triazoles (ie, itraconazole, voriconazole and posaconazole) are also recommended. Similarly, in chronic aspergillosis, the regular doses of itraconazole and voriconazole are usually prescribed by the physician. Whereas, the treatment for allergic forms of aspergillosis includes combination of corticosteroids and itraconazole.[42]

Among the pulmonary infections, nontuberculous pulmonary infections caused by NTM are very common infections, which are increasingly recognized worldwide. About 150 different species of

NTM have been described. *M. avium* complex, *M. kansasii*, and *M. abscessus* are found to be most common NTM, which cause such infections. According to Davis et al.[43] existing therapy available for the treatment of nontuberculous mycobacterial pulmonary infection by *M. avium* showed very poor clinical response rates, drug toxicities, and side effects. They demonstrated that use of aerosolized amikacin as a standard oral therapy against nontuberculous mycobacterial pulmonary infection significantly improved treatment efficacy without producing systemic toxicity. Therefore, it was proposed that aerosolized delivery of amikacin is a promising adjunct to standard therapy for pulmonary nontuberculous mycobacterial infections. Moreover, extensive experimental trials are required to define its optimal role in the therapy of this disease.

In another study, Johnson and Odell[28] reviewed that the eradication of NTM infections is very difficult by the common treatment strategies. Moreover, it requires a prolonged course of therapy with a combination of drugs. However, there are numerous challenges regarding the treatment of NTM pulmonary infections, but few drugs which can manage these infections are available. Some macrolide drugs like azithromycin and clarithromycin are the efficient and important drugs in the therapy for MAC lung infections. In the case of a resistant strain, a combination therapy with rifampin or rifabutin, and ethambutol (triple therapy) with or without an intravenous aminoglycoside is recommended for about 18 months or more. Similarly, in the case of infection by *M. kansasii*, combination of isoniazid, rifampin and ethanbutol is mostly recommended for about 12 months. In addition, macrolides, such as, clarithromycin and the fourth generation fluoroquinolone moxifloxacin demonstrated very good in vitro activity against *M. kansasii* and may be an alternative to isoniazid. Whereas, lung infections due to *M. abscessus* are very difficult to treat successfully with drug therapy alone. Chemotherapy in conjunction with surgical resection is often recommended in those who can tolerate it. Overall, different kinds of medications and therapies are available for the management of pulmonary infections, but in case of some pulmonary infections existing treatments and therapies showed certain limitations, which are briefly discussed here.

4 LIMITATIONS AND SIDE EFFECTS OF THE TREATMENT OF PULMONARY INFECTIONS

There are a number of limitations for use of conventional strategies for diagnosis and treatments of patients with pulmonary infections. Consequently, unnecessary and prolonged exposure to antimicrobial agents adversely affect patient outcomes, while inappropriate and uncontrolled use of antibiotic therapy increases chance of antibiotic resistance. According to Murdoch,[12] the role of microbiological testing methods used for the diagnosis and management of lower respiratory tract infection continues to be debated. There are many limitations of microbiology laboratories to perform the conventional diagnostic tests for pulmonary infections. The culture based methods that are currently used are slow, insensitive, may not distinguish colonization from infection and may be influenced by previous antimicrobial used. On the other hand, serological tests are also slow and poorly sensitive. Therefore, many authoritative guidelines on the management of pulmonary infection, such as, community-acquired pneumonia in adults do not support routine comprehensive microbiological testing, except in certain situations or in patients with severe disease.[44,45]

Moreover, the uncontrolled and improper use of antibiotics as a treatment for pulmonary infections leads to the development of resistance towards those antibiotics. The problem of antibiotic resistance was predominantly found in TB patients. Unfortunately, some causative agents of TB and many other pulmonary infections become resistant to multiple types of antibiotics. The loss of efficacy of antibiotics and the decrease in their ability to fight pulmonary infections in vulnerable patients is a matter of great concern.[21]

Nanotechnology is a multidisciplinary field that has recently emerged, and it looks as if it will be extremely helpful in the management of pulmonary infections. Nanotechnology based treatment methods and drug delivery strategies will help to deliver drug molecules at the specific site of infection.[46] The role of various nanomaterials in diagnosis and treatments of pulmonary infections is described here.

5 NANOTECHNOLOGY IN MEDICINE

Nanotechnology is an emerging field of science that deals with the study, synthesis and manipulation of materials at nanometer range (ie, in 1–100 nm) or the order of billionths (10^{-9}) of a meter.[47–51] Particles in the nanometer range have changed properties, such as, physicochemical properties, which are distinct from their bulk materials (the macroscopic or microscopic scale) and single atom or molecule (at the atomic scale).[48,52] Nanomaterials are referred to as engineered nanometer dimensions material, which have novel properties, such as, quantum effects and large surface area to volume ratio.[52] These materials are fabricated by a top-down approach, in which bulk materials or technologies are miniaturized, or a bottom-up approach, where assembly occurs atom by atom, from primary to larger, and towards more complex materials.[53] The synthesis of materials in the nanoscale results in magnetic, mechanical, chemical, and electronic effects that are not shown by the bulk materials. These nanoscale effects have been exploited practically in every field of technology, and include commercial applications in energy conversion, textiles, cosmetics, electronics, water purification, lubricants, computing and much more.[54,55] "Nanobiotechnology" is the study of the interaction between biological systems and nanomaterials,[56] and the related field of nanomedicine seeks to employ nanosized materials to diagnose, treat and prevent human diseases.[57–59] The changed physicochemical properties critically influenced the nature of interaction when these nanosized particles come into contact with the biological systems. Moreover, many biological processes, such as, immune recognition and passage across biological barriers are also governed by size considerations.[60] In this context, various applications of nanotechnology, such as, delivery of drug and nucleic acid-based therapeutics to particular disease site makes it a most promising technology of the era. According to Buxton[61] delivery of therapeutics by inhalation provides an opportunity for direct transport of drug to the lung epithelium of the respiratory tract.

5.1 APPLICATIONS OF DIFFERENT NANOPARTICLES FOR THE TREATMENT OF PULMONARY INFECTIONS

The multidisciplinary approach of nanotechnology plays a crucial role to find efficient solution over problems in various sectors. The application and innovation of nanotechnology in the field of medicine is known as "nanomedicine." This broad technology is capable to treat range of diseases. The use of nanoparticles in the field of medicine opens new possibilities and provides new methods for treatment of diseases. Various types of nanoparticles are used in medicine to treat diseases, such as, infectious diseases, viral, bacterial, respiratory tract diseases, and so on. The nanoparticles used for treatment of diseases include: metal nanoparticles, liposomes, polymeric nanoparticles, dendrimers, nanocomplexes, nanorods, quantum dots, nanoemulsions, and so on.[62]

Currently, most recent studies in the field of nanomedicine changed the point of view of drug delivery therapy. Scientists are using nanoparticles as a carrier for drug delivery therapy.[63] In fact, much research has not been carried out concerning the use of nanoparticles in pulmonary diseases. However, in recent years scientists may develop promising nanocarriers for treatment of diseases that affect respiratory tract. Recent studies showed that nanoparticles in pulmonary infection treatment can influence immune system, can create oxidative stress, and can cause genotoxicity.[64]

Studies on nanoparticles for treatment of pulmonary infectious diseases is an emerging field of interest, not only for the treatment of respiratory tract conditions but also for systemic administration of drug delivery for treatment of pulmonary disease. A variety of metal and metal oxide nanoparticles, such as, silver, gold, copper, titanium oxides, and so on have been successfully exploited for pulmonary drug delivery.[65]

Globally, more than 1.5 million deaths are reported annually from respiratory infections, including at least 42% of lower respiratory infections and 24% of upper respiratory infections in developing countries like China, India, Iran, Oman, Philippines, Qatar, Republic of Korea, Saudi Arabia, Thailand, and the United Arab Emirates. The frequent use of antibiotics lead to antibiotic resistance, which enables antibiotic-resistant bacteria to survive despite treatment with existing antibacterial drugs.[66,67] The growing number of multidrug resistant strains has made it imperative the development of new antibiotics and novel approaches to deliver existing agents.

The study carried out by Zhang et al.[68] revealed that polyphosphoester- based silver nanoparticles enhance in vitro antibacterial activities against pathogens associated with CF and decreases the cytotoxicity of bronchial epithelial cells in human. They developed novel degradable polyphosphoester- based polymeric nanoparticles that are able to carry silver cations toward the treatment of lung infections associated with CF through formation of silver acetylides. Another study on liposomal antibacterial targeted pulmonary infection therapy revealed that after several cycles of treatment with liposome and antibacterial drug amikacin in patients showed continuous improvement in pulmonary function and significant reduction in bacterial density.[69] These type of studies suggest that efficacy of antibacterial drugs increases when administered in combination with nanoparticles.[60]

Bhardwaj and coworkers[70] used a mixture of chemotherapeutic agent-loaded vesicular system to overcome TB by developing ligand appended liposome with dry powder inhaler. According to Barash et al.[71] the categorization and detection of specific pattern of lung cancer can be possible using gold nanoparticles sensor with a device profiles unstable organic compounds. Broza et al.[72] reported the use of nanomaterials-based sensors for the identification of breath-print of early-stage lung cancer.

5.2 NANOTECHNOLOGY FOR PULMONARY DRUG DELIVERY

Pulmonary drug delivery has many advantages compared to unusual drug delivery strategies, especially rapid drug uptake, a large surface area for solute transport, and improved drug bioavailability, as well as its noninvasive nature.[73–76] Antimicrobial agents enter into the lung by means of systemic nanoparticles administration, which is determined and potentially harmful upon systemic exposure to the drugs. On the other hand, various nanoparticles exhibit privileged accumulation in the lung—other organs have also been tried. It was reported that intratracheally administered antibiotics loaded nanoparticles were able to enter through the alveolar-capillary barrier into the systemic circulation and collect in extra-pulmonary organ containing spleen, liver, kidney, and bone.[77] Today, the drug dosage form and therapeutic efficiency is improved by "micronization" of drugs. The drug if micronized into micro-spheres with appropriate particle size can be administered directly to the lungs through the mechanical prevention of capillary bed in the lungs.[78,79]

Nanosuspensions may be an ideal approach to deliver drugs that show deprived solubility in pul-monary secretions.[80] In addition, due to the nanoparticulate nature and homogeneous size distribution of nanosuspensions, it is possible that in each aerosol droplet at least one drug nanoparticle is present leading to an even distribution of drug in lungs compared to the microparticulate form of the drug. In routine suspension aerosols, numerous droplets are free of drug and others are filled with the drug in high amounts, directing to irregular release and circulation of the drug within the lungs. Nanosuspen-sions could be utilized in all available types of nebulizer.[81]

Aggarwal et al.[80] studied antitubercular drugs, for example, pyrazinamide, isoniazid, and rifam-picin were integrated into various formulations of solid lipid ranged from 1.1 to 2.1 μm and these formulations were nebulized to guinea pigs orally for direct pulmonary delivery. Likewise, condi-tions, such as, pulmonary aspergillosis can easily be targeted by applying appropriate drug candidates like amphotericin B, in the form of pulmonary nanosuspensions as a substitute of using stealth lipo-somes.[82] Numerous respiratory diseases have been treated by using the nanocarrier systems, which can be easily transferred through the airways.[83,84] A large number of pulmonary diseases that have been searched includes- CF and some other genetic disorders, COPD, tuberculosis, pediatric diseases, and cancer.[76,85–88]

The pharmacodynamics and pharmacokinetics of a drug are exceedingly reliant on its physical and chemical features, which are influenced by the type of formulation used to deliver it. Through scaling down size of compounds, Nano-Drug Delivery System (NDDS) can transform and im-prove the performance of many drugs to an extent not reachable by conservative formulations.[89] For example, NDDS can be capitalized to encapsulate drugs and thereby (1) protect them from degrada-tion, (2) target the drugs to particular cells/tissues/organs, releasing them in a restricted behavior as a response to a precise stimulus, (3) increase their solubility, (4) enhance their epithelial absorp-tion and increase their blood circulation time, and (5) enhance their uptake by cells.[90–92] Moreover, combined NDDS can concurrently detect and treat a disease by encompassing both imaging and therapeutic compounds, which are termed as theranostics.[93] Nanomedicine could play a key role in the near future to achieve the highly desired modified medicine.[94] Over the last few decades, the usefulness of the design and development of NDDS to overcome a variety of biopharmaceutical drawbacks in the diagnosis, prevention, vaccination, and disease treatment has been intensively ex-plored by a large number of research groups globally, leading to an enormous number of scientific articles available in international journals. Moreover, it has generated a generous rational property platform. Nevertheless, and despite the fact that nanomedicine began as a discipline almost half a century ago, only a few NDDS have found their way to the market.[94,95] This experience could be explained by the lack of economic profitability, consumer mistrust and the lack of assurance be-cause of poor information or education, unproductive regulation of novel and generic products, and feeble patent protection.[96]

The respiratory system and the skin are together directly in contact with the environment, which represents a possible entrance door for the therapeutic compounds into the body. Due to the increasing frequency of pulmonary diseases with high mortality and morbidity, the pulmonary drug delivery is emerging as a noninvasive and smart approach for the treatment of a variety of pathogenic disorders.[97]

Therefore, intravenous and oral routes for disease management are acquiring an ever growing interest for systemic administration of therapeutic agents due to their various advantages.[74]

Currently, researchers have made enormous strides in the progress of pulmonary delivery technologies, both in terms of inhaler design and progresses in nanoscale carrier engineering. At present there are three main different classes of devices for pulmonary drug delivery: metered dose inhalers, nebulizers, and dry powder inhalers. These inhalers are based on diverse delivery mechanisms, and entail different types of drug formulations. Furthermore, the development of novel biologically active compounds like proteins and nucleic acids require the design of innovative delivery technologies.[98]

Bioavailability of administered drug is a major problem in pulmonary infections. Therefore, researchers have developed considerable interest in pulmonary drug delivery and also focused on enhancement of bioavailability of therapeutic biomolecules having high molecular weight.[99–100] Among the various carriers used as drug delivery systems for pulmonary infections, nanocarriers have been found to be most promising due to their significant advantages like prolonged drug release and cell-specific targeted drug delivery.[84,98]

6 CONCLUSIONS

Pulmonary infections caused by bacteria, fungi, and viruses are increasing and reemerging due to improper use of antibiotics and changing environmental conditions. The conventional methods of diagnosis and treatment of pulmonary infections have limitations. The cultural and serological methods used for identification are slow, tedious, may not distinguish colonization from infection, and may be influenced by previous antimicrobials used for the treatment. In this context, PCR methods have been useful up to a certain extent for rapid identification of the causal organism. Nano-PCR and nanobiosensors may play important role in diagnosis of pulmonary infections. The long-term treatment of pulmonary infections by antibiotics and their inappropriate use has resulted in the multidrug resistance problem. This problem is mainly evidenced by tuberculosis, which has become a global problem; therefore, there is a need to develop alternative strategies for the treatment of tuberculosis. The use of nanotechnology in diagnosis of pulmonary infections and also for delivery of drugs would be of paramount importance. Nanoparticles, particularly biodegradable nanoparticles, can be used for this purpose. The activity of the nanoparticles can also be enhanced by their use in combination with existing antibiotics. Finally, nanotechnology will provide a viable alternative for the development of a long-term strategy to tackle the problems of diagnosis and drug delivery in pulmonary infections.

REFERENCES

1. Atkinson TP, Balish MF, Waites KB. Epidemiology, clinical manifestations, pathogenesis and laboratory, detection of *Mycoplasma pneumoniae* infections. *FEMS Microbiol Rev* 2008;**32**:956–73.
2. Gupta N, Rajwanshi A. . In: Edito Amal A, editor. *Pulmonary Infections*. Intech; 2012. p. 70–84.
3. Zheng X, Zhang G. Imaging pulmonary infectious diseases in immunocompromised patients. *Radiol Infect Dis* 2014;**1**:37–41.
4. Byrne AL, Marais BJ, Mitnick CD, Lecca L, Marks GB. Tuberculosis and chronic respiratory disease: A systematic review. *Int J Infect Dis* 2015;**32**:138–46.

5. Rohilla A, Sharma V, Kumar S. Upper respiratory tract infections: an overview. *Int J Curr Pharm Res* 2013;**5**:1–3.

6. Feldman AS, Hartert TV, Gebretsadik T, Carroll KN, Minton PA, Woodward KB, et al. Respiratory severity score separates upper versus lower respiratory tract infections and predicts measures of disease severity. *Pediatr Allergy Immunol Pulmonol* 2015;**28**:117–20.

7. Pavia AT. Viral Infections of the lower respiratory tract: old viruses, new viruses, and the role of diagnosis. *Clin Infect Dis* 2011;**52**(Suppl. 4):284–9.

8. Burdon J. Adult-onset asthma. *Aust Fam Physician* 2015;**44**:554–7.

9. Cunningham TJ, Eke PI, Ford ES, Agaku IT, Wheaton AG, Croft JB. Cigarette smoking, tooth loss, and chronic obstructive pulmonary disease (COPD): Findings from the behavioral risk factor surveillance system. *J Periodontol* 2016;**87**(4):385–94.

10. Tashtoush B, Okafor NC, Ramirez JF, Smolley L. Follicular bronchiolitis: A literature review. *J Clin Diagn Res* 2015;**9**:OE01–5.

11. Murdoch DR. Impact of rapid microbiological testing on the management of lower respiratory tract infection. *Clin Infect Dis.* 2005;**1**:1445–7.

12. Rai MK, Deshmukh SD, Ingle AP, Gade AK. Silver nanoparticles: The powerful nanoweapon against multidrug-resistant bacteria. *J App Microbiol* 2012;**112**:841–52.

13. Chattopadhyay MK, Chakraborty R, Grossart HP, Reddy GS, Jagannadham MV. Antibiotic resistance of bacteria. *BioMed Res Int* 2015;**2015** Article ID 501658, 2 Pages.

14. Karchmer AW. Increased antibiotic resistance in respiratory tract pathogens: PROTEKT US-An Update. *Clin Infect Dis* 2004;**39**:S142–50.

15. Felmingham D. The need for antimicrobial resistance surveillance. *J Antimicrob Chemother* 2002;**50**(Suppl. S1):1–7.

16. Walls G, Bulifon S, Breysse S, Daneth T, Bonnet M, Hurtado N, et al. Drug-resistant tuberculosis in HIV-infected patients in a national referral hospital, Phnom Penh, Cambodia. *Glob Health Action* 2015; **8**:25964.

17. Rai M, Ingle AP, Gade A, Duran N. Synthesis of silver nanoparticles by *Phomagardeniae* and in vitro evaluation of their efficacy against human disease causing bacteria and fungi. *IET Nanobiotechnol* 2015;**9**(2):71–5.

18. Gokhan-Gozel M, Celik C, Elaldi N. *Stenotrophomonas maltophilia* infections in adults: Primary bacteremia and pneumonia. *Jundishapur J Microbiol* 2015;**8**:e23569.

19. van der Poll T, Opal SM. Pathogenesis, treatment, and prevention *of Pneumococcal pneumonia*. *Lancet* 2009;**374**:1543–56.

20. Harris M, Clark J, Coote N, Fletcher P, Harnden A, et al. British thoracic society guidelines for the management of community acquired pneumonia in children: update 2011. *Thorax* 2011;**66**(Suppl. 2):1–23.

21. Rai M, Ingle A, Bansod S, Kon K. Tackling the problem of tuberculosis by nanotechnology: Disease diagnosis and drug delivery. In: Rai M, Kon K, editors. *Nanotechnology in diagnosis, treatment and prophylaxis of infectious diseases*. USA: Elsevier Publisher; 2015. p. 133–49.

22. Lawn SD, Zumla AI. Tuberculosis. *Lancet* 2011;**378**:57–72.

23. Getahun H, Matteelli A, Chaisson RE, Raviglione M. Latent *Mycobacterium tuberculosis* Infection. *N Engl J Med* 2015;**372**:2127–35.

24. Bergeron A, Porcher R, Sulahian A, de Bazelaire C, Chagnon K, Raffoux E, et al. The strategy for the diagnosis of invasive pulmonary aspergillosis should depend on both the underlying condition and the leukocyte count of patients with hematologic malignancies. *Blood* 2012;**1198**:1831–7.

25. Tunnicliffe G, Schomberg L, Walsh S, Tinwell B, Harrison T, Chua F. Airway and parenchymal manifestations of pulmonary aspergillosis. *Respir Med* 2013;**107**(8):1113–23.

26. Kousha M, Tadi R, Soubani AO. Pulmonary aspergillosis: a clinical review. *Eur Respir Rev* 2011;**20**:156–74.

27. Restrepo-Gualteros SM, Jaramillo-Barberi LE, Rodríguez-Martinez CE, Camacho-Moreno G, Child G. Invasive pulmonary aspergillosis: a case report. *Biomedica* 2015;**35**:171–6.

28. Johnson MM, Odel JA. Nontuberculous mycobacterial pulmonary infections. *J Thorac Dis* 2014;**6**(3):210–20.

29. Griffith DE, Aksamit T, BrownElliott BA, Catanzaro A, DaleyC, Gordin F, et al. An official ATS/IDSA statement: diagnosis, treatment and prevention of nontuberculous mycobacterial diseases. *Am J Respir Crit Care Med* 2007;**175**:367.

30. Pierre Audigier C, Ferroni A, SermetGaudelus I, Le Bourgeois M, Offredo C, Vu-Thien H, et al. Agerelated prevalence and distribution of nontuberculous mycobacterial species among patients with cystic fibrosis. *J Clin Microbiol* 2005;**43**:3467.

31. Min JY, Jang YJ. Macrolide therapy in respiratory viral infections. *Mediators Inflamm* 2012;**2012** Article ID 649570, 9 Pages.

32. Henry M, Lavigne R, Debarbieuxa L. Predicting in vivo efficacy of therapeutic bacteriophages used to treat pulmonary infections. *Antimicrob Agents Chemother* 2013;**57**:5961–8.

33. Semler DD, Goudie AD, Finlay WH, Dennisa JJ. Aerosol phage therapy efficacy in *Burkholderia cepacia* complex respiratory infections. *Antimicrob Agents Chemother* 2014;**58**:4005–13.

34. http://www.medicalnewstoday.com/articles/151632.php

35. http://www.emedicine.medscape.com

36. World Health Organization, 2013. TB report http://www.who.int/tb/publications/global_report/2013/pdf/report_without_annexes.pdf

37. Jalhan S, Jindal A, Aggarwal S, Gupta A, Hemraj P. Review on current trends and advancement in drugs trends and drug targets for tuberculosis therapy. *Int J Pharm Bio Sci* 2013;**4**:320–33.

38. Dheda K, Shean K, Badri M. Extensively drug resistant tuberculosis. *N Eng J Med* 2008;**359**:2390.

39. Keshavjee S. Tuberculosis, drug resistance, and the history of modern medicine. *N Eng J Med* 2012;**367**:931–6.

40. Udwadia ZF, Amale RA, Ajbani KK, Rodrigues C. Totally drug-resistant tuberculosis in India. *Clin Infect Dis* 2012;**54**:579–81.

41. Mani V, Wang S, Inci F, De Liberoa G, Singhal A, Demirci U. Emerging technologies for monitoring drug-resistant tuberculosis at the point-of-care. *Adv Drug Deliv Rev* 2014;**78**:105–17.

42. Walsh TJ, Anaissie EJ, Denning DW, Herbrecht R, Kontoyiannis DP, Marr KA, et al. Treatment of aspergillosis: Clinical practice guidelines of the infectious diseases society of America. *Clin Infect Dis* 2008;**46**:327–60.

43. Davis KK, Kao PN, Jacobs SS, Ruoss SJ. Aerosolized amikacin for treatment of pulmonary *Mycobacterium avium* infections: an observational case series. *BMC Pulmonary Med* 2007;**7**:2.

44. Carroll KC. Laboratory diagnosis of lower respiratory tract infections: Controversy and conundrums. *J Clin Microbiol* 2002;**40**:3115–20.

45. Mandell LA, Bartlett JG, Dowell SF, File TM, Musher DM, Whitney C. Update of practice guidelines for the management of community-acquired pneumonia in immune-competent adults. *Clin Infect Dis* 2003;**37**:1405–33.

46. Farokhzad OC, Langer R. Impact of nanotechnology on drug delivery. *ACS Nano* 2009;**3**:16–20.

47. Goldberg M, Langer R, Xinqiao J. Nanostructured materials for applications in drug delivery and tissue engineering. *J Biomat Sci Polym E* 2007;**18**:241–68.

48. Sanvicens N, Marco MP. Multifunctional nanoparticles: Properties and prospects for their use in human medicine. *Trends Biotechnol* 2008;**26**:425–33.

49. Williams D. The relationship between biomaterials and nanotechnology. *Biomaterials* 2008;**29**:1737–8.

50. Ochekpe NA, Olorunfemi PO, Ngwuluka NC. Nanotechnology and drug delivery part 1: Background and applications. *Trop J Pharm Res* 2009;**8**:265–74.

51. McNeil SE. Unique benefits of nanotechnology to drug delivery and diagnostics. *Meth Mol Biol* 2011;**697**:3–8.

52. USNNI (United States National Nanotechnology Initiative). Nanotechnology 101 What is it and how it works. Available online: http://www.nano.gov/nanotech-101/what

53. Picraux T. Nanotechnology. In Encyclopaedia Britannica Deluxe Edition; Encyclopedia Britannica: Chicago: IL; USA, 2010.

54. Aitken R, Chaudhry M, Boxall A, Hull M. Manufacture and use of nanomaterials: Current status in the UK and global trends. *Occup Med* 2006;**56**:300–6.

55. Lines M. Nanomaterials for practical functional uses. *J Alloy Compd* 2008;**449**:242–5.

56. Niemeyer C, Mirkin C. *Nanobiotechnology: Concepts, applications and perspectives.* Germany: Wiley–VCH, Weinheim; 2004.

57. Medepalli KK. *Advanced Nanomaterials for biomedical applications.* UK: ProQuest, Cambridge; 2008.

58. Schulz MJ, Shanov VN. *Nanomedicine design of particles, sensors, motors, implants, robots and devices.* Boston, MA, USA: Artech House Publishers; 2009.

59. Brenner JS, Greineder C, Shuvaev V, Muzykantov V. Endothelial nanomedicine for the treatment of pulmonary disease. *Expert Opin Drug Deliv* 2015;**12**(2):239–61.

60. Parboosing R, Glenn E, Maguire M, Govender P. KrugerHG. Nanotechnology and the Treatment of HIV Infection. *Viruses* 2012;**4**:488–520.

61. Buxton DB. Nanomedicine for the management of lung and blood diseases. *Nanomedicine* 2009;**4**(3):331–9.

62. Sagadevan S, Savitha S, Preethi R. Beneficial applications of nanoparticles in medical field: A review. *Int J PharmTech Res* 2014;**6**(5):1711–7.

63. deMelo Garcia F. Nanomedicine and therapy of lung diseases. *Einstein* 2014;**12**(4):531–3.

64. Omlor AJ, Nguyen J, Bals R, Dinh QT. Nanotechnology in respiratory medicine. *Respir Res* 2015;**16**:64.

65. Card JW, Zeldin DC, Bonner JC, Nestmann ER. Pulmonary applications and toxicity of engineered nanoparticles. *Am J Physiol Lung Cell Mol Physiol* 2008;**295**(3):400–11.

66. McKenna M. Antibiotic resistance: The last resort. *Nature* 2013;**499**:7459.

67. Zumla A, Memish ZA, Maeurer M, Bates M, Mwaba P, Al-Tawfiq JA. Emerging novel and antimicrobial-resistant respiratory tract infections: New drug development and therapeutic options. *Lancet Infect Dis* 2014;**14**(11):1136–49.

68. Zhang F, Smolen JA, Zhang S, Li R, Shah PN, Cho S. Degradable polyphosphoester based silver-loaded nanoparticles as therapeutics for bacterial lung infections. *Nanoscale* 2015;**7**:2265–70.

69. Todoroff J, Vanbever R. Fate of nanomedicines in the lungs. *Curr Opin Colloid Interface Sci* 2011;**6**(3):246–54.

70. Bhardwaj A, Kumar L, Narang RK, Murthy RS. Development and characterization of ligand-appended liposomes for multiple drug therapy for pulmonary tuberculosis. *Artif Cells Nanomed Biotechnol* 2013;**41**(1):52–9.

71. Barash O, Peled N, Tisch U, Bunn Jr PA, Hirsch FR, Haick H. Classification of lung cancer histology by gold nanoparticles sensors. *Nanomedicine* 2012;**8**(5):580–9.

72. Broza YY, Kremer R, Tisch U, Gevorkyan A, Shiban A, Best LA, Haick H. A nanomaterial-based breath test for short-term follow-up after lung tumor resection. *Nanomedicine* 2013;**9**(1):15–21.

73. Laube BL. The expanding role of aerosols in systemic drug delivery, gene delivery and vaccination. *Respir Care* 2005;**50**:1161–76.

74. Patton JS, Byron PR. Inhaling medicines: delivering drugs to the body through lungs. *Nat Rev Drug Discov* 2007;**6**:67–74.

75. Nokhodchi A, Martin GP. *Pulmonary Drug Delivery: Advances and Challenges.* West Sussex, UK: John Wiley & Sons Publishers; 2015.

76. Liu WK, Qian L, Chen DH, Liang HX, Chen XK, Chen MX, et al. Epidemiology of Acute Respiratory Infections in Children in Guangzhou: A Three-Year Study. *PLOS One* 2014;**9**(5):1–9 e96674.

77. Patton JS, Fishburn CS, Weers JG. The lungs as a portal of entry for systemic drug delivery. *Proc Am Thorac Soc* 2004;**1**(4):338–44.

78. Lu B, Zhang JQ, Yang H. Lung-targeting microspheres of carboplatin. *Int J Pharm* 2003;**265**:1–11.

79. Joshi JT. A review on micronization techniques. *J Pharma Sci Technol* 2011;**3**:651–81.

80. Aggarwal P, Hall JB, McLeland CB, Dobrovolskaia MA, McNeil SE. Nanoparticle Interaction with plasma proteins as it relates to particle biodistribution, biocompatibility and therapeutic efficacy. *Adv Drug Deliv Rev* 2009;**61**:428–37.

81. Muller RH, Jacobs C. Buparvaquone muco adhesive nanosuspension: Preparation, optimization and long-term stability. *Int J Pharm* 2002;**237**:151–61.

82. Kohno S, Otsubo T, Tanaka E, Maruyama K, Hara K, Amphotericin B. encapsulated in polyethylene glycolimmunoliposomes for infectious diseases. *Adv Drug Del Rev* 1997;**24**:325–9.

83. Dames P, Gleich B, Flemmer A, Hajek K, Seidl N, Wiekhorst F, et al. Targeted delivery of magnetic aerosol droplets to the lung. *Nat Nanotechnol* 2007;**2**:495–9.

84. Garcia FM. Nanomedicine and therapy of lung diseases. *Einstein* 2014;**12**(4):531–3.

85. Jacobs REA, Gu P, Chachou A. Reactivation of pulmonary tuberculosis during cancer treatment. *Int J Mycobacteriol* 2015;**4**(4):337–40.

86. Hawn TR, Day TA, Scriba TJ, Mark H, Hanekom WA, Evans TG, et al. Tuberculosis vaccines and prevention of infection. *Microbiol MolBiol Rev* 2014;**78**(4):650–71.

87. Hartl D, Griese M. Interstitial lung disease in children-genetic background and associated phenotypes. *Resp Res* 2005;**6**:32.

88. Barnes PJ. Chronic obstructive pulmonary disease, 12: New treatments for COPD. *Thorax* 2003;**58**:803–8.

89. Andrade F, Diana R, Mafalda V, Domingos F, Alejandro S, Bruno S. Nanotechnology and pulmonary delivery to overcome resistance in infectious diseases. *Adv Drug Del Rev* 2013;**65**:1816–27.

90. Andrade F, Videira M, Ferreira D, Sarmento B. Micelle-based systems for pulmonary drug delivery and targeting. *Drug Del Lett* 2011;**1**:171–85.

91. Bailey M, Berkland C. Nanoparticle formulations in pulmonary drug delivery. *Med Res Rev* 2009;**29**:196–212.

92. Mansour HM, Rhee YS, Wu X. Nanomedicine in pulmonary delivery. *Int J Nanomed* 2009;**4**:299–319.

93. Bawarski WE, Chidlowsky E, Bharali DJ, Mousa SA. Emerging nanopharmaceuticals. *Nanomedicine* 2008;**4**:273–82.

94. Verma A. Article on latest trends in nanomedicine. *J Nanomed Res* 2015;**2**(2):00026.

95. Duncan R, Gaspar R. Nanomedicine (s) under the microscope. *Mol Pharm* 2011;**8**:2101–41.

96. Bosetti R, Vereeck L. Future of nanomedicine: obstacles and remedies. *Nanomedicine* 2011;**6**:747–55.

97. Yang W, Peters JI, Williams III RO. Inhaled nanoparticles- A current review. *Int J Pharma* 2008;**356**(1–2):239–47.

98. Marianecci C, Marzio LD, Rinaldi F, Carafa M, Alhaique F. Pulmonary delivery: Innovative approaches and perspectives. *J Biomater Nanobiotechnol* 2011;**2**:567–75.

99. Merkus FWHM, Schiepper NGM, Hermens WAJJ, Romeijin VSG, Verhoef JC. Absorption enhancers in nasal drug delivery: Efficacy and Safety. *J Cont Release* 1993;**24**:201–8.

100. Marttin E, Verhoef JC, Romeijin SG, Merkus FWHM. Effects of absorption enhancers on rat nasal epithelium in vivo: Release of marker compounds in the Nasal Cavity. *Pharma Res* 1995;**12**:1151–7.

VOLATILE OILS: POTENTIAL AGENTS FOR THE TREATMENT OF RESPIRATORY INFECTIONS

16

A. Pasdaran*,**,†, A. Pasdaran†, D. Sheikhi‡

*Guilan University of Medical Sciences, Department of Pharmacognosy, School of Pharmacy, Research and Development Center of Plants and Medicinal Chemistry, Rasht, Iran; **Shiraz University of Medical Sciences, Medicinal Plants Processing Research Center, Shiraz, Iran; †Phytochemistry Research Center, Shahid Beheshti University of Medical Sciences, Tehran, Iran; ‡Regulations (GCP/ICH), Pharmaceuticals, Denmark

1 INTRODUCTION

Referring to infectious disease, respiratory tract infections engage with all surfaces in the respiratory tract. Based on the infected zone, respiratory infections can be categorized into upper tract infection (URI or URTI) and lower tract infection (LRI or LRTI). Each involves different parts of the respiratory tract infections, which vary in type and severity of microorganisms. Although there are different types of respiratory tract infections, the acute form in the upper respiratory tract infection predominates and includes several complications, such as sinusitis, pharyngitis, epiglottitis, laryngitis, and tracheitis. On the other hand, lower respiratory tract infection (LRTI) includes both acute and chorionic types, such as pneumonia and bronchitis. Based on pathogenicity, bacterial and viral pathogens are the most common microorganisms in both types (ie, LRTI and URTI). Moreover, infection distribution leads to varieties based on the patient's age; for example, acute respiratory infections pose severe problem in childhood, which mainly occur in upper respiratory tract. Although the bacterial pathogens play a significant role in intensifying LRTIs, the major acute respiratory infections occur in upper respiratory tract, in these cases viral pathogens are the primary common pathogens, including influenza A and B, parainfluenza (type 1 and 3), adenovirus, and respiratory syncytial virus. Some of the common pathogens of the respiratory tract are listed in Table 16.1. Pathogen biodiversity, complexity, and mixed infections in many cases of respiratory tract infection have generated several problems for the treatment of respiratory infections. For example, various bacterial pathogens are encountered in several cases of viral infections. Therefore, the treatment of respiratory infections is a complex therapy which consists of several chemotherapy strategies.[1–3] Antiviral (the same as antibacterial medication) is used to control the treatment and prevention of respiratory infections.

There are several restrictive factors, such as medication resistance, recurrency, and inflammation, which will guide researchers to find new effective compounds. This will be an important field in drug development for respiratory infections. Natural compounds are considered to be one of the main sources in new drug development. Historically, numerous plants have been utilized as traditional medicines

Table 16.1 Some of the Common Pathogens Involved in Respiratory Tract Infections

Pathogen Name	Common Infected Form	Category	References
Streptococcus pneumoniae	Pneumonia/invasive pneumococcal diseases	Gram-positive	a
Haemophilus influenzae	Pneumonia, epiglottitis and sinusitis	Gram-negative	b
Chlamydophila pneumoniae	Atypical pneumonia	Obligate intracellular bacterium	c
Staphylococcus aureus	Sinusitis, pneumonia	Gram-positive	d
Pseudomonas aeruginosa	Sinusitis, pneumonia	Gram-negative	e
Legionella pneumophila	Cough with sputum or bloody sputum/pneumonia, bronchiolitis	Gram-negative	f
Moraxella catarrhalis	Bronchitis, sinusitis, laryngitis and bronchopneumonia	Gram-negative	g
Rhinoviruses	Common cold, sinusitis, pneumonia (in middle-aged adults)	Enterovirus	h
Coronaviruses	Pneumonia	Coronavirinae	i
Influenza virus	Pneumonia	Orthomyxovirus	j
Respiratory syncytial virus	Bronchiolitis, pneumonia	Pneumovirus	k
Adenovirus	Pneumonia, tonsillitis, bronchiolitis	Adenoviridae	l
Herpes simplex virus	Pneumonia, pharyngitis	Respirovirus	m
Histoplasma capsulatum	Pneumonia	Histoplasma (dimorphic fungi)	n
Cryptococcus neoformans	Pneumonia	Cryptococcus (yeast)	o
Coccidioides immitis	Pneumonia	Coccidioides (pathogenic fungus)	p
Pneumocystis jirovecii	Pneumonia	Pneumocystis (yeast-like fungus)	q

[a]*Madhi SA, Klugman KP, Group TVT. A role for* Streptococcus pneumoniae *in virus-associated pneumonia.* Nat Med *2004;* **10***:811–3. Tan TQ, Mason EO, Wald ER, Barson WJ, Schutze GE, Bradley JS, et al. Clinical characteristics of children with complicated pneumonia caused by* Streptococcus pneumoniae. Pediatrics *2002;* **110***: 1–6. Lynch JP, Martínez FJ. Clinical relevance of macrolide-resistant* Streptococcus pneumoniae *for community-acquired pneumonia.* Clin Infect Dis *2002;* **34***: S27–S46.*

[b]*Ginsburg CM, Howard JB, Nelson JD. Report of 65 cases of* Haemophilus influenzae *b pneumonia.* Pediatrics *1979;* **64***: 283–6. Peltola H. Worldwide* Haemophilus influenzae *type b disease at the beginning of the 21st century: global analysis of the disease burden 25 years after the use of the polysaccharide vaccine and a decade after the advent of conjugates.* Clin Microbiol Rev *2000;* **13***: 302–17. Cordero E, Pachón J, Rivero A, Girón JA, Gómez-Mateos J, Merino MD, et al.* Haemophilus influenzae *pneumonia in human immunodeficiency virus-infected patients.* Clin Infect Dis *2000;* **30***: 461–5. Farley MM, Stephens DS, Brachman PS, Harvey RC, Smith JD, Wenger JD. Invasive* Haemophilus influenzae *disease in adults: a prospective, population-based surveillance.* Ann Intern Med *1992;* **116***: 806–12.*

[c]*Wang X, Li H, Xia Z. Chlamydia pneumoniae* Pneumonia. Radiology of Infectious Diseases *2015;* **2** *: 69–74. Mandell LA, Wunderink RG, Anzueto A, Bartlett JG, Campbell GD, Dean NC, et al. Infectious Diseases Society of America/American Thoracic Society consensus guidelines on the management of community-acquired pneumonia in adults.* Clin Infect Dis *2007;* **44***: S27–S72. Mansel J, Rosenow E, Smith T, Martin J.* Mycoplasma pneumoniae *pneumonia.* CHEST J *1989;* **95***: 639–46. Michelow IC, Olsen K, Lozano J, Rollins NK, Duffy LB, Ziegler T, et al. Epidemiology and clinical characteristics of community-acquired pneumonia in hospitalized children.* Pediatrics *2004;* **113***: 701–7.*

[d]*Rubinstein E, Kollef MH, Nathwani D. Pneumonia caused by methicillin-resistant* Staphylococcus aureus. Clin Infect Dis *2008;* **46***: S378–S85. Rello J, Sole-Violan J, Sa-Borges M, Garnacho-Montero J, Muñoz E, Sirgo G, et al. Pneumonia caused by oxacillin-resistant* Staphylococcus aureus *treated with glycopeptides*.* Crit Care Med *2005;* **33***: 1983–7. Jiang R-S, Jang J-W, Hsu C-Y. Post-functional endoscopic sinus surgery methicillin-resistant* Staphylococcus aureus *sinusitis.* Am J Rhinol *1999;* **13***: 273–7. Solares CA, Batra PS, Hall GS, Citardi MJ. Treatment of chronic rhinosinusitis exacerbations due to methicillin-resistant* Staphylococcus aureus *with mupirocin irrigations.* Am J Otolaryngol *2006;* **27***: 161–5. Brown CA, Paisner HM, Biel MA, Levinson RM, Sigel ME, Garvis GE, et al. Evaluation of the microbiology of chronic maxillary sinusitis.* Ann Otol Rhinol Laryngol *1998;* **107***: 942–5.*

[e]Ruxana TS, Timothy S B, John WC, Alice SP. Pathogen–Host Interactions in Pseudomonas aeruginosa *Pneumonia.* Am J Respir Crit Care Med 2005; **171**: 1209–23. *Jordi R, Dolors M, Francesca M, Paola J, Ferran S, Jordi V, Pere C. Recurrent* Pseudomonas aeruginosa *Pneumonia in Ventilated Patients.* Am J Respir Crit Care Med 1998;**157**: 912–6. *Zohra B, Jean B, Walid AH, Martin D. Biofilm Formation by* Staphylococcus Aureus *and* Pseudomonas Aeruginosa *is Associated with an Unfavorable Evolution after Surgery for Chronic Sinusitis and Nasal Polyposis.* Otolaryngol Head Neck Surg 2006; **134**: 991–6.

[f]Carratala J, Gudiol F, Pallares R, Dorca J, Verdaguer R, Ariza J, et al. Risk factors for nosocomial Legionella pneumophila *pneumonia.* Am J Respir Crit Care Med 1994; **149**: 625–9. *Falco V, Fernández dST, Alegre J, Ferrer A, Martínez VJ.* Legionella pneumophila. *A cause of severe community-acquired pneumonia.* Chest 1991; **100**: 1007–11. *Beigel F, Jürgens M, Filik L, Bader L, Lück C, Göke B, et al. Severe* Legionella pneumophila *pneumonia following infliximab therapy in a patient with Crohn's disease.* Inflamm Bowel Dis 2009; **15**: 1240–4. *Sato P, Madtes DK, Thorning D, Albert RK. Bronchiolitis obliterans caused by* Legionella pneumophila. CHEST Journal 1985; **87**: 840–2.

[g]Klugman K. The clinical relevance of in vitro resistance to penicillin, ampicillin, amoxycillin and alternative agents, for the treatment of community-acquired pneumonia caused by Streptococcus pneumoniae, Haemophilus influenzae *and* Moraxella catarrhalis. *J* Antimicrob Chemother 1996; **38**: 133–40. *Karalus R, Campagnari A.* Moraxella catarrhalis: *a review of an important human mucosal pathogen.* Microbes Infect 2000; **2**: 547–59. *DiPersio JR, Jones RN, Barrett T, Doern GV, Pfaller MA. Fluoroquinolone-resistant* Moraxella catarrhalis *in a patient with pneumonia: report from the SENTRY Antimicrobial Surveillance Program (1998).* Diagn Microbiol Infect Dis 1998; **32**: 131–5.

[h]Winther B. Rhinovirus infections in the upper airway. Proc Am Thorac Soc 2011; **8**: 79–89.

[i]Peiris J, Lai S, Poon L, Guan Y, Yam L, Lim W, et al. Coronavirus as a possible cause of severe acute respiratory syndrome. Lancet 2003; **361**: 1319–25. *Peiris J, Chu C, Cheng V, Chan K, Hung I, Poon L, et al. Clinical progression and viral load in a community outbreak of coronavirus-associated SARS pneumonia: a prospective study.* Lancet 2003; **361**: 1767–72. *Woo PC, Lau SK, Chu C-m, Chan K-h, Tsoi H-w, Huang Y, et al. Characterization and complete genome sequence of a novel coronavirus, coronavirus HKU1, from patients with pneumonia.* J Virol 2005; **79**: 884–95. *Woo PC, Lau SK, Tsoi H-w, Chan K-h, Wong BH, Che X-y, et al. Relative rates of non-pneumonic SARS coronavirus infection and SARS coronavirus pneumonia.* Lancet 2004; **363**: 841–5. *Pene F, Merlat A, Vabret A, Rozenberg F, Buzyn A, Dreyfus F, et al. Coronavirus 229E-related pneumonia in immunocompromised patients.* Clin Infect Dis 2003; **37**: 929–32.

[j]Falsey AR, Walsh EE. Viral pneumonia in older adults. Clin Infect Dis 2006; **42**: 518–24.

[k]Glezen WP, Taber LH, Frank AL, Kasel JA. Risk of primary infection and reinfection with respiratory syncytial virus. Am J Dis Child 1986; **140**: 543–6. *Willson DF, Landrigan CP, Horn SD, Smout RJ. Complications in infants hospitalized for bronchiolitis or respiratory syncytial virus pneumonia.* J Pediatr 2003; **143**: 142–9. *Falsey AR. Respiratory Syncytial Virus Pneumonia.* Community-Acquired Pneumonia: *Springer; 2002: 617 28.*

[l]Siegal FP, Dikman SH, Arayata RB, Bottone EJ. Fatal disseminated adenovirus 11 pneumonia in an agammaglobulinemic patient. Am J Med 1981; **71**: 1062–7. *Castro-Rodriguez JA, Daszenies C, Garcia M, Meyer R, Gonzales R. Adenovirus pneumonia in infants and factors for developing bronchiolitis obliterans: A 5-year follow-up.* Pediatr Pulmonol 2006; **41**: 947–53. *Pichler M, Reichenbach J, Schmidt H, Herrmann G, Zielen S. Severe adenovirus bronchiolitis in children.* Acta Paediatr 2000; **89**: 1387–92. *Murtagh P, Giubergia V, Viale D, Bauer G, Pena HG. Lower respiratory infections by adenovirus in children. Clinical features and risk factors for bronchiolitis obliterans and mortality.* Pediatr Pulmonol 2009; **44**: 450–6.

[m]Luyt C-E, Combes A, Deback C, Aubriot-Lorton M-H, Nieszkowska A, Trouillet J-L, et al. Herpes simplex virus lung infection in patients undergoing prolonged mechanical ventilation. Am J Respir Crit Care Med 2007; **175**: 935–42. *Ljungman P, Ellis MN, Hackman RC, Shepp DH, Meyers JD. Acyclovir-resistant herpes simplex virus causing pneumonia after marrow transplantation.* J Infect Dis 1990; **162**: 244–8. *Ramsey PG, Fife KH, HACKMAN RC, Meyers JD, Corey L. Herpes simplex virus pneumonia: clinical, virologic, and pathologic features in 20 patients.* Ann Intern Med 1982; **97**: 813–20. *Mcmillan JA, Weiner LB, Higgins AM, Lamparella VJ. Pharyngitis associated with herpes simplex virus in college students.* Pediatr Infect Dis J 1993; **12**: 280–3.

[n]Tinelli M, Michelone G, Cavanna C. Recurrent Histoplasma capsulatum *pneumonia: a case report.* Microbiologica 1992; **15**: 89–93.

[o]Shrestha RK, Stoller JK, Honari G, Procop GW, Gordon SM. Pneumonia due to Cryptococcus neoformans *in a patient receiving infliximab: possible zoonotic transmission from a pet cockatiel.* Respir Care 2004; **49**: 606–8. *Levitz SM. The ecology of* Cryptococcus neoformans *and the epidemiology of cryptococcosis.* Rev Infect Dis 1991; **13**: 1163–9. *Subramanian S, Kherdekar S, Babu P, Christianson C. Lipoid pneumonia with* Cryptococcus neoformans *colonisation.* Thorax 1982; **37**: 319. *Jensen WA, Rose RM, Hammer SM, Karchmer AW. Serologic Diagnosis of Focal Pneumonia Caused by* Cryptococcus neoformans *1, 2.* Am Rev Respir Dis 1985; **132**: 189–91.

[p]Swartz J, Stoller JK. Acute eosinophilic pneumonia complicating Coccidioides immitis *pneumonia: a case report and literature review.* Respiration 2009; **77**: 102–6. *Standaert SM, Schaffner W, Galgiani JN, Pinner RW, Kaufman L, Durry E, et al. Coccidioidomycosis among visitors to a* Coccidioides immitis-*endemic area: an outbreak in a military reserve unit.* J Infect Dis 1995; **171**: 1672–5. *Lopez AM, Williams PL, Ampel NM. Acute pulmonary coccidioidomycosis mimicking bacterial pneumonia and septic shock: a report of two cases.* Am J Med 1993; **95**: 236–9.

[q]Kanemoto H, Morikawa R, Chambers JK, Kasahara K, Hanafusa Y, Uchida K, et al. Common variable immune deficiency in a Pomeranian with Pneumocystis carinii *pneumonia.* J Vet Med 2015. *Tasci S, Ewig S, Burghard A, Lüderitz B. Pneumocystis carinii pneumonia.* Lancet 2003; **362**: 124. *Helweg-Larsen J, Lundgren JD, Benfield T. Pneumocystis carinii pneumonia.* Curr Treat Options Infect Dis 2002; **4**: 363–75. *Mayer KH, Fisk DT, Meshnick S, Kazanjian PH. Pneumocystis carinii pneumonia in patients in the developing world who have acquired immunodeficiency syndrome.* Clin Infect Dis 2003; **36**: 70–8.

by people in many nations.[4] Many of these plants have been investigated for their antimicrobial and antiviral properties.[5–9] With regard to massive variation among natural products, chemical structure diversity causes different antimicrobial potential in natural compounds.[10]

Besides the antimicrobial activity of the essential oils in natural products, other characteristics such as high vapor pressure, low toxicity, and antiinflammatory potential create a worthwhile theme for using of these natural compounds for new drug development in respiratory infections. Parallel to the roles of the microorganisms in the pathology of respiratory infections diseases, inflammatory process also have a considerable role in the persistence and recurrence of respiratory infectious diseases.

This chapter reviews the antibacterial, antiviral, and antiinflammation effects of essential oils as effective natural compounds. It will also discuss the use of these natural compounds as traditional remedies in treatment of respiratory infections.

2 TRADITIONAL REMEDIES IN RESPIRATORY INFECTIONS

Traditional medicines utilize natural sources for the treatment of the many diseases.[11–13] Historically, infectious diseases have been the major human ailment. Natural sources are used in a variety of forms, including water extracts, tincture or alcoholic extract and incense.[14] Based on the historical uses and effective treatments that have been based on many of these traditional remedies, extensive pharmacological research of their antibacterial, antiviral, and antiinflammation activity have been performed.[15–17] Abundant information about plants and active compounds in infectious diseases and inflammation related process is available.[18–21] Aromatic and fragrant plants are a major part of traditional therapeutic remedies, and they have shown remarkable antibacterial and antiviral activity. Furthermore, many of them also have a significant antiinflammatory activity and are used as adjuvant remedies in the treatment of infection (Table 16.2 and Fig. 16.1). Some of the most active extracts of traditional herbs which have been used as antibacterial and antiviral in the treatment of respiratory infections are summarized in Table 16.3.

The use of aromatic extracts or burning plants is a common process in traditional medicine. The resultant smoke or fragrance is inhaled to treat respiratory complaints, including cough, cold, infections, and asthma.[22,23] Inhalation administration goes back to the ancient cultures and its techniques may be considered as a progressive point in respiratory complaints treatment. The direct effect of such fragrance on the respiratory tract is an advantage of this form of treatment.

Inhalation therapy often involves the aromatic extracts or burning of plant material, and the volatile fraction liberated during the process is inhaled to aid in the healing process. Inhalation of the volatile fraction from aromatic extracts or burning plant matter is a unique method of administration and has been used traditionally to treat respiratory conditions, such as, asthma, bronchitis, and other respiratory infections including the common cold.[24] In addition, aerosol delivery of such remedies is well practiced in allopathic medicine and has the advantage of being site specific, thus enhancing the therapeutic ratio for respiratory ailments.[25]

Table 16.4 and Fig. 16.2 describe several essential oils from *Achilla* species (Asteraceae family) that have demonstrated appropriate effects on some of the major respiratory infections caused by microorganisms.

Table 16.2 Some Famous Traditional Plants That Are Used as Treatment Remedies for Respiratory Diseases

Plant Species (Family)	Plant Parts Used	Indications	References
Acacia polyacantha Willd. (Forssk.) Willd. (Mimosaceae)	Stem bark	Cough	a
Andira inermis (Wright) DC. (Fabaceae)	Leaves	Cough, respiratory diseases	a
Asparagus africanus Lam. (Asparagaceae)	Whole plant	Respiratory diseases	a
Cussonia arborea Hochst. ex A. Rich. (Araliaceae)	Leaves	Cough, respiratory diseases	a
Entada africana Guill. and Perr. (Mimosaceae)	Roots	Respiratory diseases	a
Euphorbia hirta L. (Euphorbiaceae)	Whole plant	Sore throat	a
Keetia hispida (Benth.) Bridson (Rubiaceae)	Leaves	Respiratory diseases	a
Phyllanthus muellerianus (O. Ktze) Exell (Euphorbiaceae)	Leaves	Respiratory diseases	a
Terminalia schimperiana Hochst. (Combretaceae)	Leaves	Cough, respiratory diseases	a
Sophora flaescens Ait. (Fabaceae)	Roots	Respiratory diseases	b
Scutellaria baicalensis Georgi (Lamiaceae)	Root	Respiratory diseases	b
Artemisia afra (Asteraceae)	Leaves and bark	Colds, coughs, and influenza	c
Sambucus nigra L. (Caprifoliaceae)	Leaves and bark	Bronchitis	d
Anchusa italica Retz. (Boraginaceae)	Flowers	Common colds	e
Cynodon dactylon (L.) Pers. (Gramineae)	Whole plant	Coughs	e
Thymus kotschyanus Boiss. et Hoh. (Lamiaceae)	Leaves, flowers	Common colds, bronchitis	e
Glycyrrhiza echinata L. (Leguminosae)	Roots, stolons	Coughs, bronchitis	e
Trigonella foenum-graecum L. (Leguminosae)	Seeds, leaves	Cure of inflamed throat	e
Althaea officinalis L. (Malvaceae)	Flowers, leaves, roots	Coughs, bronchitis	e
Malva sylvestris L. (Malvaceae)	Whole plant	Coughs, respiratory inflammation	e
Prunus mahaleb L. (Rosaceae)	Fruits	Emollient for upper respiratory organs	f
Adiantum capillus-veneris L. (Adiantaceae)	Leaves	Respiratory ailments, cough	g
Ferula oopoda (Boiss. & Buhse.) Boiss. (Apiaceae)	Seed, latex	Cough, asthma, respiratory disorders	g
Stachys turcomica Trautv (Lamiaceae)	Whole plant	Bronchitis, influenza	g
Acacia kempeana F. Muell. (Mimosaceae)	Bark, leaves, root bark	Chest infection, severe cold	h

(Continued)

Table 16.2 Some Famous Traditional Plants That Are Used as Treatment Remedies for Respiratory Diseases (*cont.*)

Plant Species (Family)	Plant Parts Used	Indications	References
Acacia ligulata Cunn. ex Benth. (Mimosaceae)	Bark, leaves	Cough, cold, chest infection	i
Eremophila alternifolia R. Br. (Myoporaceae)	Seed, leaves	Respiratory tract infection	j
Cymbopogon ambiguus (Steudel) A. Camus (Poaceae)	Leaves	Respiratory tract infection	k

[a]*Kone W, Atindehou KK, Terreaux C, Hostettmann K, Traore D, Dosso M. Traditional medicine in North Côte-d'Ivoire: screening of 50 medicinal plants for antibacterial activity. J Ethnopharmacol 2004; **93**: 43–9.*
[b]*Ma S-C, Du J, But PP-H, Deng X-L, Zhang Y-W, Ooi VE-C, et al. Antiviral Chinese medicinal herbs against respiratory syncytial virus. J Ethnopharmacol 2002; **79**: 205–11.*
[c]*Rood B. Uit die veldapteek (Out of the field-pharmacy) Tafelberg. Cape Town 1994.*
[d]*Miraldi E, Ferri S, Mostaghimi V. Botanical drugs and preparations in the traditional medicine of West Azerbaijan (Iran). JEthnopharmacol 2001; **75**: 77–87.*
[e]*Amin GR. Popular medicinal plants of Iran: Iranian Research Institute of Medicinal Plants Tehran; 1991.*
[f]*Mir-Heidari H. Encyclopedia of Medicinal Plants of Iran. Islamic Culture Press, Tehran, Iran; 1993.*
[g]*Ghorbani A. Studies on pharmaceutical ethnobotany in the region of Turkmen Sahra, north of Iran: (Part 1): General results. J Ethnopharmacol 2005; **102**: 58–68.*
[h]*O'Connell JF, Latz PK, Barnett P. Traditional and modern plant use among the Alyawara of central Australia. Econ Bot 1983; **37**: 80–109.*
[i]*Latz PK. Bushfires & bushtucker: Iad Press; 1995.*
[j]*Territory CCotN. Traditional aboriginal medicines in the Northern Territory of Australia: Conservation Commission of the Northern Territory of Australia; 1993.*
[k]*Smith NM. Ethnobotanical field notes from the Northern Territory, Australia. J Adelaide Bot Gard 1991; 1–65.*

3 SCREENING OF THE ANTIBACTERIAL EFFECTS OF ESSENTIAL OILS

The antimicrobial effects of plants and their extracts have been recognized for a long time. Essential oil is one of the most important and wide spread secondary metabolite in plants and this class of phytochemical compounds and their activities needs attention. These phytochemicals are generally isolated from plant material by distillation methods, such as, hydrodistillation and steam distillation. They contain variable mixtures of several chemical classes, such as terpenoids, specifically monoterpenes and simple phenolic compounds. Some of the higher molecular structures with high molecular weight, such as sesquiterpenes and diterpenes, may be present. A variety of low molecular weight aliphatic hydrocarbons, acids, alcohols, esters or lactones, sulfur-containing compounds and other chemical groups may also be observed. Among the phytochemical compounds, terpenes are responsible for many therapeutic effects in medicinal plants.[26–30] Most terpenes are derived from the condensation of isoprene units and are categorized according to the number of these units present in the carbon skeleton. These compounds are responsible for aromaticity and fragrance in many of the plants. The antibacterial activity of volatile oils has been assessed by many researchers.[31–34] This potential of essential oils has been used in many pharmaceutical, cosmeceutical, and nutraceutical applications and industrials. There are many differences between the antimicrobial effects of different essential oils. Essential oils and their constituents are an attractive source in new antimicrobial compounds evaluation.[35,36]

FIGURE 16.1 Some of the Famous Edible Plants That Are Used as Traditional Antibacterial and Antiinflammations Remedies

(a) *Citrus paradise* (grapefruit), (b) *Perilla frutescens* (perilla), (c) Cymbopogon citratus (lemmon grass), (d) *Origanum vulgare* (oregano), (e) *Salvia officinalis* (sage), (f) *Thymus vulgaris* (thyme), (g) *Satureja hortensis* (savory).

Many of the essential oils have been tested for bactericidal and bacteriostatic effects against a wide range of microorganisms including food spoiling organisms, pathogenic bacteria, yeasts, fungi, and many others. The major differences in antimicrobial activity have been yielded of several distinctive parameters which identify antibacterial characters of the essential oils, some of the major parameters include: (1) bacterial membrane permeability, (2) the hydrophobicity/hydrophilicity of the bacterial membrane, (3) the metabolic characteristics of the microorganism, and (4) their Gram-positive or negative pattern. Although susceptibility of the bacteria to the essential oils is not exactly predictable, many

Table 16.3　Some Active Traditional Plants Remedies Extracts With Antibacterial and Antiviral Effects

Plant Species (Family)	Antiinfection Activity	Indications	Using Form of Plants Extracts	References
Polygonum punctatum (Polygonaceae; aerial parts)	RSV	$ED_{50} = 120$ (mg/µL) against RSV of the assayed extracts in HEp-2 cells	Aqueous extracts	a
Lithraea molleoides (Anacardiaceae; aerial parts)	RSV	$ED_{50} = 87$ (mg/µL) against RSV of the assayed extracts in HEp-2 cells	Aqueous extracts	a
Myrcianthes cisplatensis (Myrtaceae; aerial parts)	RSV	$ED_{50} = 78$ (mg/µL) against RSV of the assayed extracts in HEp-2 cells	Aqueous extracts	a
Azadirachta indica (Meliaceae; stem bark)	S.a, P.a	$MIC_{90\%} = 1$, $MBC_{90\%} = 1$ (mg/mL) for S.a; $MIC_{90\%} = 1$, $MBC_{90\%} = 2$ (mg/mL) for P.a	Methanolic extracts	b
Entada abyssinica (Leguminosae; stem bark)	S.a, P.a	$MIC_{90\%} = 0.5$, $MBC_{90\%} = 2$ (mg/mL) for S.a; $MIC_{90\%} = 0.5$, $MBC_{90\%} = 2$ (mg/mL) for P.a	Methanolic extracts	b
Eremophila duttonii (Myoporaceae; leaves)	S.a	Diameters of the zones of growth inhibition in plate-hole diffusion assays (12 mm) with 0.77 mg/mL of extract	Ethanolic extracts	c
Artemisia capillaries Thunb. (Asteraceae; aerial parts)	RSV	$IC_{50} = 13$ (µg/mL) concentration of the sample required to inhibit virus-induced	Aqueous extracts	d
Arctium lappa L. (Asteraceae; aerial parts)	RSV	$IC_{50} = 6.3$ (µg/mL) concentration of the sample required to inhibit virus-induced	Aqueous extracts	d
Prunella vulgaris L. (Lamiaceae; fruit spike)	RSV	$IC_{50} = 10.4$ (µg/mL) concentration of the sample required to inhibit virus-induced	Aqueous extracts	d
Anemone obtusiloba (Ranunculaceae; aerial parts)	HSV	Lowest concentration of extract able to partially inhibit the virus (100 µg/mL)	Methanolic extracts	e
Centipeda minima (Asteraceae; aerial parts)	HSV	Lowest concentration of extract able to partially inhibit the virus (13 µg/mL)	Methanolic extracts	e
Byrsonima verbascifolia (Malphigiaceae; aerial parts)	HSV	Minimum concentration causing complete inhibition (MIC) of viral (2.5 µg/mL)	Methanolic extracts	f
Symphonia globulifera (Clusiaceae; aerial parts)	HSV	Minimum concentration causing complete inhibition (MIC) of viral (2.5 µg/mL)	Methanolic extracts	f

Table 16.3 Some Active Traditional Plants Remedies Extracts With Antibacterial and Antiviral Effects (*cont.*)

Plant Species (Family)	Antiinfection Activity	Indications	Using Form of Plants Extracts	References
Dracaena cinnabari (Agavaceae; aerial parts)	I.A	IC_{50} = 1.5 (µg/mL) concentration of the sample required to inhibit virus-induced	Methanol extracts	g
Exacum affine (Gentianaceae; aerial parts)	I.A	IC_{50} = 0.7 (µg/mL) concentration of the sample required to inhibit virus-induced	Methanol extracts	g
Scrophularia amplexicaulis Benth. (Scrophulariaceae; aerial parts)	S.a	Diameters of the zones of growth inhibition in well-diffusion method (13 mm) with 100 mg/mL of essential oil	Essential oil	h
Cinnamomum zeylanicum (Lauraceae; bark)	H.i, S.p, S.a	The lowest concentration of oil inhibiting the growth of each organism (MIC = 0.00625, 0.00625, 0.0125 mL/mL)	Essential oil	i
Cupressus sempervirens (Cupressaceae; aerial parts)	H.i, S.p, S.a	MIC = 0.00625, 0.00625, 0.0125 mL/mL	Essential oil	i

RSV, *respiratory syncytial virus;* ADV, *adenovirus;* HSV, *herpes simplex virus 1;* I.A, *influenza virus-A;* S.a, Staphylococcus aureus; P.a, Pseudomonas aeruginosa; S.p, Streptococcus pneumonia; H.i, Haemophilus influenza.
[a]Smith NM. *Ethnobotanical field notes from the Northern Territory, Australia.* J Adelaide Bot Gard 1991; 1–65.
[b]Fabry W, Okemo PO, Ansorg R. *Antibacterial activity of East African medicinal plants.* J Ethnopharmacol 1998; **60**: 79–84.
[c]Palombo EA, Semple SJ. *Antibacterial activity of traditional Australian medicinal plants.* J Ethnopharmacol 2001; **77**: 151–7.
[d]Ma S-C, Du J, But PP-H, Deng X-L, Zhang Y-W, Ooi VE-C, et al. *Antiviral Chinese medicinal herbs against respiratory syncytial virus.* J Ethnopharmacol 2002; **79**: 205–11.
[e]Taylor R, Manandhar N, Hudson J, Towers G. *Antiviral activities of Nepalese medicinal plants.* J Ethnopharmacol 1996; **52**: 157–63.
[f]Lopez A, Hudson J, Towers G. *Antiviral and antimicrobial activities of Colombian medicinal plants.* J Ethnopharmacol 2001; **77**: 189–96.
[g]Mothana RA, Mentel R, Reiss C, Lindequist U. *Phytochemical screening and antiviral activity of some medicinal plants from the island Soqotra.* Phytother Res 2006; **20**: 298–302.
[h]Pasdaran A, Delazar A, Nazemiyeh H, Nahar L, Sarker SD. *Chemical composition, and antibacterial (against* Staphylococcus aureus) *and free-radical-scavenging activities of the essential oils of* Scrophularia amplexicaulis Benth. Rec Nat Prod 2012; **6**: 350–5.
[i]Fabio A, Cermelli C, Fabio G, Nicoletti P, Quaglio P. *Screening of the antibacterial effects of a variety of essential oils on microorganisms responsible for respiratory infections.* Phytother Res 2007; **21**: 374–7.

researchers have tried to determine the relationship between the origin of the essential oils and their compounds with their antimicrobial activity. Furthermore, the delivery of medications to the respiratory tract has become an increasingly important method for respiratory disease treatment. The use of inhaler medications has become an invaluable therapeutic in the treatment of different pulmonary disorders, including bronchitis, pneumonia, and others complications.[37] Several studies have reported the clinical efficacy of inhalation therapy for the treatment of lung disorders.[38,39] Through the effective delivery of medication to the action site, the active compounds are delivered directly into the lungs and this can result in respiratory tract local treatment. This method achieves maximum therapeutic

Table 16.4 *Achilla* Species Essential Oils, Their Major Chemical Compositions and Their Effects on Some of the Microorganisms That Cause Major Respiratory Infections

Plant Name	Tested Microorganisms	Major Compounds	References
Achillea clavennae L.	*K. pneumonia*, penicillin-susceptible and penicillinresistant *S. pneumonia*, *H. influenza* and *P. aeruginosa*	Camphor, 1,8-cineole	a
A. fragrantissima (Forssk) Sch. Bip.	*K. pneumonia, P. aeruginosa, S. faecalis, S. aureus* and *C. albicans*	Terpinen-4-ol	b
A. sintenisii Hub. Mor.	*K. pneumonia, P. aeruginosa* and *S. aureus*	Camphor, 1,8-cineole	c
A. biebersteinii Afan.	*S. pneumonia* and *S. aureus*	Piperitone, 1,8-cineole, camphor	d
A. taygetea Boiss & Heldr.	*K. pneumonia, P. aeruginosa* and *S. aureus*	Borneol, 1,8-cineole	e
A. frasii Schultz Bip.	*K. pneumonia, P.aeruginosa* and *S. aureus*	1,8-cineole, α-pinene (4), β-pinene (5)	e
A. holosericea Sibth. & Sm.	*K. pneumonia, P. aeruginosa* and *S. aureus*	Borneol, camphor	f

aBezić N, Skočibušić M, Dunkić V, Radonić A. Composition and antimicrobial activity of Achillea clavennae L. essential oil. Phytother Res *2003;* **17***: 1037–40.*

bBarel S, Segal R, Yashphe J. The antimicrobial activity of the essential oil from Achillea fragrantissima. J Ethnopharmacol *1991;* **33***: 187–91.*

cSökmen A, Vardar-Ünlü G, Polissiou M, Daferera D, Sökmen M, Dönmez E. Antimicrobial activity of essential oil and methanol extracts of Achillea sintenisii *Hub. Mor. (Asteraceae).* Phytother Res *2003;* **17***: 1005–10.*

dSökmen A, Sökmen M, Daferera D, Polissiou M, Candan F, Ünlü M, et al. The in vitro antioxidant and antimicrobial activities of the essential oil and methanol extracts of Achillea biebersteini *Afan. (Asteraceae).* Phytother Res *2004;* **18***: 451–6.*

eMagiatis P, Skaltsounis A-L, Chinou I, Haroutounian SA. Chemical composition and in vitro antimicrobial activity of the essential oils of three Greek Achillea species. Z Naturforsch C *2002;* **57***: 287–90.*

fStojanović G, Asakawa Y, Palić R, Radulović N. Composition and antimicrobial activity of Achillea clavennae *and* Achillea holosericea *essential oils.* Flavour Fragr J *2005;* **20***: 86–8.*

FIGURE 16.2 Some Plants of *Achillea* Species Whose Essential Oils Are Used in the Treatment of Respiratory Infections

(a) *Achillea fragrantissima*, (b) *A. sintenisii*, (c) *A. clavennae*, (d) *A. taygetea*, (e) *A. biebersteinii*.

effect, small dose usage, and has fewer side-effect risks compared with those associated with larger doses. Inhalation is a unique treatment with direct effects on respiratory disorder site and is based on the volatility potential of essential oils. Furthermore, there is a need to develop new therapeutic agents for respiratory infections.[40–42]

Research has been carried out on the wide spectrum of edible plants essential oils to determine the antibacterial potential of their essential oils. The role of these plants as therapeutic agents is remarkable in many cultures. Investigations have reported that thyme and oregano essential oils, based on the phenolic components [such as carvacrol (**1**) and thymol (**2**) (Fig. 16.3)] have shown a strong correlation with the inhibition of some of the pathogenic bacterial strains (eg, in *Escherichia coli*). The correlation between the antibacterial effect of the volatile oils and their chemical compounds, including high amount of the phenolic components such as carvacrol (**1**) or eugenol (**3**), has also been confirmed.[43] Other essential oils such as oregano, savory, clove, and nutmeg with high concentrations of volatile phenolic compounds inhibit Gram-positive more than Gram-negative pathogenic bacteria.[44] However, in some essential oils such as *Achillea* spp. (Yarrow) strong antibacterial activity was observed against the Gram-negative respiratory pathogens (*Haemophilus influenzae, Pseudomonas aeruginosa*) while *Streptococcus pyogenes* was the most resistant to the this oil.[45] The other essential oils such as peppermint and spearmint inhibit the methicillin-resistant type of *Staphylococcus aureus*. Previous reports have clarified that the essential oils containing aldehyde or phenol as a major component represent the highest antibacterial activity. These antibacterial potencies are lower in the essential oils that contain high amounts of terpene alcohols compared to the essential oils containing aldehyde or phenol as a major component.

Other essential oils containing terpene ketone, or ether showed much weaker activity, and oil containing terpene hydrocarbon was relatively inactive. Based on these findings, essential oils such as thyme, cinnamon, lemongrass, perilla, and peppermint have demonstrated suitable effects on respiratory tract infection.[46] The tolerance of Gram-negative bacteria to essential oils has been attributed to the presence of a hydrophilic outer membrane that blocked the penetration of hydrophobic essential oils to the target cell membrane because the Gram-positive bacteria were more exposed to the essential oils than Gram-negative bacteria, which has been reported several times.[47–50]

Lipids are one of the principal constituents for normal cell membrane function and these compounds supply many operations, such as barrier function in the bacterial cell membrane. The external capsule of some Gram-negative bacteria limits or prevents the penetration of the essential oils into the microbial cell. One of the pronounced examples of the hydrophobicity/hydrophilicity role in bacterial sensitivity to antibacterial compound is *H. influenzae*. It should be pointed out that the outer membrane of *H. influenzae* (which forms rough colonies) was more hydrophobic. Hydrophobic antibiotics, such as macrolides, are more active against *H. influenzae* than *E. coli* through their shorter oligosaccharide chains than those in *E. coli*. The effects of the cytoplasmic membrane and/or the embedded enzymes in it have demonstrated lipophilic biocide actions.[51] It is generally recognized that the antimicrobial action of essential oils depends on their hydrophilic or lipophilic character. Based on these observations, investigators are trying to indicate the relationship between structural activity of the essential oils compounds and their antibacterial activity.

Certain components of the essential oils can act as uncouplers, which interfere with proton translocation over a membrane vesicle and subsequently interrupt ADP phosphorylation pathways (primary energy metabolism). As a member of the phytochemicals, terpenoids have been observed as a model of lipid soluble agents, which have an impact on the activities of membrane catalyzed enzymes; for

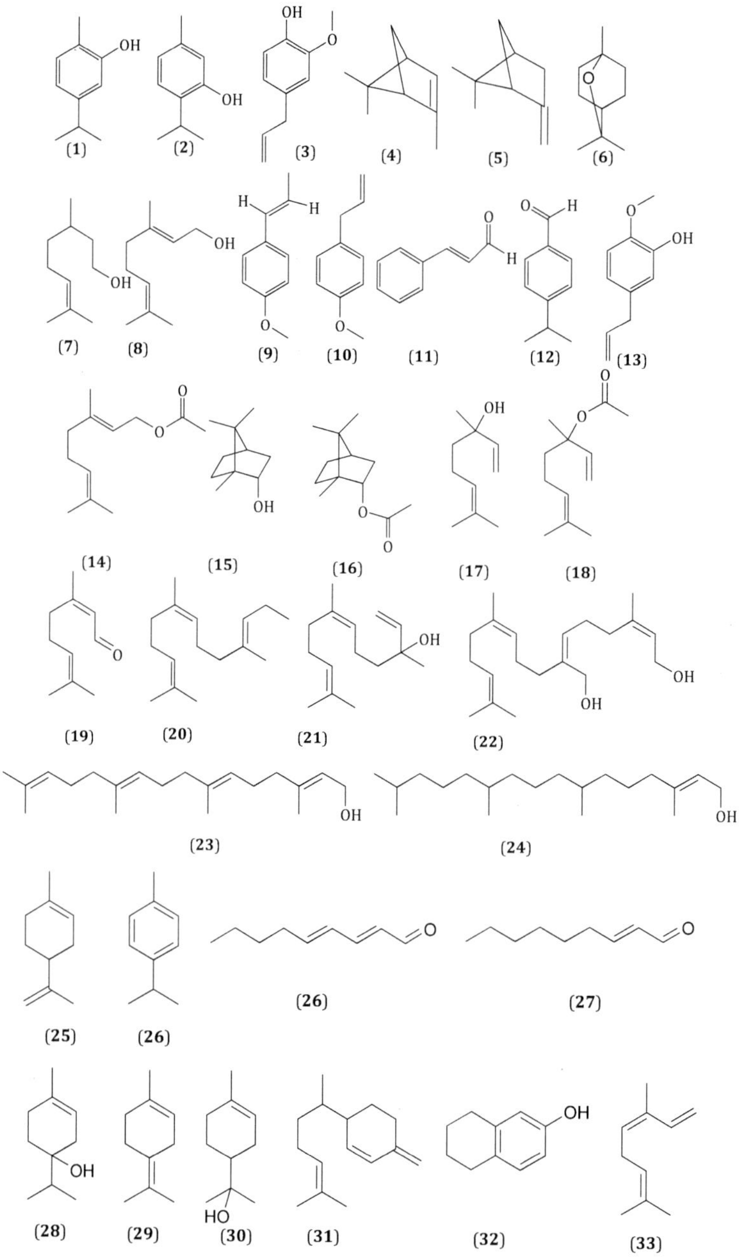

FIGURE 16.3 Structures of the Major Bioactive Chemical Compounds Isolated From Essential Oils

example, enzymes involved in respiratory pathways. Particular terpenoids with functional groups, such as phenolic alcohols or aldehydes, also interfere with membrane-integrated or associated enzyme proteins, inhibiting their production or activity. A good deal of antimicrobial compounds which act on the bacterial cytoplasmic membrane cause the loss of 260 nm absorbing material. This causes an increased susceptibility to NaCl, the lysosomes formation and loss of potassium ions, which results in inhibiting respiration and the loss of cytoplasmic material.

Investigations about the cytoplasmic membrane effects of α-pinene (**4**), β-pinene (**5**), 1,8-cineole (**6**), and electron microscopy studies have shown that the essential oils containing these compounds triggered such cytoplasmic material with losing in treated bacterial cells.[31,32] The perturbation of the lipid fraction in the plasma membrane causes antimicrobial activity of some of the phytochemicals such as α,β-unsaturated aldehydes and some of monoterpenes. Although these aldehyde compounds can elicit antibacterial effects by acting on membrane functional proteins, such antibacterial effect would be achieved with modifications of membrane permeability and intracellular materials leakage.[52–54] The membrane damage leading to whole-cell lysis has been reported by oregano and rosewood essential oils which contains major components as: carvacrol (**1**), citronellol (**7**), and geraniol (**8**).[26,55] Phenols such as carvacrol (**1**), thymol (**2**), eugenol (**3**), and other oxygenated aromatic essential oil compounds including phenol ethers [*trans*-anethole (**9**), methyl chavicol (**10**)] and aromatic aldehydes [cinnamaldehyde (**11**), cuminaldehyde (**12**)] have been reported to exert both antibacterial and antifungal activity. However, this chemical class—based on the concentration used—are known as either bactericidal or bacteriostatic agents,[56] but the phenolic component's high activity may be further explained in terms of the alkyl substitution into the phenol nucleus, which is known to increase the antimicrobial activity of phenols. The alkylation has been known to change the distribution ratio between the hydrophilic and the hydrophobic phases (including bacterial phases) by the surface tension reduction or the species selectivity mutate based on the bacteria cell wall characters.[57] This does not happen with etherified or esterified isomeric molecules, it is possible by describing their relative lack of activity.[58] As a member of these compounds carvacrol (**1**) is one of the few components that has a break apart from effect on the outer membrane of Gram-negative bacteria and causes release of lipopolysaccharide and alters cytoplasmic membrane ions transportation, Similar to carvacrol (**1**), thymol (**2**) antimicrobial activity results in structural and functional alterations in the cytoplasmic membrane.[59] Interestingly, eugenol (**3**) and isoeugenol (**13**) exhibit higher activity against Gram-negative bacteria than Gram-positive bacteria, and when cinnamaldehyde (**11**) is used against *E. coli*, its activity is similar to carvacrol (**1**) and thymol (**2**) (Fig. 16.3). These compounds alter the membrane, affect the transport of ions and ATP, and change the fatty acid profile of different bacteria.[60]

Although in some cases alcoholic form shows better potencies compared to acetate form, the presence of an acetate moiety in the structure appeared to increase the activity of the parent compound. In the case of geraniol (**8**), the geranyl acetate (**14**) demonstrated an increase in activity against the test microorganisms.[48,61,62] A similar effect was also observed in the case of borneol (**15**), bornyl acetate (**16**), linalool (**17**), and linalyl acetate (**18**) (Fig. 16.3). In addition, the effectiveness of alcoholic compounds very closely depended on the bacterial cell wall, which showed different permeability to alcohol based on chain length.[44,63] It has been suggested that an aldehyde group conjugated to a carbon double bond such as citral (**19**) is an extremely electronegative order, which may explain their activity, and an increase in electronegativity can raise up the antibacterial activity.[64] In addition, the under research of aldehydes potency seems to depend not only on the existence of the α,β-double bond but also on the chain length from the renal group and on microorganism tested. It seems that some electronegative compounds, mainly from the cell surface, are responsible for the inhibited growth of the microorganisms, which may interfere in biological processes involving electron transfer and respond with vital nitrogen components and alteration in the operation of membrane-associated proteins. Actually, a greater electronegativity of the molecule would cause a greater encounter of intermolecular hydrogen bond formation with membrane nucleophilic groups and thus a significant irregularity in the lipidic bilayer. Some studies have recommend that carbon tail length also affects the electronegativity of the aldehyde oxygen atom and thus its interaction with the nucleophilic groups of the cell membrane.[65] Comparably, the similar antimicrobial activity was detected in the series of the long-chain alcohols which is demonstrated to be resulted from the alkyl chain length.[66,67] This structural activity relationship is notable between farnesol (**20**), nerolidol (**21**), plaunotol (**22**), geranylgeraniol (**23**), phytol (**24**), geraniol (**8**), and linalool (**17**) which act on *S. aureus* with damages of the cell membranes and losing of K+ ions, while similar mode of actions can be detected by the aminoglycosides such as kanamycin and streptomycin. Farnesol (**20**) was able to damage cell membranes most effectively than other terpene alcohols. The activities of farnesol (**20**), nerolidol (**21**) (sesquiterpenes compounds) on *S. aureus* were higher than that of plaunotol (**22**) (diterpene). The effectiveness against *S. aureus* are in order as follows: farnesol (**19**) > nerolidol (**20**)> plaunotol (**22**) > geranylgeraniol (**23**), phytol (**24**) > geraniol (**8**) and linalool (**17**) (Fig. 16.3). It has been suggested that maximum activity against *S. aureus* might depend on the number of carbon atoms in the hydrophobic chain from hydrophilic hydroxyl group, which should be less than 12 but as close to 12 as possible. Neither a shorter nor a longer aliphatic carbon chain, could increase such activity.[68,69] The increased effectiveness of sesquiterpenes as enhancers of membrane permeability may stem from their structural resemblance to membrane lipids (eg, linear molecules with internal lipophilic character and a more polar terminus).[70] The bacteriostatic potential of the terpenoids was also increased when the carbonyl groups increased in structure.[63]

The type of alkyl substituent incorporated into a nonphenolic ring structure is responsible for enhancement of antibacterial activity. Such as an alkenyl substituent (1- methylethenyl) makes an increase in antibacterial activity, as seen in limonene [1-methyl-4-(1-methylethenyl)-cyclohexene] (**25**), compared to an alkyl (1-methylethyl) substituent as in p-cymene [1-methyl-4-(1-methylethyl)-benzene] (**26**). Furthermore, principally Gram-negative were the sensitive organisms that propose alkylation control of the Gram reaction sensitivity of the bacteria. An allylic side-chain seems to raise the inhibitory role of the simple phenols mainly against Gram-negative organisms. This was suggested due to the majority of the antimicrobial activity of alkylated phenols in relation to phenol which has been earlier reported (Fig. 16.3).[56]

It was observed that α-isomers are inactive relative to β-isomers in many compounds in stereochemistry, which is also effective in antibacterial activity observed from essential oils.[44] The (*E,E*)-2,4-decadienal (**26**) appears to be more toxic to bacterial cells than the correspondent monounsaturated aldehyde (*E*)-2-decenal (**27**) as another example of stereochemistry effectiveness in activity potential observed in the unsaturated aldehyde, but it is noticeable that two double bonds in the *cis* configuration in the side-chain of 2,4-decadienal (**26**) produce more bends and shorten the length of the carbon tail, such α,β-unsaturated aldehydes might be a good choice compared to other highly toxic sterilizers (Fig. 16.3).[53]

In summary, the antimicrobial activity of essential oils depends on different amounts of specific compounds. As an example, antimicrobial properties of the essential oils with high concentrations of eugenol (**3**), cinnamaldehyde (**11**), or citral (**19**), is predictable. Remarkable antimicrobial, antifungal, and antiviral activity of the monoterpenes and phenols relieve from essential oils present in thyme, sage, and rosemary. Due to the formation of polysaccharides that increase the resistance to essential oils there are some other essential oils, such as basil, sage, hyssop, rosemary, and oregano, which are active against *E. coli*, *S. aureus* but are less effective against *Pseudomonas* spp.[71–73] The typical characteristic of the essential oils is hydrophobicity, which is responsible for the disruption of bacterial structures and makes an increase in permeability due to a weakness to pull apart the essential oils from the bacterial cell membrane. Many cellular functions, including maintaining the energy status of the cell, membrane-coupled energy transducing processes, solute transport, and metabolic regulation result from the cell membrane permeability barrier. Actually, they are responsible for leakage of the cell contents, reducing the proton motive force, reducing the intracellular ATP pool via decreased ATP synthesis and augmented hydrolysis are the mechanisms of action of the essential oils including the degradation of the cell wall, damaging the cytoplasmic membrane, cytoplasm coagulation, damaging the membrane proteins, increased permeability that is different from the increased membrane permeability and reducing the membrane potential via increased membrane permeability.[33,74,75]

3.1 LABORATORY METHODS OF EVALUATION OF ANTIBACTERIAL ACTIVITY OF ESSENTIAL OILS

Unfortunately, scientific investigations on the antimicrobial activity of essential oils have been retarded by the lack of appropriate susceptibility testing methods for the essential oils and because no generally approved assay method has been established for the assessment of their antimicrobial activity. Many researchers have employed the disk assay method. However, the results of this method were not always in parallel with those of dilution assay methods. The differences were caused not only by differences in the solubility of the oils but also by interactions of the components in the concentrated solution used in the disk assay.[48] The dilution method could be more reliable than the disk method with regard to reproducibility and clinical relevance. When testing nonwater-soluble and highly volatile essential oils by the dilution method, it is necessary to obtain a homogeneous dispersion of the oils in the medium. Chemical emulsifiers such as Tween 80, Tween 20 and other have been used frequently for this homogenization, but it has been reported that emulsifiers reduced the bioactivity of the oils, probably because of the formation of micelles, which inhibit adequate contact between the oil and the test organism.[48] Evidence also shows that the minimum inhibitory concentration (MIC) values of the essential oils under open conditions of incubation caused a two- to eight-fold rise in the MICs of highly volatile oils, as compared with values obtained under sealed conditions. Sealed conditions are used to examine whether essential oils showed antibacterial activity against major respiratory tract pathogens (this method was authorized by the Japan Society of Chemotherapy to adjust for the physico-chemical properties of essential oils). The scientific information concerning the antimicrobial effectiveness of the essential oils in the vapor phase compared with direct contact case show that potential of this form of the essential oils is ambiguous, although some degree of inhibition by volatile components of the essential oils has been demonstrated in the vapor phase. In fact, more investigation into the antibacterial effectiveness of the vapor phase is required. It seems that the composition of the atmosphere generated by the essential oils is also potentially correlated with their antimicrobial behavior. This part mentioned the main antibacterial activity laboratory assay methods, as already noted. These methods do not show coordinated results in some case and this appearance is reasoned to the necessity of simultaneously running two or more methods for antibacterial activity determination of essential oils.

3.1.1 Solid diffusion assays

In this method, a Petri dish is commonly used as an assay chamber (5–12 cm diameter and filled with 10–20 mL of agar broth). The solidified medium in the Petri dish was inoculated with an appropriate colony of the microorganisms. This inoculation was accomplished by using a solution such as physiological saline solution containing adequate colony unit (commonly 100–200 (CFU)/mL). The essential oil is incorporated into the mediums in two ways: on a paper disc or into a well (hole) which is made in the agar medium. Diluted or undiluted essential oils based on the goals were added to a sterile blank filter disk (5 mm diameter) and placed on top of the cultured media in a Petri dish or added to the holes. After incubation under optimal conditions such as temperature and time, two different zones were considered in view of the average diameter of changes: (1) the zone where there is no growth of the microorganism, called total inhibition; and (2), the zone where growth of the microorganisms was significantly reduced in terms of amount of colonies, compared to blank assays. The growth of the microorganisms is recognizable with instruments or can be seen visually by one of the commonly used techniques, called turbidimetry, in which the optical density changes in the growing culture (OD) are measured. Some commercial systems have been produced for monitoring of the microorganisms growth as well as other methods were proposed (eg, bioautography).[76]

3.1.2 Vapor diffusion assays

An appropriate volume (100 µL physiological saline solution) of the microorganism's colony (10–20 CFU/mL) will be inoculated to the solidified medium. Each essential oil sample was diluted in suitable solvent such as ethyl ether to obtain serial dilutions (v/v). The required volume (10 µL) of each dilution was then added to sterile blank filter disks or cups and placed on the medium free cover of each Petri dish. Experiments were designed in two forms of Petri dish positions, including direct and invert placement. The Petri dishes were then sealed using sterile adhesive tape. Blanks were prepared by adding the same volume (10 µL) of samples of the solvent to the filter disks or cups, and this had no effect on the viability of any of the tested organisms. After the incubation period, the minimal inhibitory concentration (MIC), expressed as microliters of the essential oil per volume unit of atmosphere above the organism growing on the agar surface, that caused inhibition by comparison with control tests was measured.[77]

3.1.3 The dilution method (agar or liquid broth)

Although the modification with liquid broth is mostly practical for fungi, the serial dilution agar method is also common for bacteria and fungi. The difference is that agar broth cultures are grown in Petri dishes or tubes, but liquid broth cultures are cultivated in conical flasks filled with 100 mL medium or test tubes with 2.5–5 mL medium (bacteria and moulds). The inhibitory growth index is determined for the liquid broth in conical flasks (percent changes in mould's biomass comparing to the control culture). There is an inhibitory effect of essential oil which appears in the test tube cultures and is measured turbidimetrically or with the plate count method. The estimation of essential oil activity both in agar and liquid broth would be simplified through counting the tested microorganisms that can remain in the membrane. The lowest essential oil concentration in the broth that results in the lack of visible microorganism growth changes is known as the MIC. The microorganisms are then transferred from the lowest essential oil concentration medium (with no visible microorganism growth) into a new broth medium and incubation to determine the lethal activity of essential oil.[78,79]

4 SCREENING OF THE ANTIVIRAL EFFECTS OF ESSENTIAL OILS

Viruses as invasive microorganisms may cause serious respiratory illness and in some cases life-threatening conditions, such as acute pneumonia. Although acute respiratory infection rates are not very high, this condition has been steadily increasing in children and persons over 60 years of age. The rates of hospitalization and death increase substantially in these cases. Multiple factors, such as decline in respiratory and immune function, likely contribute to increased morbidity. Natural products in all forms including pure compounds or extracts provide massive opportunities for new antiviral-lead compounds.[4,80] At the moment, only a few effective antiviral drugs are available for the treatment of viral diseases, especially in respiratory viral infections. Therefore, finding new substances with antiviral properties is a required for medical systems.

Much evidence has been reported about antiviral potential of various essential oils and their constitutions on several genera of viruses.[9,81] In some investigations the essential oils with high hydrocarbons long-chain contents such as (*E,E*)-2,4-decadienal (**26**), showed activities against influenza virus.[82,83] On other spectrum of antiinfluenza virus compounds, sesquiterpenes and sesquiterpenes-rich essential oils showed clear effects as well as aromatic rich essential oils.[84–86] Among the terpenoids compounds, terpinen-4-ol (**28**), terpinolene (**29**), and α-terpineol (**30**), show inhibitory effect on influenza A/PR/8PR virus subtype H1N1. Also indicated that some sesquiterpenes such as β-sesquiphellandrene (**31**), and tetrahydronaphthalenol (**32**), showed potent antirhinoviral activity in a plaque reduction test.[87] In other researches that conducted about the antiviral essential oils and their active constitutions indicated that *Laurus nobilis* essential oil which was characterized by the presence of β-ocimene (**32**), 1,8-cineole (**6**), α-pinene (**4**), and β-pinene (**5**), as the main constituents, showed interesting activity against SARS-CoV.[88] This overview caused the development of commercial therapeutics based on monoterpens compounds for viral infections especially in respiratory viral infections (Fig. 16.3).[89,90]

Although, antiviral potential is an important factor for bioactivity of natural compounds, cytotoxicity effects that probably triggered host cell damaged cycles are important. Therefore, this is a limiting factor for presentation of the new useful compound in treatment or control of the viral infections. Viruses as invasive parasites are capable of using host cell organelles and mechanisms for reproduction, such perspicaciously reproducing process caused a creation of tenacious shield against many potent natural molecules used. Therefore selective activity is an important point, as are antiviral potencies.

5 ROLE OF INFLAMMATION IN RESPIRATORY TRACT INFECTIONS

The immune response to respiratory tract infection is a double-edged sword because many of the symptoms that accompany these infections are largely due to the microorganism's induction of cytokines and chemokines, which may result in protracted inflammatory responses. Phagocytic clearance of an infecting organism by inflammatory cells is an appropriate and necessary component of the host defense system.[91] At the same time, much of the evidence points to destructive effects spectrum made by products of these inflammatory cells.[92] These products can increase mucus secretion and impair ciliary clearance, thereby setting the stage for exacerbations and recurrences of infection. Primary inflammatory cell products can amplify the other steps of the inflammatory cycles and can paradoxically even impair the immune response. These observations suggest that modulating the inflammatory response may be an important aspect of definitive therapy for respiratory tract infection. The primary defense systems are the lung secretions and the mucociliary escalator, which entrap organisms and sweep them away. At the same time, the lung secretions also contain a variety of proteins that inhibit microorganisms, especially bacterial adherence for example immunoglobulin A (IgA) systems, which prevents bacterial adherence to epithelial cells, inhibit bacterial growth, and attempt to neutralize bacteria. The macrophages are a secondary defense system, which not only phagocytose microorganisms but also release biochemotactic factors that recruit other defense cells such as monocytes and neutrophils that are necessary for further phagocytosis process.[93] The inflammatory response form a key part of the secondary defense system that accompanies additional proteins, such as immunoglobulins and complement factors. Although neutrophil infestation is part of the natural defenses against an invading organism, it may have destructive effects on the pulmonary systems. One of the destructive factors is the proteolytic enzyme neutrophil elastase, which is preformed and stored within the neutrophil. This proteinase is released after the neutrophil is activated and migrates into the tissues and phagocytosis invasive microorganisms. It produces a condition that histologically simulates chronic bronchitis. The potential role of neutrophil elastase in the pathogenesis of pulmonary diseases is characterized by neutrophil infiltration of the airways and increased mucus secretion.[94] It has also been reported that "secretions from patients with acute bronchial infection cause a significant reduction in ciliary beat frequency and that the addition of a neutrophil elastase inhibitor can reverse this effect."[95] The increase in the mucus secretion and the decrease in ciliary beat have been found as an important feature of chronic lung disease frequently. Besides the phagocytic cell immediate responses, stimulated epithelial cells and macrophages produce the potent neutrophil chemoattractant, such as interleukin- 8 (IL-8), which was found to be present in high concentrations in the sputum of patients with chronic inflammatory airway disease.[96] At the same time, elastase released by activated neutrophils also stimulates IL-8 production by the epithelial cells.[97] Therefore, when these events are initiated, they become a self-amplifying cycle. This phenomenon is observed in patients with acute pneumonia that has apparently been cured by appropriate antibiotic therapy.

Natural products as inflammation inhibitors have for a long time played a key role in many traditional treatment systems. Several mechanisms including interaction with prostaglandin biosynthesis, interaction with other inflammatory mediators, and corticosteroid-like effects are involved in antiinflammatory action of natural products.[98] Among the natural chemical compounds, significant antiinflammatory activities of plant-based essential oil have been reported by many researchers, which showed the basis for folk and traditional uses of these herbs for treatment of inflammatory diseases. The essential oils and aromatic plants have had a significant place in inflammation control in many folklore medications. This evidence supports the view that appropriate investigations about the potential of essential oils should be made and their active constituents should also be evaluated for inflammation control. Monoterpene alcohols such as linalool (**17**) and linalyl acetate (**18**) and other corresponding esters which are reported as one of the major volatile components of many aromatic plants essential oil were evaluated for antiinflammatory activity. These compounds have shown promising antiinflammatory potency among the many essential oil compounds.[99] For example, 1,8-cineole (**6**) the major constituent of eucalyptus oil is another active essential oil constituent that is well tolerated in inhalation administrations. This compound can also be effective for airways inflammation in clinical trials, based on such finding 1.8-cineole (**6**) is registered as a licensed medicinal product and has been available for airways inflammation for many years.[100] Similar effects were also observed in other monoterpenes especially in carvacrol (**1**) and thymol (**2**), which are main constituents in thymus, oregano, savory, clove, and nutmeg essential oils. In other reports, similar observations have confirmed that many of the other essential oils and their constitutions can play an important role in inflammation control.

Emphasizing the structure–activity relationships is necessary to define different potential of natural compounds such as essential oils compositions. For instance, aliphatic aldehydes and aromatic aldehydes are said to predominantly have antiinflammatory and antimicrobial potentials.[101] The terpenes and sesquiterpenes evaluated appeared to be good inhibitors of 5-LOX in vitro. There is a good correlation observed between antiinflammatory activity of the limonene (**25**) and the essential oils rich in limonene (**25**) like grapefruit, lime, and celery. Other similar activity has been observed on the various inflammations pathways such as cyclooxygenase (COX) and lipoxygenases (LOX) between sesquiterpenic alcohols, aliphatic aldehydes, and some phenolic esters that caused these phytochemicals considered for more investigations in respiratory inflammations during the air ways infections (Fig. 16.3).[101,102] It has been evaluated that activity of some of the essential oils such as Western red cedar (*Thuja plicata*) to inactivate several viruses implicated in respiratory infections, and to inhibit the influenza virus-induced secretion of cytokine (IL-6) in cultured human lung cells. However, the liquid essential oil phases are generally a higher irritant and possibly toxic for nasopharyngeal or oral applications, although a few reports have indicated that the vapors of some essential oils might be useful for application in inhaled vapors for respiratory infections in low concentrations.[103] Because of the noteworthy role of inflammation in respiratory infectious disease persistence and recurrence, many laboratory models have been developed for inflammation and inflammation/infection-mixed respiratory complications research. Some of these models, such as ovalbumin-induced respiratory allergic eosinophilia arise especially for inflammation process which responds to them, including eosinophils, neutrophils, and other immune cell infiltration to the inflammation sites.[104] In other models, in addition to this cellular pathogenesis, more attention is considered on the infection's correlation with inflammation mechanisms.[105] Table 16.5 shows some of the attributes of inflammation and inflammation/infection major models that are used in screening of the new antiinflammation compounds.

Table 16.5 Typical Laboratory Models of Inflammation and Inflammation/Infection Mixed Diseases

Models	Disease	Type	Rationale	Ref
Inhaled antigen-induced tracheal constriction models	Inflammation in the upper airways	Eosinophilic inflammatory	Airway inflammation	a
Ovalbumin-induced respiratory allergic eosinophilia	Inflammation in the lower airways	Eosinophilic inflammatory	Airway inflammation	b
Neutrophil elastase (NE) inhibitory activity	Pulmonary acute injury inflammation	Neutrophils released (NE) at inflammatory sites	Pulmonary inflammation	c
Chronic respiratory infection with *Mycoplasma pneumoniae*	Chronic pulmonary infection-associated chronic inflammation	Peribronchial and perivascular mononuclear infiltrates	Pulmonary infection-inflammation	d
Antirespiratory syncytial virus (RSV)	Acute and chorionic pulmonary (lower respiratory) infection-associated with inflammation	Histopathological change including: peribronchiolar, bronchial and perivascular infiltrations	Pulmonary infection-inflammation	e

(Continued)

Table 16.5 Typical Laboratory Models of Inflammation and Inflammation/Infection Mixed Diseases (*cont.*)

Models	Disease	Type	Rationale	Ref
Chronic rat lung model of *P. aeruginosa* infection	Chronic pulmonary infection-associated chronic inflammation	Airway neutrophilic inflammation	Pulmonary infection-inflammation	f

[a]*Wills-Karp M. Immunologic basis of antigen-induced airway hyperresponsiveness.* Annu Rev Immunol 1999; **17**: 255–81. Kips J, Anderson G, Fredberg J, Herz U, Inman M, Jordana M, et al. *Murine models of asthma.* Eur Respir J 2003; **22**: 374–82.
[b]*Renz H, Smith HR, Henson JE, Ray BS, Irvin CG, Gelfand EW. Aerosolized antigen exposure without adjuvant causes increased IgE production and increased airway responsiveness in the mouse.* J Allergy Clin Immunol 1992; **89**: 1127–38. Randolph DA, Stephens R, Carruthers CJ, Chaplin DD. *Cooperation between Th1 and Th2 cells in a murine model of eosinophilic airway inflammation.* J Clin Invest 1999; **104**: 1021. Hamelmann E, Oshiba A, Loader J, Larsen G, Gleich G, Lee J, et al. *Antiinterleukin-5 antibody prevents airway hyperresponsiveness in a murine model of airway sensitization.* Am J Respir Crit Care Med 1997; **155**: 819–25.
[c]*Korkmaz B, Attucci S, Jourdan M-L, Juliano L, Gauthier F. Inhibition of neutrophil elastase by α1-protease inhibitor at the surface of human polymorphonuclear neutrophils.* J Immunol 2005; **175**: 3329–38. Vogelmeier C, Buhl R, Hoyt RF, Wilson E, Fells GA, Hubbard RC, et al. *Aerosolization of recombinant SLPI to augment antineutrophil elastase protection of pulmonary epithelium.* J Appl Physiol 1990; **69**: 1843–8.
[d]*Hardy RD, Jafri HS, Olsen K, Hatfield J, Iglehart J, Rogers BB, et al. Mycoplasma pneumoniae induces chronic respiratory infection, airway hyperreactivity, and pulmonary inflammation: a murine model of infection-associated chronic reactive airway disease.* Infect Immun 2002; **70**: 649–54. Lindsey JR, Cassell GH. *Experimental Mycoplasma pulmonis infection in pathogen-free mice: models for studying mycoplasmosis of the respiratory tract.* Am J Pathol 1973; **72**: 63.
[e]*Mejías A, Chávez-Bueno S, Ríos AM, Saavedra-Lozano J, Aten MF, Hatfield J, et al. Anti-respiratory syncytial virus (RSV) neutralizing antibody decreases lung inflammation, airway obstruction, and airway hyperresponsiveness in a murine RSV model.* Antimicrob Agents Chemother 2004; **48**: 1811–22. Jafri HS, Chávez-Bueno S, Mejías A, Gómez AM, Ríos AM, Nassi SS, et al. *Respiratory syncytial virus induces pneumonia, cytokine response, airway obstruction, and chronic inflammatory infiltrates associated with long-term airway hyperresponsiveness in mice.* J Infect Dis 2004; **189**: 1856–65. Chávez-Bueno S, Mejías A, Gómez AM, Olsen KD, Ríos AM, Fonseca-Aten M, et al. *Respiratory syncytial virus-induced acute and chronic airway disease is independent of genetic background: an experimental murine model.* Virol J 2005; **2**: 46.
[f]*Cantin AM, WOODS DE. Aerosolized prolastin suppresses bacterial proliferation in a model of chronic Pseudomonas aeruginosa lung infection.* Am J Respir Crit Care Med 1999; **160**: 1130–5. Chmiel JF, Konstan MW, Saadane A, Krenicky JE, Lester Kirchner H, Berger M. *Prolonged inflammatory response to acute Pseudomonas challenge in interleukin-10 knockout mice.* Am J Respir Crit Care Med 2002; **165**: 1176–81.

6 CONCLUSIONS

Respiratory infections are one of the most prevalent human health problems and they cause major difficulties for all age ranges.[106] Research to find new and effective therapeutic compounds for the control and treatment of these ailments is a most attractive field in natural product screening. The massive biodiversity of natural sources, such as plants, means that the selection of the suitable starting source point is a critical step for achieving the best screening results. In particular, paying attention to the main chemical constitutions of these natural sources will help to clarify the results of this research. Analysis of the relationship between the chemical structures of natural compounds has led to a better overview of the rational uses of natural compounds and natural remedies for the control and treatment of disease, and has led to the development of therapeutics. The broad diversity of the chemical makeup of essential oils has caused some uncertainty over their best selection in complementary or investigative research. Therefore, a correct understanding of the relation between the potential effects and the plant's family is a useful guide for physicians and researchers. According to this finding, the volatile oils of plant families such as

Asteraceae, Lamiaceae, Scrophulariaceae, Fabaceae, Myrtaceae, Rutaceae, Cupressaceae, and Pinaceae have good potential for use in infectious and inflammatory respiratory diseases, in both research and treatment. These plant based essential oils have a good feasibility for presentation as therapeutics in many of the respiratory infections and inflammatory diseases. Although these natural compounds have been used as complementary therapeutics, scientists must still give attention to their safety and dosage.

ACKNOWLEDGEMENTS

We thank the Medicinal Plants Processing Research Center of Shiraz University of Medical Sciences and the Phytochemistry Research Center of Shahid Beheshti University for all of the scientific support.

REFERENCES

1. Shafran SD, Singer J, Zarowny DP, Phillips P, Salit I, Walmsley SL, et al. A comparison of two regimens for the treatment of *Mycobacterium avium* complex bacteremia in AIDS: rifabutin, ethambutol, and clarithromycin versus rifampin, ethambutol, clofazimine, and ciprofloxacin. *New Engl J Med* 1996;**335**(6):377–84.
2. Gold R, Jin E, Levison H, Isles A, Fleming PC. Ceftazidime alone and in combination in patients with cystic fibrosis: lack of efficacy in treatment of severe respiratory infections caused by *Pseudomonas cepacia*. *J Antimicrob Chemother* 1983;**12**(Suppl. A):331–6.
3. Ball P, Baquero F, Cars O, File T, Garau J, Klugman K, et al. Antibiotic therapy of community respiratory tract infections: strategies for optimal outcomes and minimized resistance emergence. *J Antimicrob Chemother* 2002;**49**(1):31–40.
4. Cragg GM, Newman DJ, Snader KM. Natural products in drug discovery and development. *J Nat Prod* 1997;**60**(1):52–60.
5. Cowan MM. Plant products as antimicrobial agents. *Clin Microbiol Rev* 1999;**12**(4):564–82.
6. Iwu M, Duncan AR, Okunji CO. New antimicrobials of plant origin. *Perspectives on new crops and new uses*. Alexandria, VA: ASHS Press; 1999. p. 457–62.
7. Wiart C, Kumar K, Yusof M, Hamimah H, Fauzi Z, Sulaiman M. Antiviral properties of ent-labdene diterpenes of *Andrographis paniculata* nees, inhibitors of *herpes simplex* virus type 1. *Phytother Res* 2005;**19**(12):1069–70.
8. McCutcheon A, Roberts T, Gibbons E, Ellis S, Babiuk L, Hancock R, et al. Antiviral screening of British Columbian medicinal plants. *J Ethnopharmacol* 1995;**49**(2):101–10.
9. Reichling J, Schnitzler P, Suschke U, Saller R. Essential oils of aromatic plants with antibacterial, antifungal, antiviral, and cytotoxic properties–an overview. *Forsch Komplementarmed Klass Naturheilkd* 2009;**16**(2):79–90.
10. Butler MS, Buss AD. Natural products—the future scaffolds for novel antibiotics? *Biochem Pharmacol* 2006;**71**(7):919–29.
11. Corson TW, Crews CM. Molecular understanding and modern application of traditional medicines: triumphs and trials. *Cell* 2007;**130**(5):769–74.
12. WHO. WHO traditional medicine strategy 2002–2005; 2002.
13. Seeff LB, Lindsay KL, Bacon BR, Kresina TF, Hoofnagle JH. Complementary and alternative medicine in chronic liver disease. *Hepatology* 2001;**34**(3):595–603.
14. Pennacchio M, Jefferson L, Havens K. *Uses and abuses of plant-derived smoke: its ethnobotany as hallucinogen, perfume, incense, and medicine*. New York: Oxford University Press; 2010.
15. Atta A, Alkofahi A. Anti-nociceptive and anti-inflammatory effects of some Jordanian medicinal plant extracts. *J Ethnopharmacol* 1998;**60**(2):117–24.

16. Matu EN, Van Staden J. Antibacterial and anti-inflammatory activities of some plants used for medicinal purposes in Kenya. *J Ethnopharmacol* 2003;**87**(1):35–41.
17. Schinella G, Tournier H, Prieto J, De Buschiazzo PM, Rıos J. Antioxidant activity of anti-inflammatory plant extracts. *Life Sci* 2002;**70**(9):1023–33.
18. Eldeen I, Elgorashi E, Van Staden J. Antibacterial, anti-inflammatory, anti-cholinesterase and mutagenic effects of extracts obtained from some trees used in South African traditional medicine. *J Ethnopharmacol* 2005;**102**(3):457–64.
19. Sosa S, Balick M, Arvigo R, Esposito R, Pizza C, Altinier G, et al. Screening of the topical anti-inflammatory activity of some Central American plants. *J Ethnopharmacol* 2002;**81**(2):211–5.
20. Shale T, Stirk W, van Staden J. Screening of medicinal plants used in Lesotho for anti-bacterial and anti-inflammatory activity. *J Ethnopharmacol* 1999;**67**(3):347–54.
21. Siriwatanametanon N, Fiebich BL, Efferth T, Prieto JM, Heinrich M. Traditionally used Thai medicinal plants: in vitro anti-inflammatory, anticancer and antioxidant activities. *J Ethnopharmacol* 2010;**130**(2):196–207.
22. Hutchings A, Scott A, Lewis G, Cunningham A. *Zulu Medicinal PlantsNatal*. Pietermaritzburg: University Press; 1996.
23. Van Wyk B-E, Oudtshoorn Bv, Gericke N. *Medicinal Plants of South Africa*. Pretoria: Briza Publications; 1997.
24. Mohagheghzadeh A, Faridi P, Shams-Ardakani M, Ghasemi Y. Medicinal smokes. *J Ethnopharmacol* 2006;**108**(2):161–84.
25. Devi VK, Jain N, Valli KS. Importance of novel drug delivery systems in herbal medicines. *Pharmacogn Rev* 2010;**4**(7):27.
26. Solis C, Becerra J, Flores C, Robledo J, Silva M. Antibacterial and antifungal terpenes from *Pilgerodendron uviferum* (D. Don) Florin. *J Chil Chem Soc* 2004;**49**(2):157–61.
27. Togashi N, Inoue Y, Hamashima H, Takano A. Effects of two terpene alcohols on the antibacterial activity and the mode of action of farnesol against *Staphylococcus aureus*. *Molecules* 2008;**13**(12):3069–76.
28. Osawa K, Yasuda H, Maruyama T, Morita H, Takeya K, Itokawa H. Antibacterial trichorabdal diterpenes from *Rabdosia trichocarpa*. *Phytochemistry* 1994;**36**(5):1287–91.
29. Greay SJ, Hammer KA. Recent developments in the bioactivity of mono-and diterpenes: anticancer and antimicrobial activity. *Phytochem Rev* 2011;**14**:1–6.
30. Paiva L, Gurgel L, Silva R, Tomé A, Gramosa N, Silveira E, et al. Anti-inflammatory effect of kaurenoic acid, a diterpene from *Copaifera langsdorffii* on acetic acid-induced colitis in rats. *Vascul Pharmacol* 2002;**39**(6):303–7.
31. Knobloch K, Pauli A, Iberl B, Weigand H, Weis N. Antibacterial and antifungal properties of essential oil components. *J Essent Oil Res* 1989;**1**(3):119–28.
32. Trombetta D, Castelli F, Sarpietro MG, Venuti V, Cristani M, Daniele C, et al. Mechanisms of antibacterial action of three monoterpenes. *Antimicrob Agents Chemother* 2005;**49**(6):2474–8.
33. Juven B, Kanner J, Schved F, Weisslowicz H. Factors that interact with the antibacterial action of thyme essential oil and its active constituents. *J Appl Bacteriol* 1994;**76**(6):626–31.
34. Cristani M, D'Arrigo M, Mandalari G, Castelli F, Sarpietro MG, Micieli D, et al. Interaction of four monoterpenes contained in essential oils with model membranes: implications for their antibacterial activity. *J Agric Food Chem* 2007;**55**(15):6300–8.
35. Chen Z, He D, Deng J, Zhu J, Mao Q. Chemical composition and antibacterial activity of the essential oil from *Agathis dammara* (Lamb.) Rich fresh leaves. *Nat Prod Res* 2015;**29**:1–4.
36. Pereira MdC, Nelson DL, Stoianoff MAR, Santos SG, Nogueira LJ, Coutinho SC, et al. Chemical composition and antimicrobial activity of the essential oil from *Microlicia crenulata*. *J Essent Oil Bear Plan* 2015;**18**(1):18–28.
37. Suarez S, Hickey AJ. Drug properties affecting aerosol behavior. *Respir Care* 2000;**45**(6):652–66.
38. Neville A, Palmer J, Gaddie J, May C, Palmer K, Murchison L. Metabolic effects of salbutamol: comparison of aerosol and intravenous administration. *BMJ* 1977;**1**(6058):413–4.

39. Clark T. Effect of beclomethasone dipropionate delivered by aerosol in patients with asthma. *Lancet* 1972;**299**(7765):1361–4.

40. Federspil P, Wulkow R, Zimmermann T. Effects of standardized Myrtol in therapy of acute sinusitis–results of a double-blind, randomized multicenter study compared with placebo. *Laryngorhinootologie* 1997;**76**(1):23–7.

41. Dorow P, Weiss T, Felix R, Schmutzler H. Effect of a secretolytic and a combination of pinene, limonene and cineole on mucociliary clearance in patients with chronic obstructive pulmonary disease. *Arzneimittelforschung* 1987;**37**(12):1378–81.

42. Ferley J, Poutignat N, Zmirou D, Azzopardi Y, Balducci F. Prophylactic aromatherapy for supervening infections in patients with chronic bronchitis. Statistical evaluation conducted in clinics against a placebo. *Phytother Res* 1989;**3**(3):97–100.

43. Pei Rs, Zhou F, Ji Bp, Xu J. Evaluation of combined antibacterial effects of eugenol, cinnamaldehyde, thymol, and carvacrol against *E. coli* with an improved method. *J Food Sci* 2009;**74**(7):M379–83.

44. Dorman H, Deans S. Antimicrobial agents from plants: antibacterial activity of plant volatile oils. *J Appl Microbiol* 2000;**88**(2):308–16.

45. Ünlü M, Daferera D, Dönmez E, Polissiou M, Tepe B, Sökmen A. Compositions and the in vitro antimicrobial activities of the essential oils of *Achillea setacea* and *Achillea teretifolia* (Compositae). *J Ethnopharmacol* 2002;**83**(1):117–21.

46. Inouye S, Yamaguchi H, Takizawa T. Screening of the antibacterial effects of a variety of essential oils on respiratory tract pathogens, using a modified dilution assay method. *J Infect Chemother* 2001;**7**(4):251–4.

47. Marino M, Bersani C, Comi G. Impedance measurements to study the antimicrobial activity of essential oils from Lamiaceae and Compositae. *Int J Food Microbiol* 2001;**67**(3):187–95.

48. Inouye S, Takizawa T, Yamaguchi H. Antibacterial activity of essential oils and their major constituents against respiratory tract pathogens by gaseous contact. *J Antimicrob Chemother* 2001;**47**(5):565–73.

49. Preuss HG, Echard B, Enig M, Brook I, Elliott TB. Minimum inhibitory concentrations of herbal essential oils and monolaurin for Gram-positive and Gram-negative bacteria. *Mol Cell Biochem* 2005;**272**(1-2):29–34.

50. Elgayyar M, Draughon F, Golden D, Mount J. Antimicrobial activity of essential oils from plants against selected pathogenic and saprophytic micro-organisms. *J Food Prot* 2001;**64**(7):1019–24.

51. Sikkema J, De Bont J, Poolman B. Mechanisms of membrane toxicity of hydrocarbons. *Microbiol Rev* 1995;**59**(2):201–22.

52. Sikkema J, De Bont J, Poolman B. Interactions of cyclic hydrocarbons with biological membranes. *J Biol Chem* 1994;**269**(11):8022–8.

53. Bisignano G, Laganà MG, Trombetta D, Arena S, Nostro A, Uccella N, et al. In vitro antibacterial activity of some aliphatic aldehydes from *Olea europaea* L. *FEMS Microbiol Lett* 2001;**198**(1):9–13.

54. Trombetta D, Saija A, Bisignano G, Arena S, Caruso S, Mazzanti G, et al. Study on the mechanisms of the antibacterial action of some plant α, β-unsaturated aldehydes. *Lett Appl Microbio* 2002;**35**(4):285–90.

55. Horne D, Holm M, Oberg C, Chao S, Young DG. Antimicrobial effects of essential oils on *Streptococcus pneumoniae*. *J Essent Oil Res* 2001;**13**(5):387–92.

56. Pelczar M, Chan E, Krieg N. Control of micro-organisms, the control of micro-organisms by physical agents. *Microbiology* 1988;**469**:509.

57. Pauli A. Antimicrobial properties of essential oil constituents. *IJA* 2001;**11**(3):126–33.

58. Lee S, Buber M, Yang Q, Cerne R, Cortes R, Sprous D, et al. Thymol and related alkyl phenols activate the hTRPA1 channel. *Br J Pharmacol* 2008;**153**(8):1739–49.

59. Helander IM, Alakomi H-L, Latva-Kala K, Mattila-Sandholm T, Pol I, Smid EJ, et al. Characterization of the action of selected essential oil components on Gram-negative bacteria. *J Agric Food Chem* 1998;**46**(9):3590–4355.

60. Gill AO, Holley RA. Mechanisms of bactericidal action of cinnamaldehyde against *Listeria monocytogenes* and of eugenol against *L. monocytogenes* and *Lactobacillus sakei*. *Appl Environ Microbiol* 2004;**70**(10):5750–5.

61. Morris J, Khettry A, Seitz E. Antimicrobial activity of aroma chemicals and essential oils. *JAOCS* 1979;**56**(5):595–603.
62. Prashar A, Hili P, Veness RG, Evans CS. Antimicrobial action of palmarosa oil (*Cymbopogon martinii*) on *Saccharomyces cerevisiae*. *Phytochemistry* 2003;**63**(5):569–75.
63. Naigre R, Kalck P, Roques C, Roux I, Michel G. Comparison of antimicrobial properties of monoterpenes and their carbonylated products. *Planta Med* 1996;**62**(3):275–7.
64. Kurita N, Miyaji M, Kurane R, Takahara Y, Ichimura K. Antifungal activity and molecular orbital energies of aldehyde compounds from oils of higher plants. *Agric Biol Chem* 1979;**43**(11):2365–71.
65. Kubo A, Lunde CS, Kubo I. Antimicrobial activity of the olive oil flavor compounds. *J Agric Food Chem* 1995;**43**(6):1629–33.
66. Kubo I, Muroi H, Masaki H, Kubo A. Antibacterial activity of long-chain alcohols: the role of hydrophobic alkyl groups. *Bioorg Med Chem Lett* 1993;**3**(6):1305–8.
67. Inoue Y, Shiraishi A, Hada T, Hirose K, Hamashima H, Shimada J. The antibacterial effects of terpene alcohols on *Staphylococcus aureus* and their mode of action. *FEMS Microbiol Lett* 2004;**237**(2):325–31.
68. Togashi N, Hamashima H, Shiraishi A, Inoue Y, Takano A. Antibacterial activities against *Staphylococcus aureus* of terpene alcohols with aliphatic carbon chains. *J Essent Oil Res* 2010;**22**(3):263–9.
69. Togashi N, Shiraishi A, Nishizaka M, Matsuoka K, Endo K, Hamashima H, et al. Antibacterial activity of long-chain fatty alcohols against *Staphylococcus aureus*. *Molecules* 2007;**12**(2):139–48.
70. Brehm-Stecher BF, Johnson EA. Sensitization of *Staphylococcus aureus* and *Escherichia coli* to antibiotics by the sesquiterpenoids nerolidol, farnesol, bisabolol, and apritone. *Antimicrob Agents Chemother* 2003;**47**(10):3357–60.
71. Lis-Balchin M, Deans SG, Eaglesham E. Relationship between bioactivity and chemical composition of commercial essential oils. *Flavour Fragr J* 1998;**13**(2):98–104.
72. Davidson PM, Taylor TM. Chemical preservatives and natural antimicrobial compounds. 2007.
73. Burt S. Essential oils: their antibacterial properties and potential applications in foods—a review. *Int J Food Microbiol* 2004;**94**(3):223–53.
74. Gill A, Holley R. Disruption of Escherichia coli, Listeria monocytogenes and Lactobacillus sakei cellular membranes by plant oil aromatics. *Int J Food Microbiol* 2006;**108**(1):1–9.
75. Ultee A, Bennik M, Moezelaar R. The phenolic hydroxyl group of carvacrol is essential for action against the food-borne pathogen Bacillus cereus. *Appl Environ Microbiol* 2002;**68**(4):1561–8.
76. Das K, Tiwari R, Shrivastava D. Techniques for evaluation of medicinal plant products as antimicrobial agent: Current methods and future trends. *J Med Plants Res* 2010;**4**(2):104–11.
77. Lopez P, Sanchez C, Batlle R, Nerin C. Solid-and vapor phase antimicrobial activities of six essential oils: susceptibility of selected foodborne bacterial and fungal strains. *J Agric Food Chem* 2005;**53**(17):6939–46.
78. Hammer KA, Carson C, Riley T. Antimicrobial activity of essential oils and other plant extracts. *J Appl Microbiol* 1999;**86**(6):985–90.
79. Eloff J. A sensitive and quick microplate method to determine the minimal inhibitory concentration of plant extracts for bacteria. *Planta Med* 1998;**64**(8):711–3.
80. Cos P, Vlietinck AJ, Berghe DV, Maes L. Anti-infective potential of natural products: how to develop a stronger in vitro 'proof-of-concept'. *J Ethnopharmacol* 2006;**106**(3):290–302.
81. Bishop CD. Antiviral activity of the essential oil of *Melaleuca alternifolia* (Maiden amp; Betche) Cheel (tea tree) against Tobacco Mosaic Virus. *J Essent Oil Res* 1995;**7**(6):641–4.
82. Xia Q, Huang Z-G, Li S-P, Zhang P, Wang J, He L-N. The experiment study of the anti-virus effects of zedoary oil on influenza virus and respiratory syncytial virus. *Zhong Yao Tong Bao* 2004;**20**(3):357–8.
83. Saedisomeolia A, Wood LG, Garg ML, Gibson PG, Wark PA. Anti-inflammatory effects of long-chain n-3 PUFA in rhinovirus-infected cultured airway epithelial cells. *Br J Nutr* 2009;**101**(04):533–40.
84. Kiyohara H, Ichino C, Kawamura Y, Nagai T, Sato N, Yamada H. Patchouli alcohol: in vitro direct anti-influenza virus sesquiterpene in *Pogostemon cablin* Benth. *J Nat Med* 2012;**66**(1):55–61.

85. Wu Q-f, Wang W, Dai X-y, Wang Z-y, Shen Z-h, Ying H-z, et al. Chemical compositions and anti-influenza activities of essential oils from *Mosla dianthera. J Ethnopharmacol* 2012;**139**(2):668–71.

86. Zai-Chang Y, Bo-Chu W, Xiao-Sheng Y, Qiang W. Chemical composition of the volatile oil from *Cynanchum stauntonii* and its activities of anti-influenza virus. *Colloids Surf B Biointerfaces* 2005;**43**(3):198–202.

87. Wanga L, Jian S, Nan P, Zhong Y. Chemical composition of the essential oil of *Elephantopus scaber* from southern China. *Z Naturforsch C* 2004;**59**(5–6):327–9.

88. Loizzo MR, Saab AM, Tundis R, Statti GA, Menichini F, Lampronti I, et al. Phytochemical analysis and in vitro antiviral activities of the essential oils of seven Lebanon species. *Chem Biodivers* 2008;**5**(3):461–70.

89. Rodriguez-Gonzalez A. Broad spectrum anti-viral herbal composition. Google Patents; 2006.

90. Franklin LU. Respiratory infection prevention and treatment with terpene-containing compositions. Google Patents; 2003.

91. Green GM, Kass EH. The role of the alveolar macrophage in the clearance of bacteria from the lung. *J Exp Med* 1964;**119**(1):167–76.

92. Serhan CN, Savill J. Resolution of inflammation: the beginning programs the end. *Nat Immunol* 2005;**6**(12):1191–7.

93. Karrer H. Electron microscopic study of the phagocytosis process in lung. *J Biophys Biochem Cytol* 1960;**7**(2):357–66.

94. Sibille Y, Reynolds HY. Macrophages and polymorphonuclear neutrophils in lung defense and injury. *Am Rev Respir Dis* 1990;**141**(2):471–501.

95. Nadel JA. Role of neutrophil elastase in hypersecretion during COPD exacerbations, and proposed therapies. *CHEST Journal* 2000;**117**(5 Suppl. 2):386S–9S.

96. Garofalo R, Sabry M, Jamaluddin M, Yu RK, Casola A, Ogra PL, et al. Transcriptional activation of the interleukin-8 gene by respiratory syncytial virus infection in alveolar epithelial cells: nuclear translocation of the RelA transcription factor as a mechanism producing airway mucosal inflammation. *J Virol* 1996;**70**(12):8773–81.

97. Nakamura H, Yoshimura K, McElvaney NG, Crystal RG. Neutrophil elastase in respiratory epithelial lining fluid of individuals with cystic fibrosis induces interleukin-8 gene expression in a human bronchial epithelial cell line. *J Clin Invest* 1992;**89**(5):1478.

98. Barnes PJ, Belvisi MG, Rogers DF. Modulation of neurogenic inflammation: novel approaches to inflammatory disease. *Trends Pharmacol Sci* 1990;**11**(5):185–9.

99. Carvalho-Júnior Pd, Rodrigues R, Sawaya A, Marques M, Shimizu M. Chemical composition and antimicrobial activity of the essential oil of *Cordia verbenacea. J Ethnopharmacol* 2004;**95**:297–301.

100. Juergens U, Dethlefsen U, Steinkamp G, Gillissen A, Repges R, Vetter H. Anti-inflammatory activity of 1.8-cineol (eucalyptol) in bronchial asthma: a double-blind placebo-controlled trial. *Respir Med* 2003;**97**(3):250–6.

101. Baylac S, Racine P. Inhibition of 5-lipoxygenase by essential oils and other natural fragrant extracts. *IJA* 2003;**13**(2):138–42.

102. Salminen A, Lehtonen M, Suuronen T, Kaarniranta K, Huuskonen J. Terpenoids: natural inhibitors of NF-κB signaling with anti-inflammatory and anticancer potential. *Cell Mol Life Sci* 2008;**65**(19):2979–99.

103. Sadlon AE, Lamson DW. Immune-modifying and antimicrobial effects of Eucalyptus oil and simple inhalation devices. *Altern Med Rev* 2010;**15**(1):33–47.

104. Gavett SH, Chen X, Finkelman F, Wills-Karp M. Depletion of murine CD4+ T lymphocytes prevents antigen-induced airway hyperreactivity and pulmonary eosinophilia. *Am J Respir Cell Mol Biol* 1994;**10**(6):587–93.

105. Konstan MW, Vargo KM, Davis PB. Ibuprofen attenuates the inflammatory response to Pseudomonas aeruginosa in a rat model of chronic pulmonary infection. *Am Rev Respir Dis* 1990;**141**:186–92.

106. Treanor J, Falsey A. Respiratory viral infections in the elderly. *Antiviral Res* 1999;**44**(2):79–102.

CURRENT THERAPEUTICS AND PROPHYLACTIC APPROACHES TO TREAT PNEUMONIA

17

A. Krishnamurthy, E. Palombo

Swinburne University of Technology, Department of Chemistry and Biotechnology, Faculty of Science, Engineering and Technology, Hawthorn, Victoria, Melbourne, Australia

1 INTRODUCTION

Infections of the lower respiratory tract, such as tuberculosis, pneumonia, exacerbations of chronic obstructive pulmonary disease (COPD), and influenza are the leading infectious disease health problems that result in significant mortality and morbidity in humans. Lower respiratory tract infections (LRTI) are among the most frequent causes of mortality, as stated by the World Health Organisation, with more than three million deaths reported in children due to pneumonia.[1,2] These deaths are caused by both bacterial and viral etiological agents including *Streptococcus pneumoniae*, *Haemophilus influenzae*, *Mycoplasma pneumoniae*, Influenza A or B virus, parainfluenza virus, rhinovirus, respiratory syncytial virus (RSV) and adenovirus.[3,4] Prior to the introduction of the relevant vaccines, bacteria such as *Streptococcus pneumoniae*, *Haemophilus influenzae*, *Staphylococcus aureus*, and *Klebsiella pneumoniae* were identified as major causative agents of bacterial pneumonia.[5]

1.1 MICROBIOLOGY AND PATHOGENESIS OF *STREPTOCOCCUS PNEUMONIAE*—THE MAJOR CAUSE OF PNEUMONIA

S. pneumoniae, or pneumococcus, is a Gram-positive, nonmotile, facultative anaerobic coccus belonging to the family Streptococceae, and is classified based upon its haemolytic capability and the presence of carbohydrate antigens located in the cell wall using the Lancefield system. The most commonly recognized serotypes of *S. pneumoniae* found in children are serotypes 6, 14, 19, and 23.[6]

The various pneumococcal virulence factors associated with the pathogenesis of *S. pneumoniae* include and are not limited to; pneumolysin, autolysin, capsule and surface adhesins, enzymes and surface proteins such as neuraminidase, SpxB (pyruvate oxidase), and choline-binding proteins.[7] The choline-binding proteins are known to bind to techoic or lipotechoic acid, and function as an anchoring device for most Gram-positive bacteria, including *S. pneumoniae*. It is believed that there are approximately 10–15 different choline-binding proteins including PspA, PspC, and LytA that are encoded by *S. pneumoniae*[8]. After adherence of pneumococcus to a cell surface, further tissue invasion is facilitated by enzymes such as hyaluronidase and neuraminidase. Hyaluronic acid is one of the most abundant glycans in the extracellular matrix of connective tissues,[9] while neuraminidase comprising of NanA and NanB, of which NanA is shown to enhance intracellular survival and replication of the bacteria in the lung.[10] Being a nasopharyngeal commensal, *S. pneumoniae* can gain access to the eustachian tube and lungs via the nasopharynx, and can cause middle ear infection and pneumonia, respectively.[11] During this process, there is a massive influx of neutrophils driven by proinflammatory cytokines, such as IL-6, IL-1 and TNF-α. The progression to pneumonia is associated with a viral infection, which can enhance the adherence of *S. pneumoniae* to the respiratory epithelia and is recognized due to the interactions between the phosphorylcholine of the pneumococcus cell wall and platelet-activating factor receptors that are believed to be upregulated upon cytokine stimulation.[12] In addition, various cell wall degradation products such as peptidoglycan, techoic acid, and toxins like pneumolysin are also released following autolysis which can induce inflammatory responses in the host. The various actions of pneumolysin, which include induction of cytokines like TNFα and IL-1β, disruption of the epithelial cell integrity, decreased bactericidal activity, inhibition of neutrophil migration, and inhibition of lymphocyte proliferation and antibody synthesis, highlight its important role in the pathogenesis of *S. pneumoniae*.[13]

1.2 BIOLOGY OF PNEUMOCOCCAL PNEUMONIA

The mucosal epithelium of the nasopharynx is a well-recognized primary site of bacterial colonisation including the opaque and transparent phenotypes of the pneumococcus.[14] The pneumococcus can traverse down to the lung upon aspiration and start adhering to the alveolar type II cells to initiate bacterial infection.[15] The progress to pneumonia can occur more rapidly if there is a preexisting respiratory viral infection or increased bacterial adherence facilitated by viruses or cytokines.[16] The different stages of pneumococcal pneumonia are well known. The first stage is characterized by a bulge or engorgement due to the bacterial presence and serous exudate in the alveoli which provides nutrients to the bacteria and facilitates further infection in the lung.[17] The next stage is the intense inflammatory reaction involving leakage of erythrocytes into the alveoli (red hepatisation), followed by the migration of leukocytes into the consolidated area (gray hepatisation), and surface phagocytosis by the leukocytes.[18] Due to the intact immune system of the host, normally the type-specific antibodies and the polymorphonuclear leukocytes phagocytise the pneumococci and the lung returns to its normal state. However, in patients with a compromised immune system having certain complement deficiencies or hypogammaglobulinemia (absence of type-specific antibodies), bacteremia can occur and the pneumococcus with its virulence factors (mostly cell wall components such as peptidoglycan, techoic acid, and proteins such as pneumolysin) induce inflammation causing subsequent tissue damage.[19]

2 CHILDHOOD PNEUMONIA

Since the early 1980s, the LRTI caused by *S. pneumoniae* in certain developing countries, such as Ghana and South Africa, have been reported at >60–90% of all cases in children less than 5 years of age, and 100 per 1000 population of adults per year, and has remained at equivalent rates even a decade later.[20–22] The incidence of pneumonia in children less than 5 years of age was higher in regions such as South-East Asia, Africa and Western Pacific countries in comparison with those in the developed countries like the Americas and Europe.[22] More recently, the incidence of childhood pneumonia has been estimated to be >120 million globally, of which ~12% progressed to severe disease with 1%

mortality in children <2 years of age.[23] The risk factors for childhood pneumonia, especially in the developing countries, include nutrition deficiencies, lack of breastfeeding, indoor air pollution due to passive smoking, HIV infection, and substandard housing and living conditions.[24]

The etiological agents causing pneumonia include both bacteria and respiratory viruses. Certain vaccine studies have indicated the predominant bacterial agent causing pneumonia to be *S. pneumoniae* resulting in almost 18–25% severe cases and 30–35% mortality, followed by *H. influenzae* accounting for 4% of severe cases and 16% of deaths. Influenza virus remains the dominant viral etiological agent responsible for 7% of the severe cases and 11% of deaths.[23,25] *S. pneumoniae* is also well-recognized to cause community-acquired pneumonia (CAP) in children with lower fatality rates of 1.5%.[26] In addition to these etiological agents, bacteria such as *Staphylococcus aureus*, *Klebsiella pneumoniae*, and respiratory viruses such as RSV, rhinovirus, human metapneumovirus, human bocavirus and para-influenza viruses are the other commonly identified agents that contribute to the burden of childhood pneumonia.[25,27]

3 ADULT PNEUMONIA AND COMMUNITY-ACQUIRED PNEUMONIA

Community-acquired pneumonia (CAP) is an increasing health problem and the third most common reason for hospitalization for adults, especially the elderly aged >65 years. The prevalence of CAP has been reported as 18–20 cases per 1000 population with an increase from 9% in 65–74 year olds to 17% in 75–84 year olds, and 30% in >85 year olds.[28,29] Several predisposing factors such as impaired immunity and lung function, dysfunctional nasal mucociliary clearance, lung and heart diseases, smoking have been identified as independent predictors for CAP in adults and the elderly.[24]

Certain studies have reported *S. pneumoniae*, *Legionella* species, *H. influenzae*, *Moraxella catarrhalis*, and *S. aureus* as the predominant pathogens in CAP.[30] Although the role of respiratory viruses has been well-recognized in CAP in children and infants, it is not well understood in adults and the elderly. It is still unclear whether a virus by itself can cause pneumonia or whether the virus can act in conjunction with other respiratory pathogens. One study has reported that respiratory viruses such as influenza virus, RSV, adenovirus, and rhinovirus were commonly isolated as part of a co-infection, especially with *S. pneumoniae*.[31] Thus, viral agents in adults with CAP most often seem to be part of a mixed infection, usually with *S. pneumoniae* as the co-pathogen.

4 VACCINATION

The World Health Organisation (WHO) recommends routine childhood immunization programs that include vaccines that offer protection from various respiratory disease such as pneumonia, influenza (flu), measles, and pertussis. The *Haemophilus influenzae* type b (Hib) vaccine and the pneumococcal conjugate vaccines are increasingly available in both developed as well as developing countries, especially the 7- and 13-valent pneumococcal conjugate vaccines which have shown effectiveness in reducing the incidence and severity of pneumonia and other lower respiratory infections in children. Currently, there are three vaccines in the childhood routine immunization schedule; measles, Hib, and the pneumococcal conjugate vaccine that is well-recognized to reduce childhood mortality from and related to pneumonia.

4.1 HIB VACCINE

Vaccination remains the primary preventative strategy for pneumonia, including CAP in the elderly. The Hib vaccine has a proven efficiency of >90% against invasive meningitis and bacteremia and noninvasive pneumonia caused by *H. influenzae* type b.[32,33] This impressive efficiency has resulted in the introduction and addition of the Hib vaccine worldwide into National Immunization Programs, and has resulted in a significant reduction in the vaccination gap between developed and developing countries.[34] A recent review from several randomized controlled trials (from 1970s to 2008) from different developing countries indicated a significant reduction of severe pneumonia by 6%, pneumonia-associated mortality by 7%, and reduction of radiological confirmed cases of pneumonia by 18%.[35] Based on the preventative approach of pneumonia and pneumonia-related mortality with effective vaccination, a certain modeling-based study has estimated that if implemented at present annual rates of increase in developing countries, the vaccine could save up to 51% of pneumonia deaths by 2025 at a cost saving of US$3.8 billion.[36] High coverage of the Hib vaccine immunization in children less than 5 years of age could reduce childhood pneumonia resulting in decreased incidence of severe pneumonia.

4.2 PNEUMOCOCCAL VACCINES

The 23-valent polysaccharide vaccine (PPV23) and the 13-valent protein-conjugated polysaccharide vaccine (PCV13) are the two vaccines that offer protection against pneumococcal disease, and have replaced the 7-valent conjugate vaccine (PCV7). As the polysaccharide vaccine (PPV) is T-cell independent, it does not boost immunological memory and the immunity offered may not last for a long time. For this reason, this vaccine is not offered to infants aged <2 years of age, but is provided to children aged >2 years and to elderly people who are at risk (>50 years of age) for developing pneumonia. In contrast, the conjugate vaccines stimulate a T-cell dependent response and are more effective in infants and children <2 years of age.[37]

The different pneumococcal vaccines, serotypes covered, and the conjugate protein used are mentioned in Table 17.1 PCV7 and PCV10 are offered for children aged from 6 weeks to 5 years of age, whereas PCV13 is given to children aged between 6 weeks and 17 years, and to adults aged >50 years of age. Since the introduction of the PCV7 vaccine in 2000, its efficacy against invasive meningitis, pneumonia, and otitis media is well documented.[38–40] The subsequent vaccines, PCV10 and PCV13,

Table 17.1 Current Pneumococcal Conjugate Vaccines			
Pneumococcal Vaccine	**Serotypes Covered**	**Conjugate Protein Used**	**Trade Name**
PCV7	4, 6B, 9V, 14, 18C, 19F, and 23F	Mutant diphtheria toxoid	Prev(e)nar (Pfizer)
PCV10	4, 6B, 9V, 14, 18C, 19F, 23F, 1, 5, and 7F	Protein D from nontypeable *H. influenzae*, tetanus toxoid, and diphtheria toxoid	Synflorix (GlaxoSmithKline)
PCV13	4, 6B, 9V, 14, 18C, 19F, 23F, 1, 5, 7F, 3, 6A, 19A	Mutant diphtheria toxoid	Prev(e)nar-13 (Pfizer)

The table shows the currently available pneumococcal vaccines and the conjugate protein used in its making, and the coverage of various serotypes of S. pneumoniae *by each vaccine.*

have also demonstrated comparable immunogenicity to PCV7 in several clinical trials.[41–43] Although PPV23 covers 23 serotypes of *S. pneumoniae*, and is recommended for adults aged >65 years of age, its effectiveness in reducing invasive pneumococcal disease remains uncertain.[44,45] As PCV13 has equal or greater immunogenicity than PPV23, and has greater immunological memory, the use of PCV13 is now recommended for adults in addition to PPV23, particularly the elderly and high risk individuals. In addition to this, newer PCV's have been shown to reduce the number of healthy carriers of the pathogen in the community because of "herd immunity" where unvaccinated people are protected from the pathogen. Since the introduction of PCV's as a part of "herd immunity", the incidence of invasive pneumococcal diseases was shown to decline by almost 70% among vaccinated children <2 years of age, and by 32% in adults aged 20-39 years, and by 18% in the elderly aged >65 years of age, who were not previously vaccinated.[46]

5 CURRENT ANTIINFECTIVE TREATMENTS AGAINST BACTERIAL PATHOGENS

The classification of antibiotics is based on the cellular component that they affect, as well as on whether they can induce cell death (bactericidal) or inhibit cell growth (bacteriostatic). Antibiotic-mediated cell death is a complex process involving physical interaction between a drug molecule and its bacterial-specific target, and/or modulation of the affected bacterium at the biochemical, molecular, and ultrastructural levels.[47] Fig. 17.1 summarizes the different drug targets and mechanism of actions of various antimicrobials such as; inhibition of cell wall synthesis, inhibition of protein synthesis, injury to cytoplasmic membrane, and inhibition of nucleic acid synthesis and replication. DNA synthesis and cell division are well-recognized processes that involve modulation of chromosomal supercoiling through topoisomerase-catalyzed strand breakage and rejoining reactions. Antimicrobials such as quinolone

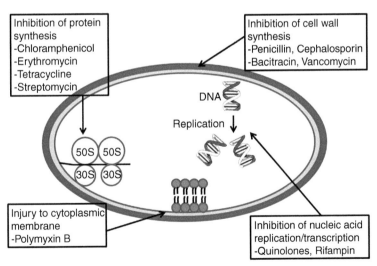

FIGURE 17.1 Summary of Target and Mechanisms of Actions of Various Antimicrobials

This figure summarizes the various targets and mechanism of actions of different antimicrobials along with relevant examples of antibiotics.

target enzymes like DNA gyrase (topoisomerase II) and topoisomerase IV (topoIV) that are required for bacterial DNA synthesis and replication, and prevent DNA strand rejoining.[48] Peptidoglycan, a co-valently cross-linked polymer matrix composed of peptide-linked β-(1–4)-N-acetyl hexosamine, is the main component of bacterial cell walls that contributes towards the structural integrity of the bacterial cell. The peptidoglycan layer is maintained through the activity of transglycosylase and transpeptidase enzymes, which add disaccharide pentapeptides to extend the glycan strands of existing peptidoglycan molecules and cross-link adjacent peptide strands of immature peptidoglycan units, respectively.[49] β-lactams and glycopeptides are among the classes of antibiotics that interfere with cell wall biosynthesis resulting in changes to bacterial cell shape and size and induction of cellular stress responses that leads to bacterial cell lysis.[50] The process of protein synthesis via mRNA translation involves the ribo-some that is composed of two major components, the 50S and 30S subunits. Drugs that inhibit protein synthesis are divided into two subclasses: the 50S inhibitors and 30S inhibitors. The 50S ribosome inhibitors include the macrolide, lincosamide, streptogramin, amphenicol, and oxazolidinone classes of antibiotics.[51] The 30S inhibitors include the tetracycline and aminocyclitol families of antibiotics.[52]

5.1 CURRENT ANTIINFECTIVE ANTIMICROBIALS

The mortality caused due to pneumonia can be avoided through cost-effective and life-saving treat-ment from antibiotics for bacterial pneumonia, thereby significantly increasing the patient's chances of survival. The pneumonia management strategy with the use of appropriate antibiotics and supportive care including oxygen systems remains an effective cornerstone in the treatment and management of children suffering from pneumonia. The WHO Integrated Management of Childhood Illness program has consistently reported a reduction of childhood mortality rates by approximately 20%, while certain community based management strategies have reported a decrease in 70% mortality due to the usage of oral antibiotics such as amoxicillin.[53–55] The four types of antibiotics recommended for children <5 years of age for the treatment of pneumonia are; cotrimoxazole, amoxicillin, cephalosporins, and macrolides, with oral amoxicillin (40 mg/kg/dose) used for 3 days (nonsevere pneumonia) and 5 days for children with severe pneumonia.[56] During severe pneumonia, the first line of treatment is often parenteral ampicillin (penicillin) and gentamicin, followed by ceftriaxone if the first line of treatment is not effective.[56] Various randomized controlled studies from the Cochrane database of Systemic Re-views have shown a multitude of available treatments for pneumonia in children with (1) cefpodoxime proving to be more effective than amoxicillin, (2) amoxicillin more effective than chloramphenicol, (3) amoxicillin being an effective alternative to cotrimoxazole for CAP patients, (4) coamoxyclavu-lanic acid and cefpodoxime as alternative second-line drugs of choice, and (5) penicillin/ampicillin plus gentamicin more effective than chloramphenicol for children hospitalized with severe CAP.[57] A 3-year pediatric study of the susceptibility of 208 *S. pneumoniae* isolates, including serotype 19A, us-ing antibiotics such as second- and third-generation cephalosporins showed significant efficacy against 60–70% of the isolates, with clindamycin susceptibility of 60–85%, levofloxacin 95%, and ceftriaxone >95%.[58] The American Thoracic Society and the European Respiratory Society recommend that hos-pitalized patients with CAP are preferably treated with a respiratory fluoroquinolone or combination therapy with a β-lactam and a macrolide.[59] The success rates of incorporating the fluoroquininolone or combination with a macrolide based on the clinical, bacteriological, or radiological examinations ranged from 87–96%.[60] Vancomycin or clindamycin (based on local susceptibility data) should also be provided in addition to β-lactam therapy if clinical, laboratory, or radiological characteristics are consistent with infection caused by *S. aureus*.[61] Although nonsevere and severe CAP have been managed by many antimicrobials as a result of various studies from developing countries that compared different types of antibiotics, there is need for more studies and larger clinical trials for better management of pneumonia in developed countries. Another major health concern is the continual rise in antibiotic resistance with approximately 30% of the isolates being resistant to macrolides including erythromycin, azithromycin, and clarithromycin.[62]

The introduction and inclusion of the Hib vaccine over the past 25 years has resulted in almost complete elimination of *H. influenzae* in children, therefore it is not considered as a pathogen in CAP. Nontypeable *H. influenzae* is also not considered as a pathogen in pediatric pneumonia unless detected in lung disease or COPD. When detected as a true pathogen in CAP, oral amoxicillin is considered effective against β-lactamase negative strains, and for β-lactamase producing strains, amoxicillin-clavulanate, cefuroxime, cefdinir, cefixime, cefpodoxime are all considered effective therapies, while children allergic to oral β-lactam agents are only given fluoroquinolones.[61]

Although an infrequent cause of CAP, group A *Streptococcus* may cause severe necrotizing pneumonia. Penicillin G at the dosage of 100,000–200,000 U/kg/day in 4–5 divided doses is used to treat patients suffering from CAP due to group A *Streptococcus*. As macrolide resistance is greater in streptococcal infections, along with lower tolerability by tissues, erythromycin and other macrolides are not administered.[63]

S. aureus capable of causing pneumonia are usually methicillin-sensitive and are treated with either a β-lactamase stable penicillin (oxacillin or nafcillin) or a first-generation cephalosporin, like cefazolin. Community-associated methicillin-resistant *S. aureus* (MRSA) represents >50–70% of the clinical isolates in some region of the United States,[64] but are shown to be susceptible to vancomycin, clindamycin, and linezolid. However, children intolerant to vancomycin and clindamycin could be treated with linezolid. However, severe adverse effects, including suppression of platelets and neutrophils, nerve injury, mean that this drug should be used with caution.

In situations where *Mycoplasma pneumoniae* and *Chlamydophila pneumoniae* are of significant consideration upon diagnostic evaluation, empiric combination therapy with a macrolide and a β-lactam antibiotic is considered.

6 CURRENT ANTIINFECTIVE TREATMENTS AGAINST VIRAL PATHOGENS

Children with moderate to severe CAP consistent with influenza virus infection during widespread local circulation of influenza viruses should be administered with influenza antiviral therapy. The susceptible strains of influenza A virus are commonly treated with adamantanes and neuraminidase inhibitors. As the occurrence of genetic variations is highly substantial among influenza strains, resistance to either class of antiviral agents may develop quickly. However, the dosages of antiviral agents currently recommended for seasonal influenza are developed for fully susceptible strains and evaluated in clinical trials mandating the requirement of treatment within 3 days of the onset of symptoms.[65] Early antiviral treatment has been shown to provide maximal benefit, and treatment should not be delayed until confirmation of positive influenza test results. Negative results of influenza diagnostic tests, especially the rapid antigen tests, do not conclusively exclude influenza disease. Therefore, treatment after 48 h of symptomatic infection may still provide some clinical benefit to those with more severe disease.[61]

The efficacy of ribavirin for the treatment of RSV CAP in infants is debatable, as certain in vitro studies have shown activity of ribavirin against RSV, but its usage for RSV infection is not routinely recommended in the management of lower respiratory tract disease because of the high cost, aerosol administration, and possible toxic effects among healthcare providers. Palivizumab (Synagis), a humanised murine monoclonal antibody is another effective prophylaxis for RSV infection that is administered intramuscularly.[66]

Although parainfluenza virus, adenovirus, metapneumovirus, rhinovirus, coronavirus, and bocavirus are associated with pediatric CAP, there are no prospective, controlled studies for antiviral therapy against these viruses.

7 ANTIBIOTIC RESISTANCE AND ITS IMPACT

Since the introduction of penicillin in 1950s, it has been the first choice for treating pneumococcal pneumonia. During the early 1970s, infants and children were successfully treated with amoxicillin (40–45 mg/kg/day divided into 3 equal doses) because of the susceptible nature of the strains at that time. Resistance to the commonly used antibiotics poses a major problem and concern for health practitioners while choosing an empirical therapy against bacterial pneumonia, and there are large geographical variations indicating different resistance patterns. In the 1990s with the widespread pneumococcal resistance to penicillin, the dosage was increased to ~90 mg/kg/day given twice daily for treating children with acute otitis media.[67] A recent review has highlighted advances in the understanding of the various mechanisms by which bacteria acquire resistance to antibiotics, how they prevent access to different drug targets, and modulate or inactivate antibiotics.[68]

The introduction of pneumococcal conjugate vaccine and the changes in antimicrobial usage have both significantly altered the resistance patterns of *S. pneumoniae*. The decreased degree of penicillin resistance further prompted a decrease in amoxicillin dosage compared to that administered in the prevaccine era.[69] Over the last decade, a certain multicentre clinical trial study has reported a significant decrease in the susceptibility rates of the commonly used antibiotics such as amoxicillin/clavulante, penicillin, and ceftriaxone from 93.8% to 82.7%, 94.7% to 84.1%, and 97.4% to 87.5%, respectively.[70] The susceptibility rates of macrolides such as erythromycin and clindamycin were also reported to be decreased from 82.2% to 60.8% and from 96.2% to 79.1%, respectively. Recently, increasing resistance against macrolides has been reported in several European countries, including the United Kingdom. A 3-year surveillance study involving 1545 clinical isolates reported around 26% and 20% increase in the rates of ampicillin and trimethoprim/sulfamethoxazole resistance against *H. influenzae*, respectively, while antibiotics such as ceftaroline, ceftriaxone, amoxicillin/clavulante, and levofloxacin showed 99–100% susceptibility.[71] This study also showed increasing resistance of penicillin (96.4%) towards *M. catarrhalis*, because of the prevalence of β-lactamase that is known to reduce the susceptibility to penicillins. The resistance to macrolides against *M. pneumoniae* in children and adults with CAP has been increasingly emerging in countries like Japan, France, Denmark, United States, and China, with rates as high as 40% in Japan, 85% in China, and 3–10% in the Europe and the United States.[72,73] Community-acquired MRSA, although primarily associated with skin and soft tissue infections, are now being recognized to cause invasive infections including CAP, with almost 50% mortality rates reported in the United States and Europe.[74,75]

There are certain ways by which resistance to antimicrobials can be minimized, such as: limiting the exposure to any antibiotic, whenever possible; limiting the spectrum of usage of antimicrobials to that specifically required to treat the identified pathogen; using the proper dosage of antimicrobial to achieve a minimal effective concentration at the site of infection; treatment for the shortest effective duration that will minimize the exposure of both pathogens and normal microbiota to antimicrobials and further minimize the selection for resistance.[61]

8 ADVANCES IN ANTIBIOTIC TREATMENT FOR PNEUMONIA

The increasing incidence of antimicrobial resistance remains one of greatest challenges against emerging bacterial infections and has resulted in some bacteria being essentially untreatable with the current available treatment options. As a result, newer antimicrobials with novel modes of action against multidrug resistant strains are being developed. A recent review has highlighted how combinations of drugs can offer synergistic and antagonistic drug interactions, and how these drug interactions can provide opportunities for discovery of newer drugs.[76] In recent years, the availability of new antimicrobials for human consumption has been lower than in the recent past, with no new classes of antimicrobials developed since the introduction of nalidixic acid (1962) and linezolid (2000). The availability of antimicrobials in recent years has mostly been the result of modification of existing molecules. One of the reasons for this is that the development of any new antimicrobial agent is very expensive and time consuming, with research and development of infective drugs taking around 15–20 years, and costing around US$1000 million, with further additional costs for bringing the new drug into the market.[77] There is a strong need for newer unexploited targets and strategies for the next generation of antimicrobial drugs against drug resistant and emerging pathogens. Some of the new antimicrobial agents that are in the clinical development stage are listed in Table 17.2.

Table 17.2 Newer Antimicrobial Agents Against Bacterial Pneumonia

Drug	Class	Year of Approval/Trial	Activity Against
Daptomycin	Lipopeptide	2003	Gram+ve bacteria
Telithromycin	Ketolide	2004	Gram+ve and −ve bacteria
Ceftaroline	Cephalosporin	2010	Gram+ve bacteria
Fidaxomicin	Macrocyclic	2011	Gram+ve bacteria
Retapamulin	Pleuromutilin	2007	Gram+ve bacteria
Under Research and Development			
Torezolid	Oxazolidinones	Phase II	Gram+ve bacteria
Cethromycin	Ketolides	Phase III	Gram+ve bacteria
Oritavancin	Glycopeptides	Phase III	Gram+ve bacteria
Still Needing FDA Approval			
Ceftobiprole	Cephalosporin		Gram+ve bacteria
Iclaprim	Dihydrofolatereductase		Gram+ve bacteria

The table shows the availability of the newer antimicrobial agents, future promising antiinfective antimicrobials, and those awaiting FDA approval for treating bacterial pneumonia.

Some of the new antibiotics that have shown promising results in the treatment of pneumonia and CAP are as follows:

Ceftaroline: a fifth generation cephalosporin known to bind to penicillin-binding proteins and preventing synthesis of bacterial cell walls. It is a novel broad-spectrum antibiotic effective against MRSA, penicillin and cephalosporin resistant *S. pneumoniae*, vancomycin-intermediate *S. aureus* (VISA), and vancomycin-resistant *S. aureus* (VRSA).[78] It is also active against many Gram-negative pathogens but inactive against extended-spectrum β-lactamase (ESBL) producing bacteria. It has been approved for the treatment of CAP. Different randomized, double-blind, multicentre trials have demonstrated the efficacy (>82% clinical cure) and safety of ceftaroline (intravenous, 600 mg twice daily) for the treatment of CAP.[79]

Ceftobiprole: another newer cephalosporin that has a broad-spectrum activity against MRSA, penicillin-resistant *S. pneumoniae*, *P. aeruginosa* and enterococci.[80] A randomized trial consisting of 706 hospitalized adults with severe CAP who were administered ceftobiprole (intravenous, 500 mg over 120 min every 8 h) showed no significant differences between the treatment groups but found adverse events including nausea and vomiting in 36% of the patients.[81]

Telavancin: a semi-synthetic lipoglycopeptide derivative of vancomycin known to disrupt peptidoglycan synthesis and alter cell membrane function. It has been in use for treating complicated skin infections caused by *S. aureus*, and hospital-acquired bacterial pneumonia, including ventilator-associated bacterial pneumonia caused by susceptible isolates of *S. aureus*.[82]

Telithromycin: the first ketolide to enter clinical use for the treatment of CAP, chronic bronchitis and acute sinusitis. Telithromycin is a protein synthesis inhibitor blocking the progression of the growing polypeptide chain by binding to the 50S subunit of the bacterial ribosome. It exhibits 10 times higher affinity to the subunit 50S subunit than erythromycin. In addition, telithromycin strongly binds simultaneously to two domains of 23S subunit of the 50S ribosomal subunit; older macrolides bind to only to one domain and weakly to the second domain. An in vitro study showed activity of telithromycin against *S. pneumoniae* and, compared with clarithromycin and azithromycin, was found to maintain its activity against macrolide-resistant strains of *S. pneumoniae* and *S. pyogenes*.[83] It is formulated as 400 mg tablet for oral administration with good absorption and bioavailability.[84] However, the FDA withdrew its approval in the treatment of CAP in 2007 due to its safety concerns involving hepatotoxicity, myasthenia gravis exacerbation, and visual disturbances.

Cethromycin: a 3-keto,11,12carbamate derivative of erythromycin A with an O-6 linked aromatic ring. It binds strongly to the 50S ribosomal subunit and inhibits bacterial protein synthesis.[85] Cethromycin displays in vitro activity against streptococci, including strains of *S. pneumoniae* that are resistant to penicillins and macrolides.[86] Its activity was greater than telithromycin against macrolide-resistant streptococci and is more potent than macrolides and fluoroquinolones against penicillin-resistant streptococci. It also displays comparable in vitro activity to azithromycin against respiratory Gram-negative organisms including β-lactamase-producing *H. influenzae* and *M. catarrhalis*. It was shown to be more potent than erythromycin and clarithromycin but less potent than fluoroquinolones against β-lactamase-producing *H. influenzae*.[86,87] It showed similar potency against β-lactamase-producing *M. catarrhalis*.

Solithromycin: a new macrolide, and the first fluoroketolide in clinical development, with proven activity against macrolide-resistant bacteria. Solithromycin is being developed in both intravenous and oral formulations for the treatment of CAP, which should allow both oral therapy and i.v.-to-oral step-down therapy in appropriate patients. A recent multicentre, double-blind, randomized phase II study consisting of 132 patients with moderate to severe CAP administered with oral solithromycin (800-mg loading dose and 400 mg maintenance dose/5 days) showed efficacy comparable to that of levofloxacin in the treatment of CAP, with a favorable safety and tolerability profile.[88]

Nemonoxacin: a novel nonfluorinated quinolone with proven in vitro and in vivo activity against CAP pathogens including multidrug resistant *S. pneumoniae*. A randomized multicentre trial consisting of 265 CAP patients treated with an oral administration of nemonoxacin (750 and 500 mg/7 days) showed a remarkable 80–85% clinical and bacteriological success rate, which was comparable tolevofloxacin therapy.[89] A recent comprehensive review has well documented all the data available on the pharmacodynamics, the pharmacokinetics, and the clinical treatment studies of this antimicrobial agent.[90]

Zabofloxacin: is being developed as a new fluoroquinolone antibiotic that is a potent and selective inhibitor of the essential bacterial type II topoisomerases and topoisomerase IV and is indicated for community-acquired respiratory infections due to Gram-positive bacteria. Two dosing regimens of zabofloxacin (zabofloxacin hydrochloride 400 mg capsule andzabofloxacin aspartate 488 mg tablet) were well-tolerated with no adverse effects.[91]

JNJ-Q2 and KPI-10: two novel fluoroquinolones that are being developed for the treatment of bacterial pathogens responsible for respiratory infections including CAP, and other skin infections. Both agents have demonstrated increased potency when compared with the marketed fluoroquinolones, thus encouraging further clinical development.[92,93]

BC-3781: a recent semisynthetic pleuromutilin antibiotic with excellent antibacterial activity against skin pathogens such as *S. aureus*, β-haemolytic streptococci, viridans streptococci, and *Enterococcus faecium* as well as against respiratory pathogens. Its activity against respiratory pathogens has also been confirmed in various murine models of infection using *S. pneumoniae*, *H. influenzae*, *S. aureus*, and MRSA (nosocomial and community-associated), with better drug penetration, strongly supporting its potential use in the treatment of bacterial respiratory tract infections.[94]

9 NEWER TARGETS FOR THE NEXT GENERATION ANTIMICROBIALS FOR COMBATING DRUG RESISTANCE

Although there are a wide variety of clinically efficacious antibiotics in use today, the development of bacterial resistance has rendered them less effective, with most being bacteriostatic, and acting by either protein or cell wall synthesis inhibition. Further research is needed in the design of novel antibacterial agents with new targets.

9.1 TARGETING BACTERIAL PROTEINS

One approach could be to design antibiotics that can be used against novel drug targets such as the bacterial enzymes β-ketoacyl-acyl-carrier-protein synthase I/II which are required for fatty acid biosynthesis. Platensimycin is one such drug undergoing preclinical trials and is known to block these enzymes that are involved in the biosynthesis of essential fatty acids by Gram-positive bacteria.[95] It has potent antibacterial activity against Gram-positive bacteria including multidrug resistant staphylococci and enterococci.

9.2 COMBINING β-LACTAMASE ENZYME WITH β-LACTAM ANTIBACTERIAL DRUGS

Another approach worth investigating could be to combine β-lactam antibiotics with naturally occurring β-lactamase enzymes in the gastrointestinal microbiota. These enzymes are shown to hydrolyse various antibiotics including penicillin, ampicillin, and piperacillin. P1A protein (29 kDa) is one such example of having both structural and functional similarities to the β-lactamase enzyme. The emergence of resistant microbes can be significantly reduced by taking advantage of combining this naturally occurring hydrolysis of the antibacterial drug with currently available β-lactam drugs. A Phase II trial for the treatment of serious respiratory infections which incorporated treatment with P1A (β-lactamase product) and ampicillin showed only a 20% change in gut microbiota compared to 50% change in patients treated with ampicillin alone.[96]

9.3 IMMUNOMODULATORY STRATEGIES

Apart from antimicrobials, strategies involving immunomodulation of inflammatory responses (targeting pattern recognition receptor signaling, corticosteroids, complement inhibitors etc.), improving pulmonary barrier function (using adrenomedullin, angiopoietin etc.) during pneumonia and its associated complications could add a new dimension in providing better therapeutics for patients.[97]

10 CONCLUSIONS AND FUTURE PERSPECTIVES

Despite great advances in management and preventative approaches, pneumonia still remains a major burden of mortality and morbidity in young children and the elderly, especially in the developing and under-developed countries. Prevention by means of vaccination is critical for reducing pneumonia mortality in children <5 years of age, and an effective antibiotic therapy is important for the elderly. The widespread emergence of antimicrobial resistance is a well-recognized cause of the ineffectiveness of the large number of the currently used antimicrobials. Although numerous efforts have been made to combat this, newer targets need to be identified for the generation of the next level of effective antimicrobials. In addition, a complete understanding of the various aspects of drug resistance in microbes is essential to assist us in designing better targets and help us discover new antibacterial drugs. In the near future, the next challenge will be to identify newer agents for the treatment of multidrug resistant pathogens which are emerging at a rapid rate. The constant and unpredictable nature of pneumococcal pathogens can outpace technological and drug development strategies. Therefore, it is critical for researchers, pharmaceutical companies, and governments and other funding bodies to continue making progress in developing new strategies and antimicrobials towards respiratory infections, including pneumonia.

REFERENCES

1. Moellering RC. The continuing challenge of lower respiratory tract infections. *Clin Infect Dis* 2002;**34**:S1–3.
2. Leowski J. Mortality from acute respiratory infections in children under five years of age: global estimates. *Wld Hlth Statist Quart* 1986;**39**:138–44.
3. Garbino J, Gerbase MW, Wunderli W, Kolarova L, Nicod LP, Rochat T, et al. Respiratory viruses and severe lower respiratory tract complications in hospitalized patients. *Chest* 2004;**125**(3):1033–9.

4. Kais M, Spindler C, Kalin M, Ortqvist A, Giske CG. Quantitative detection of *Streptococcus pneumoniae.* Haemophilus influenzae and Moraxella catarrhalis in lower respiratory tract samples by real-time PCR. *Diag Microbiol Infect Dis* 2006;**55**:169–78.

5. Shann F. Etiology of severe pneumonia in children in developing countries. *Pediatr Infect Dis J* 1986;**5**:247–52.

6. Bogaert D, Engelen MN, Timmers-Reker AJM, Elzenaar KP, Peerbooms PGH, Coutinho RA, et al. Pneumococcal carriage in children in The Netherlands: a molecular epidemiological study. *J Clin Microbiol* 2001;**39**(9):3316–20.

7. Brooks-Walter A, Briles DE, Hollingshead SK. The *pspC* gene of *Streptococcus pneumoniae* encodes a polymorphic protein, PspC, which elicits cross-reactive antibodies to PspA and provides immunity to pneumococcal bacteremia. *Infect Immun* 1999;**67**:6533–42.

8. Bergmann S, Hammerschmidt S. Versality of pneumococcal surface proteins. *Microbiol* 2006;**152**:295–303.

9. Yang B, Yang BL, Savani RC, Turley EA. Identification of a common hyaluronan binding motif in the hyaluronan binding proteins RHAMM, CD44 and link protein. *EMBO J* 1994;**13**:288–98.

10. Mitchell AM, Mitchell TJ. *Streptococcus pneumoniae*: virulence factors and variation. *Clin Microbiol Infect* 2010;**16**(5):411–8.

11. Tuomanen EI. Pathogenesis of pneumococcal inflammation: otitis media. *Vaccine* 2001;**19**:S38–40.

12. Cundell DR, Gerard NP, Gerard C, Idanpaan-Heikkila I, Tuomanen EI. *Streptococcus pneumoniae* anchor to activated human cells by the receptor for platelet-activating factor. *Nature* 1995;**377**(6548):435–8.

13. AlonsoDeVelasco E, Verheul AFM, Verhoef J, Snippe H. *Streptococcus pneumoniae*: virulence factors, pathogenesis and vaccines. *Microbiol Rev* 1995;**59**(4):591–603.

14. Weiser JN, Austrian R, Sreenivasan PK, Masure HR. Phase variation in pneumococcal opacity: relationship between colonial morphology and nasopahryngeal colonisation. *Infect Immun* 1994;**62**:2582–9.

15. Cundell DR, Tuomanen EI. Receptor specificity of adherence of *Streptococcus pneumoniae* to human type II pneumocytes and ascular endothelial cells in vitro. *Microb Pathog* 1994;**17**:361–74.

16. Hakansson A, Kidd A, Wadell G, Sabharwal H, Svanborg C. Adenovirus infection enhances in vitro adherence of *Streptococcus pneumoniae. Infect Immun* 1994;**62**(7):2707–14.

17. Harford CG, Hara M. Technical Assistance of Alice H. Pulmonary Edema in Influenzal Pneumonia of the Mouse and the Relation of Fluid in the Lung to the Inception of Pneumococcal Pneumonia. *J Exp Med* 1950;**91**(3):245–60.

18. Kline BS, Winternitz MC. Studies Upon Experimental Pneumonia in Rabbits : Viii. Intra Vitam Staining in Experimental Pneumonia, and the Circulation in the Pneumonic Lung. *J Exp Med.* 1915;**21**(4):311–9.

19. Chiou CC, Yu VL. Severe pneumococcal pneumonia: new strategies for management. *Curr Opin Crit Care* 2006;**12**(5):470–6.

20. Cashat-Cruz M, Morales-Aguirre JJ, Mendoza-Azpiri M. Respiratory tract infections in children in developing countries. *Semin Pediatr Infect Dis* 2005;**16**(2):84–92.

21. Obaro SK, Monteil MA, Henderson DC. Fortnightly review: the pneumococcal problem. *BMJ* 1996;**312**(7045):1521–5.

22. Rudan I, Boschi-Pinto C, Biloglav Z, Mulholland K, Campbell H. Epidemiology and etiology of childhood pneumonia. *Bull World Health Organ* 2008;**86**(5):408–16.

23. Walker CL, Rudan I, Liu L, Nair H, Theodoratou E, Bhutta ZA, et al. Global burden of childhood pneumonia and diarrhoea. *Lancet* 2013;**381**(9875):1405–16.

24. Zar HJ, Madhi SA, Aston SJ, Gordon SB. Pneumonia in low and middle income countries: progress and challenges. *Thorax* 2013;**68**(11):1052–6.

25. Gilani Z, Kwong YD, Levine OS, Deloria-Knoll M, Scott JA, O'Brien KL, et al. A literature review and survey of childhood pneumonia etiology studies: 2000-2010. *Clin Infect Dis* 2012;**54**(Suppl. 2):S102–8.

26. Kaplan SL, Mason Jr EO, Wald E, Tan TQ, Schutze GE, Bradley JS, et al. Six year multicentre surveillance of invasive pneumococcal infections in children. *Pediatr Infect Dis J* 2002;**21**(2):141–7.

27. Ruuskanen O, Lahti E, Jennings LC, Murdoch DR. Viral pneumonia. *Lancet* 2011;**377**(9773):1264–75.
28. May DS, Kelly JJ, Mendlein JM, Garbe PL. Surveillance of major causes of hospitalization among the elderly, 1988. MMWR. CDC surveillance summaries: Morbidity and mortality weekly report. CDC surveillance summaries/Centers for Disease Control 1991;40(1):7–21.
29. Ochoa-Gondar O, Vila-Corcoles A, de Diego C, Arija V, Maxenchs M, Grive M, et al. The burden of community-acquired pneumonia in the elderly: the Spanish EVAN-65 study. *BMC public health* 2008;**8**:222.
30. El-Solh AA, Sikka P, Ramadan F, Davies J. Etiology of severe pneumonia in the very elderly. *Am J Respir Crit Care Med* 2001;**163**(3 Pt 1):645–51.
31. Johansson N, Kalin M, Tiveljung-Lindell A, Giske CG, Hedlund J. Etiology of community-acquired pneumonia: increased microbiological yield with new diagnostic methods. *Clin Infect Dis* 2010;**50**(2):202–9.
32. Swingler G, Fransman D, Hussey G. Conjugate vaccines for preventing *Haemophilus influenzae* type B infections. *Cochrane Database Syst Rev* 2007;(2):CD001729.
33. Watt JP, Wolfson LJ, O'Brien KL, Henkle E, Deloria-Knoll M, McCall N, et al. Burden of disease caused by *Haemophilus influenzae* type b in children younger than 5 years: global estimates. *Lancet* 2009;**374**(9693):903–11.
34. Centers for Disease C. Prevention. Global routine vaccination coverage, 2011. *MMWR. Morbidity and mortality weekly report.* 2012;**61**(43):883–5.
35. Theodoratou E, Johnson S, Jhass A, Madhi SA, Clark A, Boschi-Pinto C, et al. The effect of *Haemophilus influenzae* type b and pneumococcal conjugate vaccines on childhood pneumonia incidence, severe morbidity and mortality. *International journal of epidemiology* 2010;**39**(Suppl. 1):i172–185.
36. Bhutta ZA, Das JK, Walker N, Rizvi A, Campbell H, Rudan I, et al. Interventions to address deaths from childhood pneumonia and diarrhoea equitably: what works and at what cost? *Lancet* 2013;**381**(9875):1417–29.
37. Zimmerman RK, Middleton DB. Vaccines for persons at high risk, 2007. *J Fam Pract* 2007;**56**(2 Suppl. Vaccines):S38–46 C34–C35.
38. Aljunid S, Abuduxike G, Ahmed Z, Sulong S, Nur AM, Goh A. Impact of routine PCV7 (Prevenar) vaccination of infants on the clinical and economic burden of pneumococcal disease in Malaysia. *BMC Infect Dis* 2011;**11**:248.
39. Picon T, Alonso L, Garcia Gabarrot G, Speranza N, Casas M, Arrieta F, et al. Effectiveness of the 7-valent pneumococcal conjugate vaccine against vaccine-type invasive disease among children in Uruguay: an evaluation using existing data. *Vaccine.* 2013;**31**(Suppl. 3):C109–13.
40. Ekstrom N, Ahman H, Palmu A, Gronholm S, Kilpi T, Kayhty H, et al. Concentration and high avidity of pneumococcal antibodies persist at least 4 years after immunization with pneumococcal conjugate vaccine in infancy. *Clin Vaccine Immunol* 2013;**20**(7):1034–40.
41. Ordonez JE, Orozco JJ. Cost-effectiveness analysis of the available pneumococcal conjugated vaccines for children under five years in Colombia. *Cost Eff Resour Alloc* 2015;**13**:6.
42. Fletcher MA, Fritzell B. Pneumococcal conjugate vaccines and otitis media: an appraisal of the clinical trials. *International J Otolaryngol* 2012;**2012**:312935.
43. Andrews NJ, Waight PA, Burbidge P, Pearce E, Roalfe L, Zancolli M, et al. Serotype-specific effectiveness and correlates of protection for the 13-valent pneumococcal conjugate vaccine: a postlicensure indirect cohort study. *Lancet Infect Dis* 2014;**14**(9):839–46.
44. Ochoa-Gondar O, Vila-Corcoles A, Rodriguez-Blanco T, Gomez-Bertomeu F, Figuerola-Massana E, Raga-Luria X, et al. Effectiveness of the 23-valent pneumococcal polysaccharide vaccine against community-acquired pneumonia in the general population aged >/= 60 years: 3 years of follow-up in the CAPAMIS study. *Clin Infect Dis* 2014;**58**(7):909–17.
45. Cadeddu C, De Waure C, Gualano MR, Di Nardo F, Ricciardi W. 23-valent pneumococcal polysaccharide vaccine (PPV23) for the prevention of invasive pneumococcal diseases (IPDs) in the elderly: is it really effective? *J Prev Med Hyg* 2012;**53**(2):101–3.

46. Whitney CG, Farley MM, Hadler J, Harrison LH, Bennett NM, Lynfield R, et al. Decline in invasive pneumococcal disease after the introduction of protein-polysaccharide conjugate vaccine. *NEJM* 2003;**348**(18):1737–46.
47. Kohanski MA, Dwyer DJ, Collins JJ. How antibiotics kill bacteria: from targets to networks. *Nat Rev Microbiol* 2010;**8**(6):423–35.
48. Drlica K, Malik M, Kerns RJ, Zhao X. Quinolone-mediated bacterial death. *Antimicrob Agents Chemother* 2008;**52**(2):385–92.
49. Bugg TD, Walsh CT. Intracellular steps of bacterial cell wall peptidoglycan biosynthesis: enzymology, antibiotics, and antibiotic resistance. *Nat Prod Rep* 1992;**9**(3):199–215.
50. Tomasz A. The mechanism of the irreversible antimicrobial effects of penicillins: how the beta-lactam antibiotics kill and lyse bacteria. *Annu Rev Microbiol* 1979;**33**:113–37.
51. Katz L, Ashley GW. Translation and protein synthesis: macrolides. *Chemical Rev* 2005;**105**(2):499–528.
52. Chopra I, Roberts M. Tetracycline antibiotics: mode of action, applications, molecular biology, and epidemiology of bacterial resistance. *Microbiol Mol Biol Rev* 2001;**65**(2):232–60 second page, table of contents.
53. Principi N, Esposito S. Management of severe community-acquired pneumonia of children in developing and developed countries. *Thorax* 2011;**66**(9):815–22.
54. Sazawal S, Black RE. Pneumonia Case Management Trials G. Effect of pneumonia case management on mortality in neonates, infants, and preschool children: a meta-analysis of community-based trials. *Lancet Infect Dis* 2003;**3**(9):547–56.
55. Soofi S, Ahmed S, Fox MP, MacLeod WB, Thea DM, Qazi SA, et al. Effectiveness of community case management of severe pneumonia with oral amoxicillin in children aged 2-59 months in Matiari district, rural Pakistan: a cluster-randomized controlled trial. *Lancet* 2012;**379**(9817):729–37.
56. Recommendations for Management of Common Childhood Conditions: Evidence for Technical Update of Pocket Book Recommendations: Newborn Conditions, Dysentery, Pneumonia, Oxygen Use and Delivery, Common Causes of Fever, Severe Acute Malnutrition and Supportive Care. GenevaWHO 2012.
57. Kabra SK, Lodha R, Pandey RM. Antibiotics for community-acquired pneumonia in children. *Cochrane Database Syst Rev Cochrane Database Syst Rev* 2010;(3):CD004874.
58. Harrison CJ, Woods C, Stout G, Martin B, Selvarangan R. Susceptibilities of *Haemophilus influenzae*, *Streptococcus pneumoniae*, including serotype 19A, and *Moraxella catarrhalis* paediatric isolates from 2005 to 2007 to commonly used antibiotics. *J Antimicrob Chemother* 2009;**63**(3):511–9.
59. Mandell LA, Bartlett JG, Dowell SF, File Jr TM, Musher DM, Whitney C, et al. Update of practice guidelines for the management of community-acquired pneumonia in immunocompetent adults. *Clin Infect Dis* 2003;**37**(11):1405–33.
60. Bjerre LM, Verheij TJ, Kochen MM. Antibiotics for community-acquired pneumonia in adult outpatients. *Cochrane Database Syst Rev* 2009;(4):CD002109.
61. Bradley JS, Byington CL, Shah SS, Alverson B, Carter ER, Harrison C, et al. The management of community-acquired pneumonia in infants and children older than 3 months of age: clinical practice guidelines by the Pediatric Infectious Diseases Society and the Infectious Diseases Society of America. *Clin Infect Dis* 2011;**53**(7):e25–76.
62. Farrell DJ, Couturier C, Hryniewicz W. Distribution and antibacterial susceptibility of macrolide resistance genotypes in Streptococcus pneumoniae: PROTEKT Year 5 (2003–2004). *Int J Antimicrob Agents* 2008;**31**(3):245–9.
63. Lappin E, Ferguson AJ. Gram-positive toxic shock syndromes. *Lancet Infect Dis* 2009;**9**(5):281–90.
64. Como-Sabetti K, Harriman KH, Buck JM, Glennen A, Boxrud DJ, Lynfield R. Community-associated methicillin-resistant *Staphylococcus aureus*: trends in case and isolate characteristics from six years of prospective surveillance. *Public Health Rep* 2009;**124**(3):427–35.

65. Heinonen S, Silvennoinen H, Lehtinen P, Vainionpaa R, Vahlberg T, Ziegler T, et al. Early oseltamivir treatment of influenza in children 1-3 years of age: a randomized controlled trial. *Clin Infect Dis* 2010;**51**(8):887–94.

66. Committee on Infectious D. From the American Academy of Pediatrics: Policy statements--Modified recommendations for use of palivizumab for prevention of respiratory syncytial virus infections. *Pediatrics* 2009;**124**(6):1694–701.

67. Dagan R, Hoberman A, Johnson C, Leibovitz EL, Arguedas A, Rose FV, et al. Bacteriologic and clinical efficacy of high dose amoxicillin/clavulanate in children with acute otitis media. *Pediatr Infect Dis J* 2001;**20**(9):829–37.

68. Blair JM, Webber MA, Baylay AJ, Ogbolu DO, Piddock LJ. Molecular mechanisms of antibiotic resistance. *Nat Rev Microbiol* 2015;**13**(1):42–51.

69. Kyaw MH, Lynfield R, Schaffner W, Craig AS, Hadler J, Reingold A, et al. Effect of introduction of the pneumococcal conjugate vaccine on drug-resistant *Streptococcus pneumoniae*. *N Engl J Med* 2006;**354**(14):1455–63.

70. Jones RN, Sader HS, Moet GJ, Farrell DJ. Declining antimicrobial susceptibility of Streptococcus pneumoniae in the United States: report from the SENTRY Antimicrobial Surveillance Program (1998–2009). *Diagn Microbiol Infect Dis* 2010;**68**(3):334–6.

71. Pfaller MA, Farrell DJ, Sader HS, Jones RN. AWARE Ceftaroline Surveillance Program (2008-2010): trends in resistance patterns among Streptococcus pneumoniae, Haemophilus influenzae, and Moraxella catarrhalis in the United States. *Clin Infect Dis* 2012;**55 Suppl. 3**:S187–93.

72. Bebear C, Pereyre S, Peuchant O. *Mycoplasma pneumoniae*: susceptibility and resistance to antibiotics. *Future Microbiol* 2011;**6**(4):423–31.

73. Morozumi M, Takahashi T, Ubukata K. Macrolide-resistant *Mycoplasma pneumoniae*: characteristics of isolates and clinical aspects of community-acquired pneumonia. *J Infect Chemother* 2010;**16**(2):78–86.

74. David MZ, Daum RS. Community-associated methicillin-resistant *Staphylococcus aureus*: epidemiology and clinical consequences of an emerging epidemic. *Clin Microbiol Rev* 2010;**23**(3):616–87.

75. Kwong JC, Chua K, Charles PG. Managing Severe Community-Acquired Pneumonia Due to Community Methicillin-Resistant Staphylococcus aureus (MRSA). *Curr Infect Dis Rep* 2012;**14**(3):330–8.

76. Bollenbach T. Antimicrobial interactions: mechanisms and implications for drug discovery and resistance evolution. *Curr Opin Microbiol* 2015;**27**:1–9.

77. Birkett D, Brosen K, Cascorbi I, Gustafsson LL, Maxwell S, Rago L, et al. Clinical pharmacology in research, teaching and health care: Considerations by IUPHAR, the International Union of Basic and Clinical Pharmacology. *Basic Clin Pharmacol Toxicol* 2010;**107**(1):531–59.

78. Ge Y, Biek D, Talbot GH, Sahm DF. In vitro profiling of ceftaroline against a collection of recent bacterial clinical isolates from across the United States. *Antimicrob Agents Chemother* 2008;**52**(9):3398–407.

79. Saravolatz LD, Stein GE, Johnson LB, Ceftaroline:. a novel cephalosporin with activity against methicillin-resistant *Staphylococcus aureus*. *Clin Infect Dis* 2011;**52**(9):1156–63.

80. Widmer AF, Ceftobiprole:. a new option for treatment of skin and soft-tissue infections due to methicillin-resistant *Staphylococcus aureus*. *Clin Infect Dis* 2008;**46**(5):656–8.

81. Nicholson SC, Welte T, File Jr TM, Strauss RS, Michiels B, Kaul P, et al. A randomized, double-blind trial comparing ceftobiprole medocaril with ceftriaxone with or without linezolid for the treatment of patients with community-acquired pneumonia requiring hospitalization. *Int J Antimicrob Agents* 2012;**39**(3):240–6.

82. Karlowsky JA, Nichol K, Zhanel GG. Telavancin: mechanisms of action, in vitro activity, and mechanisms of resistance. *Clin Infect Dis* 2015;**61**(Suppl. 2):S58–68.

83. Zuckerman JM. Macrolides and ketolides: azithromycin, clarithromycin, telithromycin. *Infect Dis Clin North Am* 2004;**18**(3):621–49.

84. Namour F, Wessels DH, Pascual MH, Reynolds D, Sultan E, Lenfant B. Pharmacokinetics of the new ketolide telithromycin (HMR 3647) administered in ascending single and multiple doses. *Antimicrob Agents Chemother* 2001;**45**(1):170–5.

85. Champney WS, Pelt J. The ketolide antibiotic ABT-773 is a specific inhibitor of translation and 50S ribosomal subunit formation in *Streptococcus pneumoniae* cells. *Curr Microbiol* 2002;**45**(3):155–60.

86. Brueggemann AB, Doern GV, Huynh HK, Wingert EM, Rhomberg PR. In vitro activity of ABT-773, a new ketolide, against recent clinical isolates of *Streptococcus pneumoniae, Haemophilus influenzae*, and *Moraxella catarrhalis. Antimicrob Agents Chemother* 2000;**44**(2):447–9.

87. Dubois J, St -Pierre C. In vitro activity of ABT-773 versus macrolides and quinolones against resistant respiratory tract pathogens. *Diagn Microbiol Infect Dis* 2001;**40**(1–2):35–40.

88. Oldach D, Clark K, Schranz J, Das A, Craft JC, Scott D, et al. Randomized, double-blind, multicentre phase 2 study comparing the efficacy and safety of oral solithromycin (CEM-101) to those of oral levofloxacin in the treatment of patients with community-acquired bacterial pneumonia. *Antimicrob Agents Chemother* 2013;**57**(6):2526–34.

89. van Rensburg DJ, Perng RP, Mitha IH, Bester AJ, Kasumba J, Wu RG, et al. Efficacy and safety of nemonoxacin versus levofloxacin for community-acquired pneumonia. *Antimicrob Agents Chemother* 2010;**54**(10):4098–106.

90. Qin X, Huang H. Review of nemonoxacin with special focus on clinical development. *Drug Des Dev Ther* 2014;**8**:765–74.

91. Han H, Kim SE, Shin KH, Lim C, Lim KS, Yu KS, et al. Comparison of pharmacokinetics between new quinolone antibiotics: the zabofloxacin hydrochloride capsule and the zabofloxacin aspartate tablet. *Curr Med Res Opin* 2013;**29**(10):1349–55.

92. Biedenbach DJ, Farrell DJ, Flamm RK, Liverman LC, McIntyre G, Jones RN. Activity of JNJ-Q2, a new fluoroquinolone, tested against contemporary pathogens isolated from patients with community-acquired bacterial pneumonia. *Int J Antimicrob Agents.* 2012;**39**(4):321–5.

93. Viasus D, Garcia-Vidal C, Carratala J. Advances in antibiotic therapy for community-acquired pneumonia. *Curr Opin Pulm Med* 2013;**19**(3):209–15.

94. Sader HS, Paukner S, Ivezic-Schoenfeld Z, Biedenbach DJ, Schmitz FJ, Jones RN. Antimicrobial activity of the novel pleuromutilin antibiotic BC-3781 against organisms responsible for community-acquired respiratory tract infections (CARTIs). *J Antimicrob Chemother* 2012;**67**(5):1170–5.

95. Habich D, von Nussbaum F. Platensimycin, a new antibiotic and "superbug challenger" from nature. *ChemMedChem Sep* 2006;**1**(9):951–4.

96. Tarkkanen AM, Heinonen T, Jogi R, Mentula S, van der Rest ME, Donskey CJ, et al. P1A recombinant beta-lactamase prevents emergence of antimicrobial resistance in gut microflora of healthy subjects during intravenous administration of ampicillin. *Antimicrob Agents Chemother* 2009;**53**(6):2455–62.

97. Muller-Redetzky H, Lienau J, Suttorp N, Witzenrath M. Therapeutic strategies in pneumonia: going beyond antibiotics. *Eur Respir Rev* 2015;**24**(137):516–24.

Subject Index